A Study through Skull Morphology on the Diversity of Holocene African Populations in a Historical Perspective

Isabelle Ribot

BAR International Series 2215
2011

Published in 2016 by
BAR Publishing, Oxford

BAR International Series 2215

A Study through Skull Morphology on the Diversity of Holocene African Populations in a Historical Perspective

ISBN 978 1 4073 0773 2

BAR Publishing is the trading name of British Archaeological Reports (Oxford) Ltd. British Archaeological Reports was first incorporated in 1974 to publish the BAR Series, International and British. In 1992 Hadrian Books Ltd became part of the BAR group. This volume was originally published by Archaeopress in conjunction with British Archaeological Reports (Oxford) Ltd / Hadrian Books Ltd, the Series principal publisher, in 2011. This present volume is published by BAR Publishing, 2016.

Printed in England

BAR
PUBLISHING

BAR titles are available from:

BAR Publishing
122 Banbury Rd, Oxford, OX2 7BP, UK
EMAIL info@barpublishing.com
PHONE +44 (0)1865 310431
FAX +44 (0)1865 316916
www.barpublishing.com

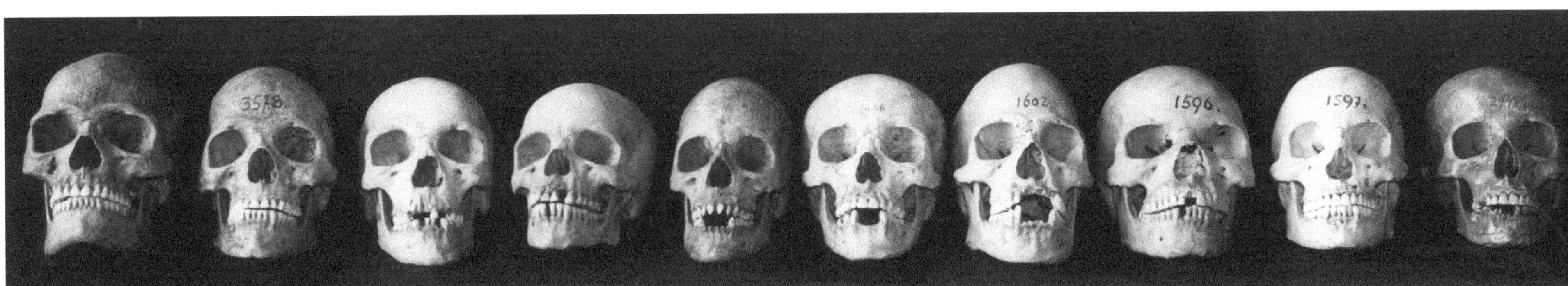

This photograph was taken at the American Museum of Natural History of New York (with the permission of Gary Sawyer), showing ten skulls (anterior view) from different African countries. Although all the specimens are extremely diverse morphologically, it shows only a small percentage of the wide variation that is present within the African continent. Four specimens (99/7189, VL/363, VL/2458 and VL/1602) have been included in the present analyses.

From left to right: VL3575 (male, Zulu, South Africa); VL3578 (female, Zulu, South Africa); 99/8428 (unknown sex, KhoiSan, Murraysburg, South Africa); 99/7189 (unknown sex, Pygmy, Medje, D.R.C.); VL/363 (female, Elmina, Gold Coast, Ghana); VL/2458 (female, Ibea, Cameroon); VL/1602 (male, Ashanti, Ghana); VL/1596 (male, Schasi, Tanzania); VL1597 (female, Ussandawi, Tanzania); VL/2947 (unknown sex, El Hesa, Egypt).

CONTENTS

LIST OF FIGURES

LIST OF TABLES

ABSTRACT

The main objective of the present research (**Chapter 1**) was to explore through skull morphology some potential sources of biological diversity within sub-Saharan Africa, such as: geography and especially history, in relation to large-scale population movements (expansion of Bantu-speakers). Therefore, through various statistical analyses, morphological variation was re-evaluated within modern sub-Saharan African populations, using a very large modern human sample (N=2508) and a set of metric variables related to the cranium and mandible. In the same way, morphological patterns through time were also traced, focusing on various Later Stone Age (LSA) and Iron Age (IA) populations, originating in particular from strategic areas of various influences. For example, temporal trends of differentiation and continuity in West African populations were explored to find to what extent they reflected microevolutionary mechanisms of isolation and gene flow.

In **Chapter 2**, after having briefly introduced both geographical and historical backgrounds of sub-Saharan Africa, the dispersal of Bantu-speakers, a very long-term and large-scale phenomenon, which initiated since the Early Iron Age (*c.* 1,000 BC) was presented in more depth. Its effects were discussed in the light of various results obtained from different fields of research. In fact, biological traces of the dispersal of Bantu-speakers appeared to be difficult to distinguish from many confounding factors (geography, selection and micro-scale historical events). This literature review strongly supported the choice of a multi-layered approach, thereby considering each factor in relative isolation. In this perspective, a series of microevolutionary models were proposed for explaining biological diversity within sub-Saharan Africa.

In **Chapter 3**, after a detailed presentation of the populations and variables under study, a preliminary analysis about inter- and intra-observer errors was achieved.

In **Chapter 4**, various factors (geography, sex and ecology) were tested as a source of modern diversity. On a geographical level, inter-regional craniometric differences on a worldwide scale showed similar results to previous studies. However, additional information about variation within Africa was provided on an inter-regional level with new combinations of variables in relation to both the face and mandible. Although the localization of the geographical differences did not change when sexes were considered separately, a slightly more marked geographical differentiation was observed within African females. Most of the ecological variables (distance from equator, temperature, rainfall, vegetation and humidity) were shown to be interrelated to each other, as well as, to geography. It was not possible to determine a causal relationship between selective pressures and skull morphology, but various morphological trends within sub-Saharan Africa were observed in relation to climate.

In **Chapter 5**, the effects of historical factors on skull morphology were explored through both modern and past African diversity. The morphological similarities observed between most of the modern sub-Saharan Africans (except for the KhoiSan) supported a homogenization process due to a common origin and/or founder effect phenomenon (in relation to the expansion of Bantu-speakers). However, as a diversification was also noted, the results could reflect other factors related to both geography and history (two streams of dispersal of Bantu-speakers). A morphological continuity was observed between various Holocene specimens and modern West Africans, but affinities were also identified with Central Africans. These facts supported the presence of a biological input originating mainly from sub-Saharan Africa in Senegal and Mali. Focusing on an Iron Age Senegalese coastal population (Dioronboumak), social factors relating to high levels of admixture could explain the morphological heterogeneity that contrasted with the cultural continuity.

In conclusion (**Chapter 6**), although the effects (homogenization and/or diversification) of historical events such as large-scale and long-term expansions that occurred since the Mid-Holocene (Bantu-speakers expansion) were visible within sub-Saharan Africa, they were difficult to distinguish from many confounding factors. In fact, the morphological patterns found on both inter- and intra-regional levels, could also reflect other long-term processes (geography and/or selection), as well as short-term ones (micro-scale historical events).

ACKNOWLEDGEMENTS

I would like to thank first the Foundation Philippe Wiener-Maurice Anspach (Brussels) who financially supported my entire Ph.D. research. Additional institutions also supported financially my fieldwork such as the Leakey Trust (London), the Downing College (Cambridge), the NERC (Dr M. Lahr, Cambridge) and the American Museum of Natural History (New York). Philippe De Bruycker and Mrs Bosmans of the Foundation Philippe Wiener-Maurice Anspach are also especially thanked for their help and encouragement, as well as all the people who supported me for grant applications (Rosine Orban, Pierre de Maret, Marta Lahr and Rob Foley).

My thanks also go first to my supervisor Marta Lahr who supported my research through her advice and encouragement, as well as Rob Foley who also initially supervised this research during the M.Phil. dissertation. Many thanks also goes to all the Staff of the Department of Biological Anthropology (University of Cambridge) for their assistance such as Nick Mascie-Taylor for his patience about my statistical problems, Theresa Parker and David Jordan for their various technical and administrative support, and Maguy Bellatti for her major help in the skeletal collections of the Duckworth Laboratory.

Alain Froment (Musée de l'Homme, Paris, France) has been often very helpful with his advice through e-mails, and his work in collaboration with Jean Hiernaux has been for me a great source of inspiration.

Many thanks also to William Howells for his worldwide data that are currently available online, and to John Relethford for his software (also available online) and his advices on RMET analyses.

Thanks to my Belgian friends who helped me during my fieldwork in Belgium (IRScNB, Bruxelles) such as Rosine Orban for the access to Rwandese and Congolese cranial collections, Patrick Semal and André Leguebe for their statistical advice. Staff at the IRScNB was also very helpful to me such as Freddy Vande Meulebroeke. In the Musée Royal de l'Afrique Centrale, Els Cornelissen (MRAC, Tervuren) was also very helpful when I was looking for archival information about the origin of certain skeletal collections.

Many thanks to various people for access to various collections. In London, Louise Humphrey and Robert Krusinsky (NHM, London) allowed me to measure the African skeletal collections of the Natural History Museum, and they have been always very helpful. In Leiden, Roger Bedaux and Annette Schmidt also allowed me to study the Malian collections of the Volkenkunde Museum of Leiden. In Senegal, Hamady Bocoum and Guy Thilmans (IFAN, Dakar) allowed me to study the skeletal collections of the Laboratory of Prehistory at the Institute Cheikh Anta Diop. They were always very helpful as well as Abu Deme, who allowed me to study Walalde materials. In New York, Ian Tattersall and Kenneth Mowbray allowed me to study the cranial collections of the American Museum of Natural History and they supported me in obtaining a small research grant from their institution. Many thanks go to them for their help, as well as to Gary Sawyer who introduced me to the collections.

I did not mention my numerous friends and family so far as this piece of work is dedicated to them.

Lastly, I would like to thank John Hedges for his continued support over the years to publish this work.

PREFACE

This book is essentially, a shorter and updated version of my Ph.D. dissertation I wrote at the University of Cambridge (Ribot, 2002), "Craniomandibular variation in sub-Saharan Africa: sexual dimorphism, geography, ecology and history". It presents the main results and interpretations, which have been used as a basis for various publications that I did in the following years (e.g. Ribot, 2003, 2004; Ribot *et al.*, 2002; Morris and Ribot, 2006; Ribot *et al.*, 2006a, 2006b). Despite updates with new references and additional analyses, it has been left as much as possible similar to the original piece of work achieved in 2002. Michael Petraglia (University of Cambridge) and Alain Froment (Musée de l'Homme, Paris, France) who had been my Ph.D. examiners, also greatly improved the layout of the present book with their various comments and suggestions on the original thesis. The results that have been presented originally in four chapters have been summarised here in only two (**4 and 5**). A re-evaluation of the modern African diversity is presented in **Chapter 4** relating to geography and ecology. Sexual dimorphism was addressed to a lesser extent, as there were too many methodological biases involved (e.g. standardization of sexing methods). RMET analyses of John Relethford's method have also been added, as they also provided some information. After these preliminary analyses, history was then better explored in **Chapter 5** through both modern and past African diversity. Iron Age populations from West Africa were still under focus, as sample size was more reliable for interpretations than other regional archaeological samples.

All the figures have been drawn with Coreldraw X3. Some of them were adapted from other sources (always cited in brackets) and re-drawn adding more information.

All the text has been proof-read by Ericka L'Abbe (University of Pretoria, RSA), Donna Ellender (Paris, France), Alain Froment (Musée de l'Homme, Paris, France) and Otto Graf (Montreal, Canada).

CHAPTER 1

INTRODUCTION

Africa, cradle of human origins, reflects probably the longest history, as it has been continuously populated through time, although with varying intensity. African populations occupied a wide range of habitats and relied on various types of subsistence strategies. During the last 10,000 years, microevolutionary processes became increasingly complex, especially in high density populations (agriculturists), who replaced and/or assimilated low density populations (foragers). As new ways of subsistence were appearing especially along the savannah belt, later human groups began to reach over-population and expanded from core areas through natural corridors. Therefore, both West Africa and Central Africa have probably been peopled by a mosaic of populations that either differentiated or homogenized successively.

As a consequence of these past processes, sub-Saharan Africa, which is nowadays inhabited by more than three thousands ethnic groups (Murdock, 1959), has often been chosen to study human diversity. Although it corresponds mainly to Bantu-speaking populations, it provides on a worldwide scale the highest levels of "bio-cultural" heterogeneity. Recent biological studies provided new research perspectives, as they took into account more closely both historical and environmental aspects (Hiernaux, 1953-1977; Cavalli-Sforza *et al.*, 1994; Froment, 1995). For example, Hiernaux (1974) showed that the marked differences observed among African populations are the result of both phenotypic and genetic adaptations and that they have probably been enhanced by a high rate of ethnic endogamy.

As history and environment are key-elements in shaping human morphology, the present study explores through craniomandibular variation the differentiation of a large African sample, especially from the Western and Central regions. Are differences and/or similarities in skull morphology interpretable in relation to aspects such as sexual dimorphism, geography, ecology (climate), culture (ethno-linguistic background), and especially history (e.g. Bantu-speakers expansion)?

These factors cited above are to be approached as a potential source of biological diversity, and focuses on the possible biological effects of the Bantu-speakers expansion (Figure 1, p. 4). In order to analyse the latter, the population samples under study will be defined on different geographical and cultural scales (e.g. worldwide, continental, regional, ethnic).

According to numerous anthropometrical studies (Hiernaux, 1974; Howells, 1989; Lahr, 1996; Froment, 1998; Lieberman *et al.*, 2002; Viðarsdóttir *et al.*, 2002; Ousley *et al.*, 2009), geography often appears to be the key aspect in the morphological differentiation of populations. In addition, this fact is also supported by genetic data (Cavalli-Sforza *et al.*, 1994; Biondi *et al.*, 1996; Salas *et al.*, 2002; Ramachandran *et al.*, 2005). However, geography and sex still need to be explored on both worldwide and regional scales, as they could provide different pictures of morphological variation.

Morphological evolution is also essential to link to several other aspects such as ecology and history. As suggested previously, climate might have had a greater effect on the face than other parts of the skull (Hiernaux and Froment, 1976; Harvati and Weaver, 2006; Roseman and Weaver, 2004). However, the (causal) relationship between skull morphology and various ecological parameters is difficult to determine, as selection is often the result of complex factors.

Historical aspects probably also affected present population diversity. For example, biological effects of large-scale population movements (such as the Bantu-speakers expansion) are often one of the key questions for paleoanthropological research. As suggested by linguistics and archaeology, the dispersal of Bantu languages, has been related to the emergence and diversification of several food production methods in sub-Saharan Africa in the last 5,000 years (Vansina, 1995). In biological terms, this rather complex cultural phenomenon could have accelerated genetic admixture with different groups, decreased the effects of genetic drift and isolation, and therefore facilitated the process of homogenization. However, according to Hiernaux and Basu (1980), these microevolutionary mechanisms also interplayed with other factors such as the selective pressures, as the latter in some circumstances can mimic gene flow. Furthermore, additional historical factors not related to the Bantu-speakers expansion (e.g. Sudanic migrations along the savanna belt during the 17th–18th centuries AD) might also have accelerated or attenuated the processes of biological homogenization in some areas such as in West-Central Africa.

In fact, previous studies cited above that tried to approach the Bantu-speakers expansion, could observe that its resulting effect (homogenization process) was difficult to distinguish from other factors (geography, ecology, ethno-history). As a consequence, the present work will focus on different levels of analysis, in order to confront various types of results.

Although historical processes and selective pressures are difficult to distinguish between each other, it is assumed as according to previous literature (Howells, 1989; Lahr, 1996; Froment, 1998; Lieberman *et al.*, 2002; Pietrusewsky, 2008; Smith 2009; Ousley *et al.*, 2009) that skull morphology is a good indicator of biological affinities. It is a kind of "paleogenetic" tool that allows us to extend our study on past populations. In the present case, the (re-)evaluation of the affinities of a few West and Central Africans dated to the Later Stone Age and/or Iron Age periods will provide us the opportunity to explore some microevolutionary processes related to sedentism and various other historical events. During the Iron Age (from first millennium BC), population movements and changes accelerated in relation to, not only the Bantu-speakers expansion in sub-Saharan Africa, but also the appearance and/or disappearance of various West African kingdoms.

In order to introduce all these microevolutionary processes within a historical framework and biological context, **Chapter 2** summarises the major aspects related to the Holocene peopling (human settlement) of West and Central Africa. It is an extremely complicated puzzle where several historical and/or environmental factors have played an important role in shaping present diversity. First, a brief overview is presented about the major events that occurred during the Holocene, both before and during the Bantu-speakers expansion. The latter is also expanded using linguistic and archaeological data. Thereafter, one will focus on the evidence obtained from various aspects of biological anthropology (physical anthropology and genetics).

Chapter 3 summarises the populations and the variables under study, in relation to various aspects (skeletal preservation, archaeology, history and environment) that will be later useful for the analysis. A very large sample of modern African populations (N=1907) are compared to various archaeological specimens (N=601) dated mainly to the Iron Age period (last 3,000 years). For this purpose, a set of 61 metric traits (and 4 non-metric traits) of the skull are used as the main features for the statistical morphological analysis. The cranium and mandible are analysed together or separately, in order to compare the results obtained from different parts of the skull.

At the onset, a selection of the variables related to skull morphology is achieved according to results obtained from both intra- and inter-observer errors. It was important to do this procedure in order to validate the composition of our large corpus of data, which consists of both personal and compiled data.

The following two chapters (**4 and 5**) correspond to the focal part of our statistical analysis that is shaped as a "multi-layered" approach. Various univariate and multivariate statistics are performed here to test a series of questions mainly related to processes involved in the differentiation of West and Central African populations. The present study will focus in depth on historical populations dated to around the last three centuries, before making comparisons with ancient populations.

In **Chapter 4**, various factors (geography, sex and ecology) will be tested as potential sources of the modern morphological diversity. Geographical differences will be re-evaluated on both worldwide and continental scales, focusing mainly on Africa. A very large data set for sub-Saharan Africa and new series of variables (especially in relation to the mandible) will be used, in comparison to previous work (Howells, 1989; Froment, 1992a, 1992b; Lahr, 1996; Hájek *et al.*, 2008). It will allow us to see whether additional information about between- and within- population variation can be extracted from the multivariate analyses.

Levels of sexual dimorphism will be explored both on a worldwide context and within sub-Saharan Africa. As their variability will show, the results will suggest to us what are the most dimorphic variables and populations.

Concerning the ecological aspect, the following issues will be tested: Are the environmental parameters related to climate, indicators of morphological evolution as was suggested previously by Hiernaux and Froment (1976) and others (e.g. Roseman, 2004; Roseman and Weaver, 2004)? For example, does the nose shape reflect trends of population differentiation on both inter- and intra-regional levels?

In **Chapter 5**, the effects of historical factors (e.g. the Bantu-speakers expansion) on skull morphology will be explored through both modern and past African diversity.

In relation to the ethno-linguistic background of the modern population under study, two main questions will be tested in relation to the pattern of the Bantu-speakers expansion, such as:

1. Are Bantu-speakers morphologically more homogeneous than other linguistic groups in sub-Saharan Africa?
2. Are the Central Bantu-speakers (e.g. Basuku) morphologically closer to derived Eastern ones (e.g. Bahutu) rather than a different linguistic group such as the Kwa-speakers (e.g. Ashanti)?

The first issue will be approached on a continental scale, and the second one on an ethnic scale, in order to compare different types of results. However, it will not be possible to test the source of the Bantu-speakers dispersal on the basis of only modern data. In fact, additional recent expansions, not related to the Bantu-speakers, occurred successively in West-Central Africa. Therefore, they probably diversified and/or mainly homogenized the present populations such as from Cameroon, as was suggested by genetics (Spedini *et al.*, 1999; Hájek *et al.*, 2008).

However, the analysis of a few Later Stone Age and/or Iron Age populations from West and Central Africa will allow us to test the following two issues:

1. Are there morphological continuity and/or discontinuity during the Holocene in West and Central Africa?
2. Is the position of ancient "populations" within the modern variation, indicative of recent historical processes such as large- (and/or micro-) scale population movements along the savanna belt and/or into the sub-Saharan region (e.g. Bantu-speakers expansion)?

This last question in particular is tested on Iron Age Senegalese specimens, as the latter could reflect the high frequency of population movements that occurred between sub-Saharan Africa and the Sahelian zone. Again, it was not possible to test the source of the Bantu-speakers dispersal, in the present case mainly due to the lack of human remains found in equatorial areas. In fact, there are less than 30 individuals dated to the Later Stone Age and Iron Age periods (around last nine millennia), who, in addition, also originate from the hypothetical cradle of "proto"-Bantu-speakers region (Nigeria-Cameroon).

Finally (**Chapter 6**) several microevolutionary models will be re-assessed on the basis of the results that will be obtained from the "multi-layered" approach. The key morphological features responsible for variation will be explored in relation to different aspects (geography, sex, ecology and time). They might provide meaningful results in terms of population differentiation and additional evidence supporting various hypotheses on the large-scale and long-term expansion of Bantu-speakers within sub-Saharan Africa. In particular, was the latter either a simple process of homogenization, or a more complex phenomenon in relation to various confounding factors (e.g. population admixture, selection)?

In order to introduce the present study, various aspects, such as geography, history, archaeology, linguistics and biology, are reviewed here, especially in relation to the Bantu-speakers expansion.

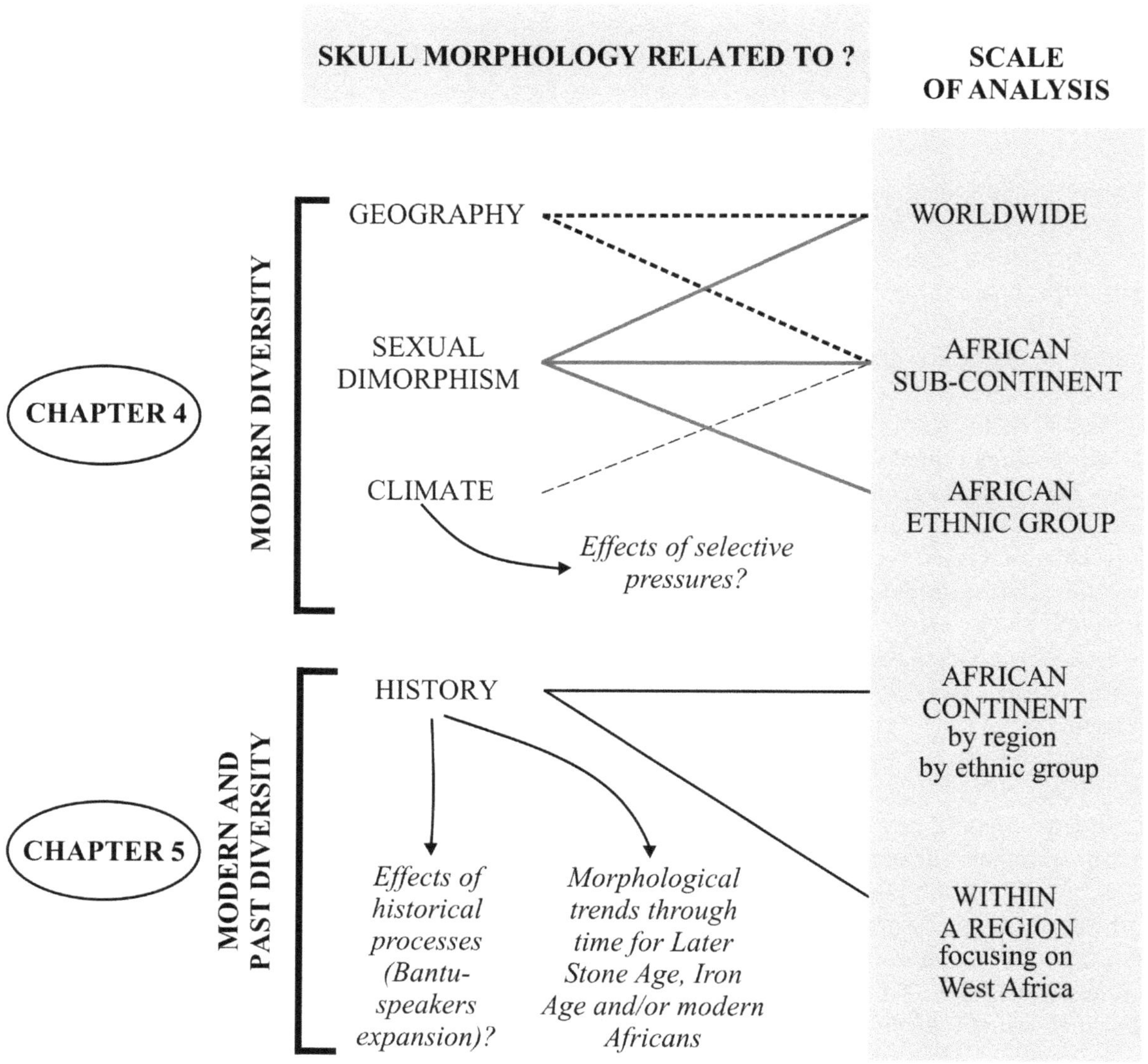

Figure 1. Main objectives of the "multi-layered" approach undertaken through two chapters of analysis, before a final discussion on the biological effects of the Bantu-speakers expansion and its confounding factors.

CHAPTER 2

THE HOLOCENE PEOPLING OF WEST AND CENTRAL AFRICA: A COMPLICATED PUZZLE

The Holocene period (approximately the last 10,000 years) is a key phase to explore the origins and evolution of both modern populations and societies. Various environmental factors and historical processes probably contributed to shape the present cultural and biological differences present nowadays among African populations.

In order to place this study and its objectives in the context of what is known about the population history of Africa in the Holocene, this chapter is subdivided into four sections.

Firstly, broad chronological and spatial frameworks of Africa are summarised, as they provide information on key historical events that affected the peopling (human settlement) of the continent. They will help us to better understand the current linguistic and ethnic diversity, especially in West and Central Africa.

Secondly, the Bantu-speakers expansion is approached in detail not as an isolated phenomenon but rather as one aspect within a broader historical context. Both linguistic and archaeological data in relation to this phenomenon are also presented.

Thirdly, hypotheses from biology about the peopling of West and Central Africa, based on both anthropometric and genetic data, are discussed. The question of the Bantu-speakers expansion is raised again as a testing framework. Two problems, identifying selective pressures and linking physical anthropology to genetics, are also developed, as they can affect our interpretations on biological diversity.

Finally, various hypotheses about microevolutionary processes are summarised for the Holocene period in Africa, and in a discussion, an attempt is made to integrate several aspects (geography, demography and morphological diversity). After underlining interactions between culture, languages and biology, the importance of large-scale population movements in shaping present biological diversity is discussed, especially around the phenomenon of the Bantu-speakers expansion. The question of the relationship between expanding and indigenous populations is also added to the discussion, as it is a determining factor in creating patterns of cultural and biological diversity.

2.1. Overview of major events

In order to understand the biological diversity of present African populations in historical terms, it is necessary first to introduce the major events that shaped the history of Africa during the last 10,000 years. As Vansina (1990) suggested, sociocultural differences between peoples are directly shaped by past and present history. The documentation of these differences in a diachronic manner is necessary, in order to explain various evolutionary processes.

2.1.1. Environmental factors affecting human settlements, climate and geography

During the Holocene in Africa, climate (temperature, rainfall and humidity) was a determining factor in influencing direction and duration of population movements. After a very long dry period (15,000-10,000 BC), the appearance of three wet phases provided favourable temperatures for human settlements especially in the Sahara. Their duration ranging from a millennium to about five hundred years, probably determined the limits of population expansion (Muzzolini, 1993).

Figure 2 (p. 7) shows for the Sahara a climatic curve that was obtained from various paleoclimatological data compiled by Muzzolini (1993), and its width reflects divergences between regions, contradictory evidence or variations in dating. More recently, climatic oscillations have been explored further by various inter- and intra-regional studies; most of them supporting the broad lines of Muzzolini's curve (e.g. Hoelzmann *et al.*, 2001; Salzmann and Hoelzmann, 2005).

The Sahara and West Africa provide much more paleoclimatological data than Central Africa, as they have many paleo-lake deposits dated to the Holocene period (Giresse *et al.*, 2005; Grove, 1993; Maley, 1993, 1998). During the Pleistocene, North Africa was often cut off from sub-Saharan Africa, apart from few possible corridors such as the Nile Valley. During the Holocene, the Sahara became a key area of cultural crossroads, as well as the Sahel and the Middle Niger valley that provided a long occupational sequence since 2,500 BC (MacDonald, 1999).

Despite the occurrence of wet phases during the last twelve millennia, processes of aridification in Africa also accelerated progressively, with increasing numbers of droughts (Maley, 1993; Jousse, 2006). Although there is regional variation, the trend on a continental scale follows the pattern of increasing dry phases and decreasing length of wetter phases. During the Early Holocene, temperature and moisture patterns rose progressively in sub-Saharan Africa. During dry phases, Africa south of the equator was regularly cut off from the northern part. The pollen and lake-level data obtained from the core of Lake Bosumtwi in Ghana support all these facts, showing that several regressions occurred during the Holocene *circa* 1,000 BC (Coetze and Van Zinderen, 1984).

The position of the Sahara-Sahel limit also varied enormously during the last 22,000 years, migrating northwards during the wet phases (Goudie, 1996; Ballouche and Neumann 1995): it was localized at a latitude of 15° north during a very dry phase of the Late Pleistocene (or Late Glacial Maximum; 16,000 BC), at 24° north during a wet phase of the Early Holocene (*c.* 7,000 BC), and presently at 18° north, corresponding to increasing dryness.

According to Hassan (1997), when we examine both long-term records for climate and human response, we can observe interactions between climate changes and culture. Climatic fluctuations seemed to have favoured adaptation to new food resources (e.g. roots, wild plants, various types of foraging). For example, during a long wet period interrupted by small droughts (10,000-9,500 BC; 6,500 BC; 5,500 BC; 2,500 BC), the Sahara became much more inhabited by pastoral populations coming from the East. A very dry phase (*c.* 2,000-1,700 BC) probably encouraged expansions, along savanna corridors, of both pastoralists in the Paleo-Sahel northern fringe and Bantu-speaking agriculturists in sub-Saharan Africa. The equatorial forest reduced to refuge areas, and the Dahomey gap that corresponds to an interruption of the West African rain-forest belt in Togo and Benin, was most probably formed during this very dry phase.

Geography is also an important aspect in understanding the settlement of past and present populations. Environmental conditions, either bad or good, are often associated with various kinds of habitats (very diffuse habitat in bad conditions; grouped habitat in good conditions), resulting in different demographical patterns, which are not detailed for each country (Adiwas Kouerey *et al.*, 1989) (Figure 3, p. 8).

Still from present high densities of populations (Nigeria, Cameroon and the East African highlands), it is possible to localize core areas, which attracted populations because of their environments with abundant food resources. For present Africa, these facts can be illustrated by the comparison of geographical features, vegetation and population density distributions. Unfortunately, from the Early to Mid-Holocene, there is not enough data for reconstructing both vegetational and demographical maps of the whole continent. However, observations from present-day data can suggest key elements of the geography and/or ecology, such as both “repulsive” and “attractive” areas, that could have influenced human settlement (Cornelissen, 2002; Mercader, 2002).

“Repulsive” areas for settlement (e.g. forest, desert) are nowadays characterized by a very low population density: 1 to 10 inhabitants per km^2 (e.g. Gabon, Namibia, Central Africa, Botswana and Congo). Their present populations tend to have very low birth rates and fecundity, and low life expectancy (Adiwas Kouerey *et al.,* 1989). The occupation of forest areas is in general organized along axes of penetration, such as roads and rivers, leaving the remaining areas completely empty. More than 40% of the forest populations moved to cities, and they still tend to settle in urban centres.

“Attractive” areas for settlement (e.g. highland plateaux, lowland savannas) provide a better demographic picture than the previous one, as their environment is favourable to agriculture and/or herding. They have however a very variable population density, ranging from 10 to 250 inhabitants per km^2.

Highest concentrations are reached, for example, in the highlands of Rwanda and Burundi, where 80% of population is rural. For example, the Cameroonian highlands (the Grassfields) may also have attracted many populations through time mainly because of its fertile soil conditions and few parasitic diseases (Warnier, 1984).

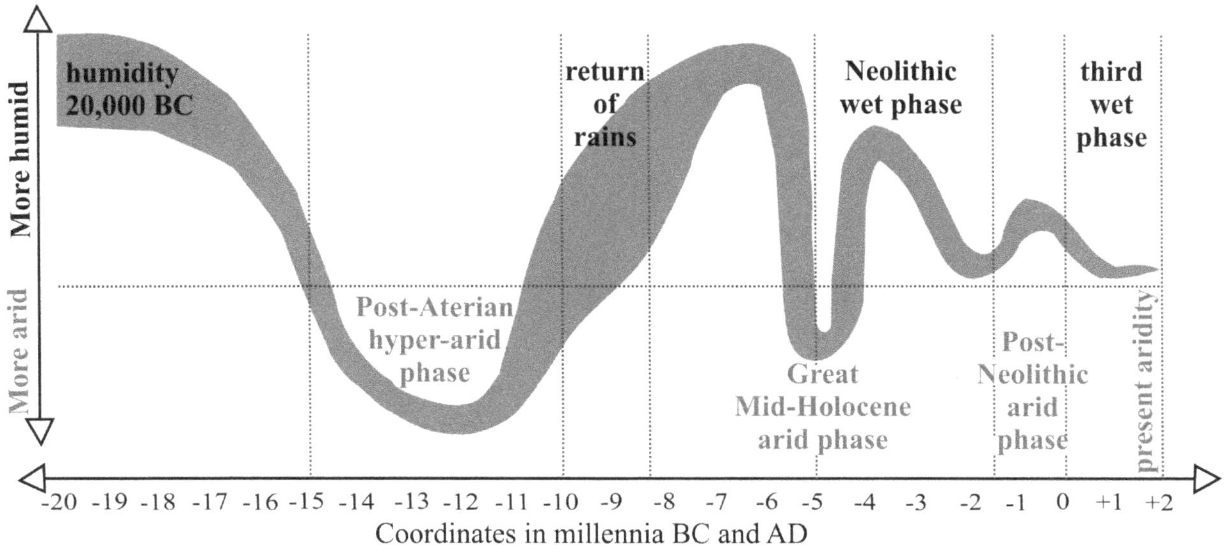

Figure 2. Paleoclimatological data for the Sahara (last 22,000 years): occurrence of humid/arid phases (line drawing after Muzzolini, 1993, p. 228).

However, human occupation of lowland savanna is more diffuse than in the highlands, as it can spread more easily. Hydrographic systems are also key elements of geography, as they may have determined the direction of a migration. Four large basins (e.g. Chad, Nile, Congo and Zambezi) could have been potential corridors for population movements. Most of them have very ramified secondary systems that probably helped people to spread in a diffuse manner. For example, in sub-Saharan Africa, the Congo basin was probably used by the Bantu-speaking populations in order to expand southwards (Vansina, 1995).

2.1.2. The Holocene period in a broad chronological context

The Holocene period is characterized by a very short chronological scale (last 10,000 years) but intensively interrupted by various events that probably shaped present population diversity. Terminal Pleistocene and Early Holocene populations within Africa adapted much more than previously to a wider variety of environments. In fact, according to the paleoclimatological records, peoples re-populated the Sahara and experimented the earliest stages of several new ways of subsistence, especially between 10,000 BC and 6,000 BC (Muzzolini, 1993).

Up to the Mid-Holocene (*c.* 3,000 BC), some "cultural" traditions of the Upper Paleolithic (e.g. microlithic industries) reflecting hunting-gathering activities, were still practised extensively. However, this broad cultural continuity is only apparent, as through time there is regional variation.

The Mid-Holocene (6,000-2,000 BC) is a key period, as several new food production methods appear and spread throughout the continent, especially during the transitional period of "Stone to Metal Age" in West Central Africa (3,000-1,000 BC) (de Maret, 1996). The Late Holocene period (since 2,000 BC) illustrates all these previous historical developments in an accelerated manner, leading to the appearance and disappearance of various African states (Oliver, 1991; Curtin *et al.*, 1995).

2.1.3. From the Early Holocene to Mid-Holocene

2.1.3.1. Population expansion, cultural diversity and diffusion of ceramic technology

Before the Early Holocene, population distribution was very scattered on the African continent, as a hyper-arid period (18,000-10,000 BC) probably affected human expansion (Muzzolini, 1993; Brooks 2006). Then, *c.* 10,000-6,000 BC, it changed completely due to a wetter phase affecting especially the Sahara. According to the radiocarbon dates of lake high strandlines, the culmination of the last pluvial period (when all closed basins between tropics were occupied by lakes) was dated between 7,500 and 6,500 BC (Grove, 1993).

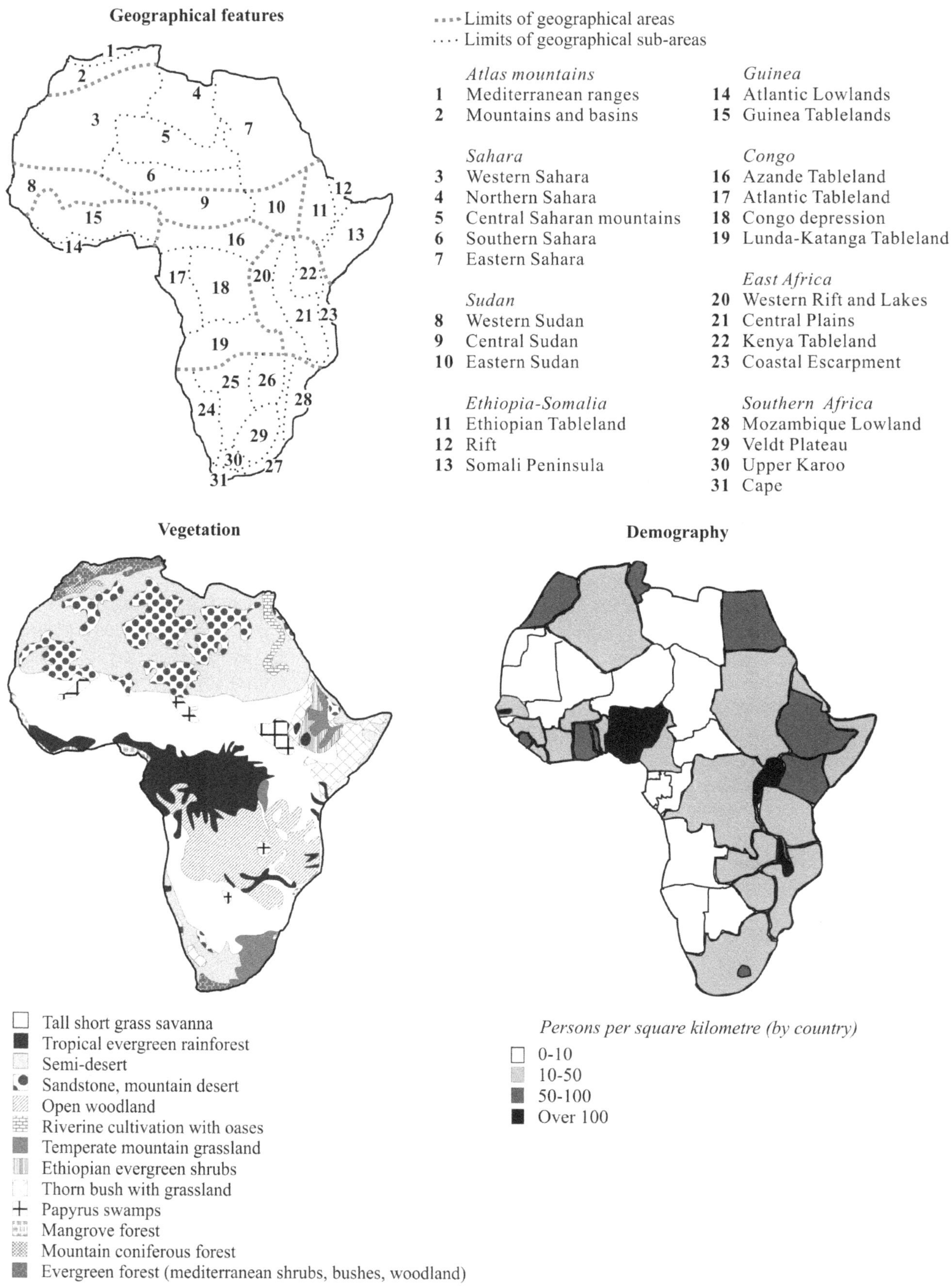

Figure 3. Distribution of geographical features, vegetation and population density in Africa according to recent data (line drawings, after Philip, 1998, p. 18-24).

First, the West African littoral could have been peopled by hunter-gatherer groups, who settled in refugia areas (MacDonald, 1999; Jousse, 2006). For example, two sites with microlithic industries such as Iwo Eleru (9,250±200 BC; Brothwell and Shaw, 1971) and Shum Laka (7,150±70 BP; de Maret *et al.*, 1987, 1993) could be considered as settlement refugia close to the coast and forest zone (Figure 4, p. 10).

After 10,000 BC, an expansion in a northern direction was probably initiated during a wet phase, and this fact is attested by the increasing density of microlithic industries from the West African coast up to Mali such as the site of Korounkorokale (5,185±95 BP; MacDonald, 1994, 1997).

Since 8,500 BC, according to archaeological evidence (e.g. bone harpoons), an "aquatic" life style developed in the Sahara especially close to lakes and rivers (Chad, Middle Niger Valley) (Figure 5, p. 10). As suggested by various sites, hunting-gathering peoples used both ceramic technology and produced epi-paleolithic industries. For example, at Adrar Bous 10 in the Aïr Massif, microlithic artefacts are associated with both grinding stones and ceramics (9,130±65 BP; Roset, 1987).

The diffusion of ceramic technology (*c.* 7,500 BC) into the Sahara cannot really be directly related to the southward expansion of robust populations or so-called "Mechtoid". The latter probably originated in the Maghreb (Dutour, 1989): they are represented in sites like Afalou-bou-Rhummel in Algeria (10,500-8,500 BC; Vallois, 1969) and Taforalt in Morocco (13,750 BC; Chamla, 1978), where they produced one of the various Maghrebian epi-paleolithic industries (Iberomaurusian).

South of the Sahara, the West Central African woodland traditions are characterized by other industries with regional and chronological variations. For example, in Central Africa, the Tshitolian which is a large-core tradition of the Sangoan and Lupemban, carried on from the Early Holocene up to the first millennium BC (e.g. Matupi Cave; 10,000-1st century BC) (Van Noten, 1977, 1982; Mercader, 2002).

2.1.3.2. Spread of new ways of subsistence

The spread of new ways of subsistence within Africa reflects a wide range of adaptations to environmental diversity. Linked to climatic changes, new economic strategies occurred mainly along the savanna belt.

During the Mid-Holocene (6,000-2,000 BC), episodes of dryness, which led to a depletion of hunting resources, compelled populations to turn to new ways of food subsistence (Clark, 1980). In fact, increasing population density has probably been a factor associated with the intensification of food procurement, especially in a deteriorating environment (Jousse, 2006; Brooks, 2006).

2.1.3.2.1. Complexity of the phenomenon related to environmental diversity

The appearance of new subsistence strategies was a complex phenomenon involving various factors and influences. For example, the origins of agriculture are related to both indigenous and imported species.

Furthermore demographic events related to population movements (e.g. Bantu-speakers expansion) also reshaped the distribution of human settlements on a continental scale.

All these factors are different from those in Europe, as the local African contribution was greater than the local one in Central Europe (Renfrew, 1999). Experimentations occurred locally with several different indigenous domesticates leading to different kinds of subsistence strategies (Harlan, 1993). Therefore, the "neolithic" in Africa does not imply a single "culture", as it is characterized by a very high diversity of environmental adaptations (Brooks, 2006; Jousse, 2006; Smith, 2005).

When both agriculture and sedentism spread gradually, the economy of the African continent differentiated into two main groups - cereal and root-crop cultivations, characteristic of temperate and tropical regions, respectively. However, this geographical and economical dichotomy did not imply a well-defined limit between the two types of "agricultural" regions. In fact, there was probably a greater degree of flexibility between various types of agriculture and other ways of subsistence (e.g. fishing, hunting and gathering, and pastoralism) (Garcea, 2004).

In various parts of Africa, well before the appearance of agriculture, specialist activities such as fishing linked to permanent water sources encouraged a (semi-) sedentary way of life. For

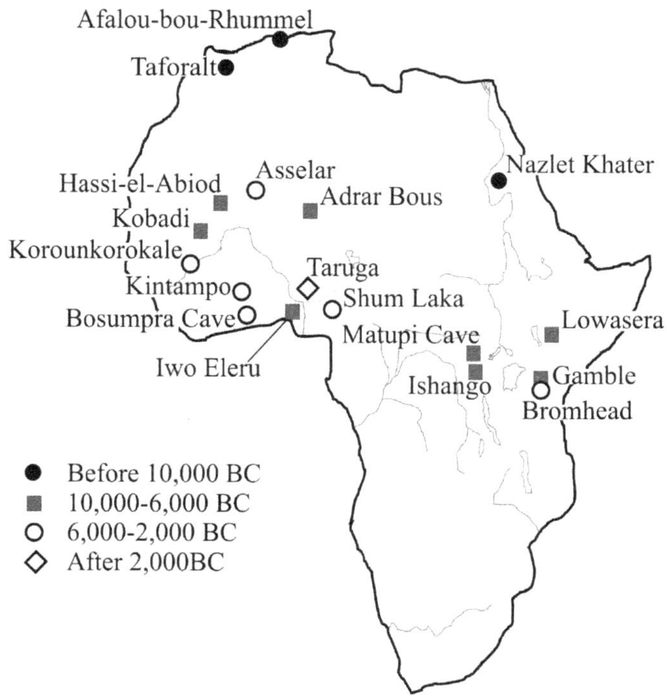

Figure 4. Distribution of a few Holocene and earlier sites within Africa.

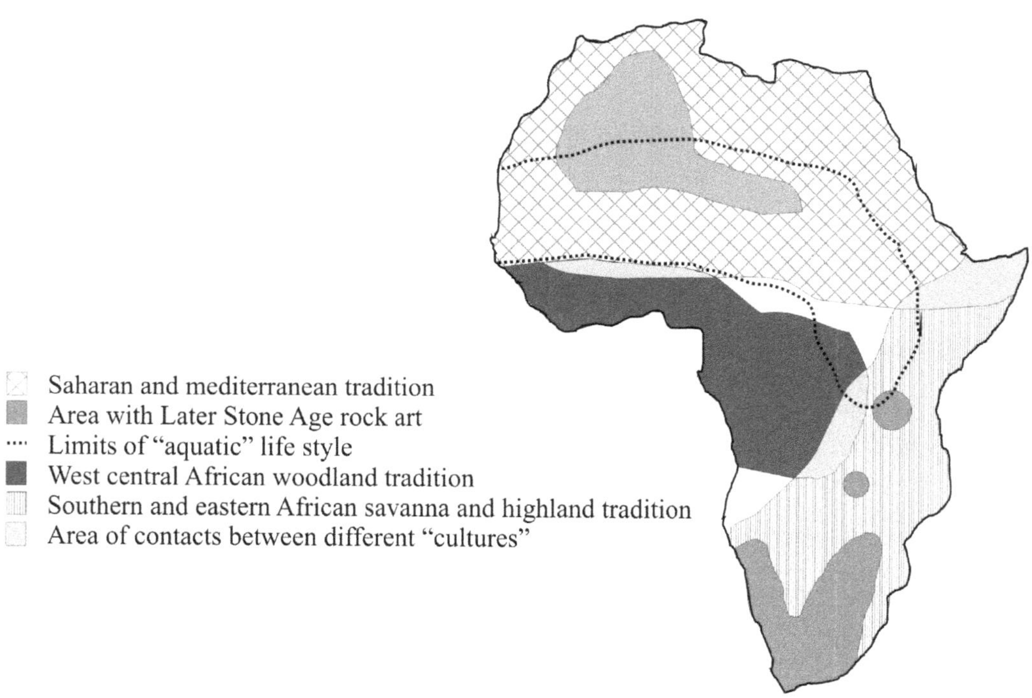

Figure 5. Distribution of key Later Stone Age "cultures" in Africa (after Newman, 1995, p. 32).

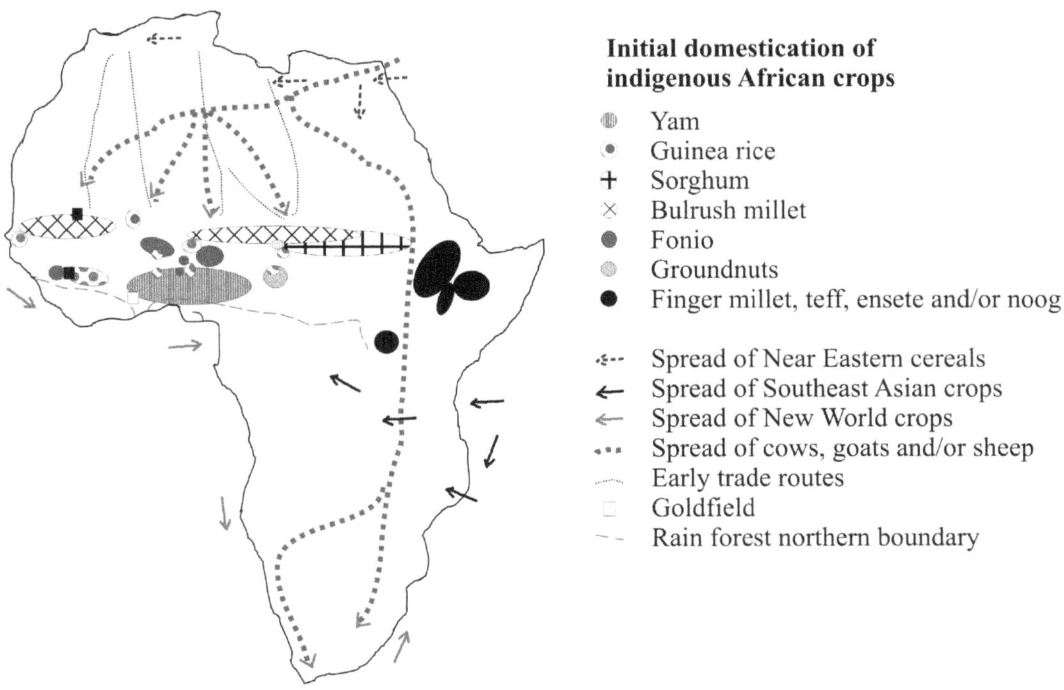

Figure 6. Spread of new ways of subsistence since *c.* 5,000 BC in Africa (line drawing after Harlan, 1971; Murray, 1998, p. 43).

example, the "aquatic" life style reached its peak in the Western Sahara between 6,000 and 2,000 BC (e.g. Asselar, Kobadi, Gobero) (Dutour, 1992; Georgeon *et al.,* 1992; Mayor *et al*., 2005; Sereno *et al.,* 2008). At *c.* 7,000 BC, it also appeared in East and Central Africa's Lake Region in sites such as Ishango (Brooks and Smith, 1987) and Lowasera (Phillipson, 1977a).

2.1.3.2.2. Importance of the savanna belt and persistence of hunter-gatherers

Domestication of both plants and animals began more or less simultaneously in the savanna belt (Sahel) between the Sahara and the equator (Figure 6, p. 11). During the seventh millennium, domesticated cattle was already present in the Sahara (Gautier, 1987; Muzzolini, 1993), and slightly later (6,000 BC) domesticated barley, wheat, sheep and goats were introduced from Asia into Northern Africa through the Nile Valley, and into Ethiopia (Wendorf and Schild, 1994; Bower, 1995; Marshall and Hildebrand, 2002).

As indigenous plant species (e.g. millet, sorghum and African rice) became more adapted to tropical environments, they began to appear along the savanna belt. Millet was first domesticated in the Sahara (6,000 BC), and both sorghum and African rice in Sudan (4,000 BC) (Andrea and Casey, 2002; Ballouche and Neumann, 1995; Oumar *et al.*, 2008).

Closer to the equator, root crops (African yam, palm oil tree) were better suited to the forest than grain crops, and these crops could also have been domesticated in the savanna according to Vansina (1990). However, as they have left no botanical evidence through time, no dates are available. Up to the Late Holocene, pastoralism coupled with grain gathering persisted as the main way of subsistence in West Africa.

Despite these new subsistence strategies, specialized fishing peoples and hunter-gatherers persisted in West and Central Africa throughout the Mid-Holocene (MacDonald, 1999; Mayor *et al.*, 2005). This fact is suggested by the presence of coastal settlements with a specific microlithic industry. As the latter was completely absent north of the savanna belt, it could suggest some demarcation of different ecotones reflecting in turn different ways of subsistence.

Furthermore, only between 3,500 and 2,000 BC, ceramic technology spread southward throughout the coastal and forested areas of West Africa, as is found in Bosumpra Cave (Ghana) (5,303±100 BP; Stahl, 1993). At the same time, horticulture (oil palm and tubers) may have intensified.

2.1.3.2.3. From Later Stone Age to Iron Age: fusion and/or expansion of cultures

Between the Later Stone Age and the Iron Age periods (*c.* 3,000 to 1,000 BC), episodes of dryness increased as well as demographic density in the peripheral zones of the Sahara (Muzzolini, 1993). These factors were probably in part responsible for various population movements. For example, south of the Sahara, Sahelian populations probably expanded southward and came into contact with indigenous hunter-gatherers and early horticulturalists.

According to MacDonald (1999), this "cultural" fusion and adaptation of populations from both the savanna and forest ecotones could be illustrated by the "Kintampo Complex". The latter is characterized by projectile points, microlithic quartz assemblages, the presence of livestock (Kintampo: 3,600-3,000 BC) (Stahl, 1985) and since the fourth millennium BP, cultivated pearl millet (D'Andrea and Casey, 2002).

Further south, the "proto"-Bantu-speaking agriculturists expanded into sub-Saharan Africa (Vansina, 1995). This phenomenon, which lasted more than 2,000 years, will be discussed later. At around 3,000-2,000 BC, populations corresponding hypothetically to "proto"-Berbers entered into the Central Sahara (Paris, 1990; Muzzolini, 1992). According to rock art depictions of rituals, they may share common features of nomadic pastoral life with modern Fulani groups (Smith, 1993). They might also have been influenced and/or absorbed by the latter.

In comparison to previous periods, the Iron Age corresponds to an intensification of fusion and/or expansion of cultures (Figure 7, p. 14). For example, the spread of metal working accelerated the spread of cultivation and pastoralism, and it also increased the trans-Saharan trade, especially after the introduction of the camel from Asia (1st century BC) (Newman, 1995).

As suggested by rock art, iron-smelting techniques obtained from North African pastoralists may have spread southward along trade routes. Iron south of the Sahara was first discovered in relation to the Nok culture in Nigeria (Taruga; 5th century BC) (Holl, 1993).

In West-Central Africa, food production was strengthened not only by the appearance of iron, but also the later adoption of other imported species originating from Asia (e.g. rice, sugar cane and lemon) and the Americas (corn and cassava) (Curtin *et al.*, 1995). As a consequence, a demographic explosion occurred, especially in sub-Saharan Africa. Large settlements of Bantu-speaking agriculturists displaced hunter-gatherer populations, who were driven back sometimes into remoter areas (Vansina, 1995).

2.1.4. The recent Holocene: an increasingly complex history

During the last two millennia, several indigenous chiefdoms and/or states emerged in West and Central Africa. Social changes were related to both increasing wealth and an appearance of social stratification (Curtin *et al.*, 1995).

2.1.4.1. Appearance and disappearance of earliest African "desert" states

Since the first millennium BC, especially along the western area of the savanna belt (Figure 8, p. 14), various states (e.g. Ghana, Songhay, Mali, Bornu-Kanem) appeared and disappeared successively. For example, Djenne-Djeno city (3rd century BC - 14th century AD) and the Ghana state (from 8th century AD) were both localized between the Sahara and the Niger river bend, as they were established in key areas for both food production (fishing, agriculture) and trans-Saharan trading (Kasule, 1998). In particular, the Ghana state that resulted from the coalition of both iron working and agricultural communities (Soninke) is also one of the earliest Iron Age communities of Western Africa, and its influence extended eastward up to Sudan.

The Late Iron Age economy of these "desert" states (12th-16th centuries AD), were closely linked to the development and control of trans-Saharan trade routes. Trading activities also increased significantly when Arabs arrived in sub-Saharan Africa (*c.* 8th century AD) (Newman, 1995). The coalition of several groups (e.g. Mali Empire) that were bound together by the adoption of Islam controlled numerous trading routes. Between the 12th and 16th centuries AD, they conquered other empires at the edge of the Sahara (Ghana and Songhay). Since the 12th century AD, Muslim pastoralists, such as the Fulani, also began an eastward expansion from their original home between the Senegal River Valley and eastern Guinea (Curtin, 1995; Newman, 1995).

2.1.4.2. "Forest" states in West and Central Africa

In West and Central Africa, "forest" states (Yoruba, Ibo, Benin, Asante and Dahomey) reflect different historical and economic backgrounds because of their geographical situation closely related to the equatorial forest and the Atlantic coast (Kasule, 1998; MacDonald, 1999) (Figure 8, p. 14). In Nigeria, they practised metallurgy and traded with other African and European states. For example, the Yoruba state (since the 9th century AD) probably corresponded to the earliest settlements of Ibo populations, who specialized in the production of bronze and copper artefacts. The centralized city-state of Benin, founded by Edo groups (14th-15th centuries AD), was also very successful in trade, exchanging products and slaves with both Portuguese and other Africans. After the 16th century AD, as the Atlantic coast became more accessible, the Asante and Dahomey states increased in both economic importance and political power. The power of these states was also affected by the decline of the Songhay state.

2.1.4.3. Appearance of other kingdoms in sub-Saharan Africa: expansion and/or intermingling of populations

In Central and East Africa, during and after the southward expansion of Bantu-speaking agriculturists, several kingdoms emerged as the result of social hierarchization of settled communities.

Early Iron Age settlements spread across equatorial Africa in areas particularly rich in mineral deposits. In the Katanga Province of D.R.C., the Luba kingdom (*c.* 15th century AD) developed its wealth by trading copper and salt (de Maret, 1977). In Southern Africa, the Great Zimbabwe kingdom whose origins are dated to the 10th century AD, reached greatest wealth around the 12th century AD because of its gold mines (Huffman, 2007; Kim and Kusimba, 2008).

In East Africa and the Horn of Africa, the Late Iron Age period (10th-16th centuries AD) corresponded to an increase in population movements not just related to the Bantu-speakers expansion. For example, the spread of various Nilo-Saharan-speakers from southern Sudan and Ethiopia was triggered by the arrival of displaced pastoralists. Some of these, such as the Tutsi and the Maasai, moved southward into the Great Lake region and northern Tanzania respectively. Other pastoralists such as Afro-Asiatic speakers (Somali) also moved southward from the northern part of the Horn to the coast of Somalia (Newman, 1995).

At the beginning of the 19th century AD, in both East and Central Africa, several centralized kingdoms and smaller chiefdoms were present. In the Great Lakes, they were based on either agriculture (Buganda, Bunyoro) or pastoralism (Haya, Ankore, Rwanda, and Burundi). These differed from each other not only in their economy but also in their political structure (Murdock, 1959).

2.1.4.4. Slave trade and European colonialism

During the last five centuries, two major processes took place in Africa - the slave trade and European colonialism.

Even before the Trans-Atlantic trade, the slave trade was already a secondary economical resource within Africa (Kasule, 1998; Alexander, 2001). However, it became an increasing source of profit when European traders settled various posts along the West African coast in the 15th century AD (Gore Island, Gold Coast). As a result, a dramatic demographic loss took place, as 15 million Africans were shipped aboard during four centuries (Eltis, 2001) (Figure 9, p. 15). The economy of the whole continent slowed down, and populations (Azande) were displaced on a regional level, as they tried to resist slavery (Lovejoy, 1983). The 19th century AD put an end to these "depopulations" with the abolition of the slave trade, and two freed-slave colonies (Sierra Leone and Liberia) were formed.

2.2. The Bantu-speakers expansion

The Bantu-speakers dispersal is one of the greatest population movements in the history of the world and a complex historical event which impacted all of sub-Saharan Africa (Diamond and Bellwood 2003). It is also part of much wider phenomena, both linguistic and cultural - the establishment of the whole linguistic phylum to which the "Bantu" groups belong, and the emergence of several food production strategies.

However, a single approach is not sufficient to understand the microevolutionary processes, which

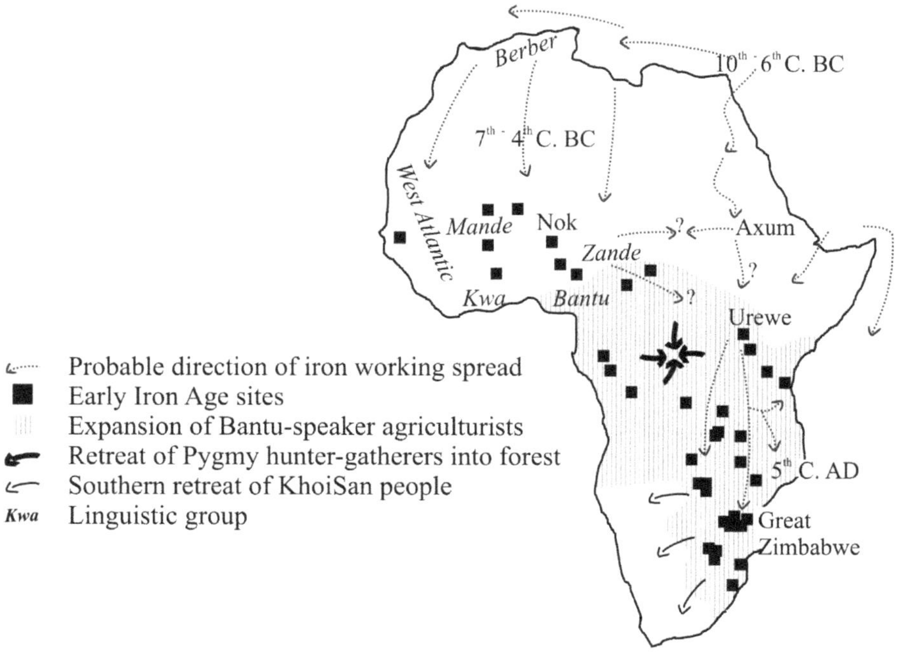

Figure 7. Spread of iron-working (since first millennium BC) in Africa (line drawing after Harlan, 1971; Murray, 1998, p. 43).

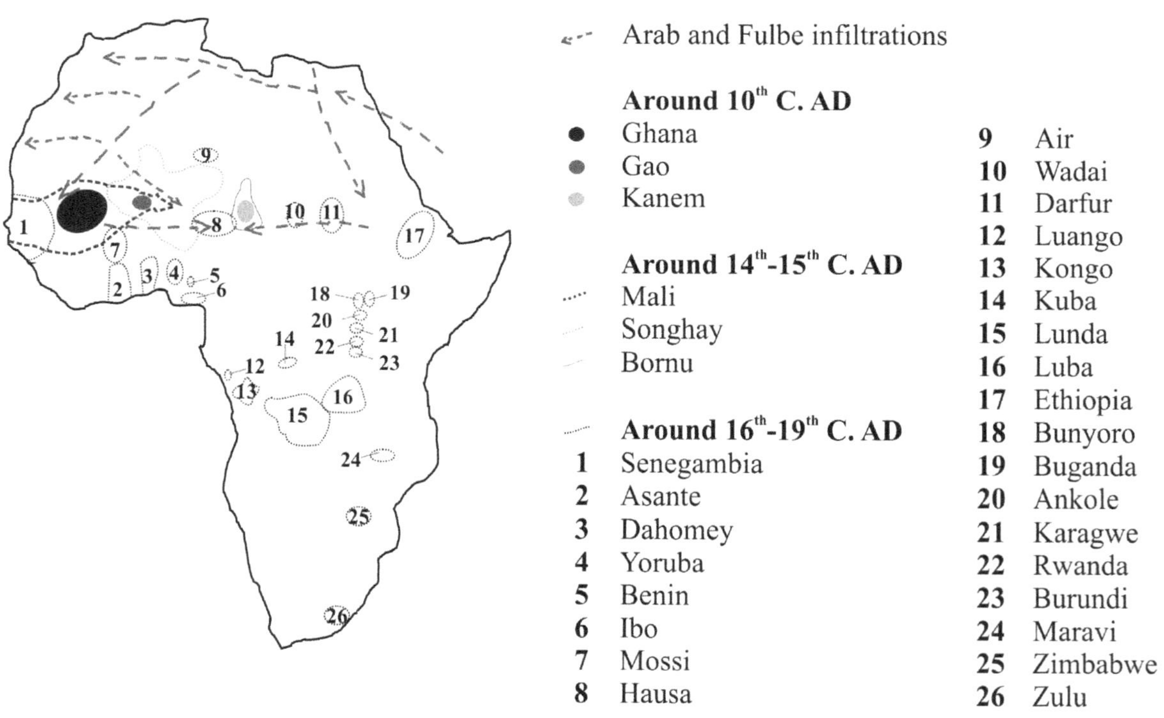

Figure 8. Distribution of various African states for the last two millennia (line drawing after Newman, 1995; Kasule, 1998, p. 79).

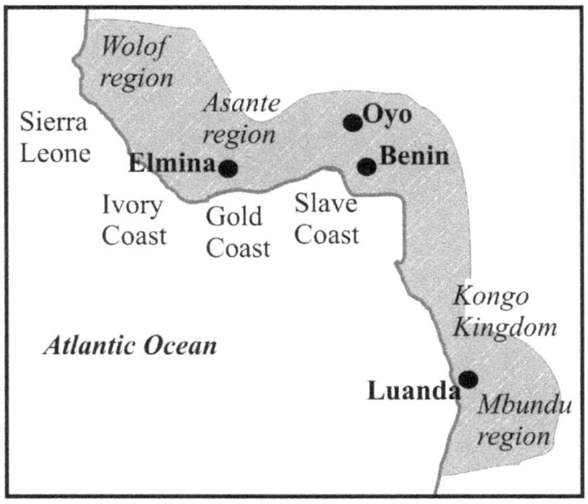

Figure 9. Regions affected by slave trade in West and Central Africa (16th-18th century AD) (line drawing after Oliver, 1991, p. 124).

shaped diversity of African populations today. This is why the "problematic" of the Bantu-speakers dispersal has been and still is presently an "arena", where Linguistics, Archaeology, Physical Anthropology and Genetics try to reconcile their findings (Beleza *et al.*, 2005; Berniell-Lee *et al.*, 2009; Phillipson, 1985; Robertson and Bradley, 2000).

The context in which the interests for the Bantu-speakers expansion arose is related to both historical and political factors (Pan-Africanism, Cheikh Anta Diop ideas) (Mandela, 1996; Holl, 1997a). As African identity was completely alienated culturally and economically by colonialism, it re-emerged even stronger in order to resist and argue against colonial systems.

During the 1970s, a new line of thinking also emerged, encouraging a multidisciplinary approach. Linguistics initially played an important role, as a first impulse to encourage links with other disciplines (e.g. Ehret, 1984, 1997, 1998). In the context of New Archaeology, the African continent appeared to be at the centre of the debate (Andah and Bagodo, 1993).

For a long time, African Archaeology was influenced by Eurocentric ideas and fascinated only with the history of Pharaonic Egypt. Early archaeologists completely neglected historical aspects of sub-Saharan Africa, such as the Bantu-speakers expansion. Since the 1970s, the cultural and biological diversity of sub-Saharan Africa, as well as its origins, have been intensively analysed (e.g. Hiernaux, 1974; Reed and Tishkoff, 2006).

2.2.1. A linguistic phenomenon

The term "bantu" is an entirely linguistic concept, grouping peoples into a cluster of closely related languages. Due to its linguistic diversity, a category (or single entity) of people cannot explain the phenomenon of the "Bantu" expansion or modern groups associated with "Bantu" speaking groups. While the term 'Bantu' has negative social connotations in Southern Africa, in this work, the word is used to describe the linguistic association of a variety of groups of people in Sub-Saharan Africa (Vansina, 1995; Diamond, 1997).

2.2.1.1. Present distribution of African languages

The Bantu languages belong to the largest linguistic phylum (Niger-Congo), as well as to the largest linguistic sub-set in Africa (Guthrie, 1962; Greenberg, 1964; Heine, 1980; Blench, 1997; Ehret, 2002a, 2002b) (Figure 10, p. 17). The Niger-Congo phylum is spoken from West to South Africa throughout most of two thirds of sub-Saharan Africa, and it has been ramified into several branches. The eastern group (Benue-Congo) includes the Bantoid languages from which the Bantu originated. The Kwa and Atlantic languages belong to the western group. In comparison to other branches, Bantu languages (Gikuyu, Rwanda, Rundi, Kongo, Luba Lulua, Swahili, Makua, Chewa Nyanja, Shona, Zulu and Xhosa) are widely distributed across the southern half of the continent. The Shona (Zimbabwe and Mozambique) represent the largest sub-group of native Bantu-speakers.

2.2.1.2. Diversity and relatively recent origin of Bantu languages

The Bantu languages are also relatively young in evolutionary terms (Blench, 1997). This has been shown through Lexicostatistics that use lexical variability to reconstruct the affinities of languages. For example, Bastin (1980), Bastin *et al.* (1983) and Coupez (1989) found that 300 to 600 Bantu languages correspond to a large continuum of dialects, which display small differences between them. Their studies suggest that language differentiation was a process of splits between dialects occurring simultaneously in a relatively short time (around the last 5,000 years). The meaning itself of the word "bantu" supports this fact: these languages were probably called "bantu" meaning "people", because of the frequent occurrence of this word in the whole group (Comrie *et al.*, 1997).

The nature of this linguistic phenomenon is closely linked to a "wave model" (Vansina, 1995). The process of language differentiation operated from an ancestral mother language, the so-called "proto"-Bantu that has been localized in North West Cameroon (Grassfields). Looking at the geographical coherence of the Bantu-speakers "world", it can be admitted intuitively that the spread of migrants was mainly southwards. The Bantu-speakers expansion was also probably at least a two stream phenomenon, the eastern one being more difficult to identify than the western one.

2.2.1.3. Reconstructing language history: lexicostatistics and various methodologies

Linguistic methodologies are mainly derived from lexicostatistics, as the latter developed quantitative methods. The pioneering work of Johnston (1919/1922) already reconstructed the Bantu-speakers migration from lexical variability, but its methodology was not well defined (Coupez, 1977).

In fact, a more systematic comparison of grammatical and lexical structures from language to language provided a quantifiable measure of linguistic affinity through an obtained percentage of total number of general roots. Using this method, Greenberg (1964) established a common index for the Bantu. He also determined a nuclear area or centre of origin like Guthrie (1962), who used the same methodology but on a smaller scale.

However, both Greenberg (1964) and Guthrie (1962) have been criticized by more recent studies. According to Ruhlen (1991), Greenberg used the misleading relationship of sound correspondences and genetic affiliation: word similarity was identified, but no distinction was made between basic words and borrowed words, the latter being evidence of contact between populations but not of linguistic affinity.

According to Heine (1980), Guthrie confused language groupings with genetic units, as linguistic data were too quickly extrapolated to groupings obtained from other fields. Nevertheless, Greenberg's classification of African languages, although still debated, is often used as a reference and starting point for various hypotheses.

The work of Bastin (1980) is an important contribution to African Linguistics, as it applied the lexicostatistics approach in a more critical manner and on a much larger sample of very well described languages. Basal vocabulary was used to establish genealogical links between Bantu languages. After having discarded borrowed elements, the proportions of conservation/innovation were analysed, using especially shared innovations to define sub-groups and date their splits (e.g. *c.* 86% conservation and 14% innovation within one millennium).

However, there are various methodological problems in Lexicostatistics, especially in relation to the "historical depth" of language classification. Firstly, the occurrence of a parallel innovation is possible and it does not imply genealogical relationship. Secondly, there is also no absolute dating (glottochronology), as the rate of linguistic change is not always constant (Coupez, 1989). Finally, although lexicostatistics can yield a "preliminary genetic classification" (Vansina, 1995) of the Bantu languages, the language trees should not be confused with population trees (Maho, 2001).

In addition, various methods (onomastic, vocabulary cross-comparisons) have also been used for the study of the Bantu languages, but without any statistical treatment (N'Goie-N'Galla, 1989). According to Blench (1993) and Williamson (1993), the most promising technique is the cross-comparison of words of cultural significance. Word roots, reconstructed from

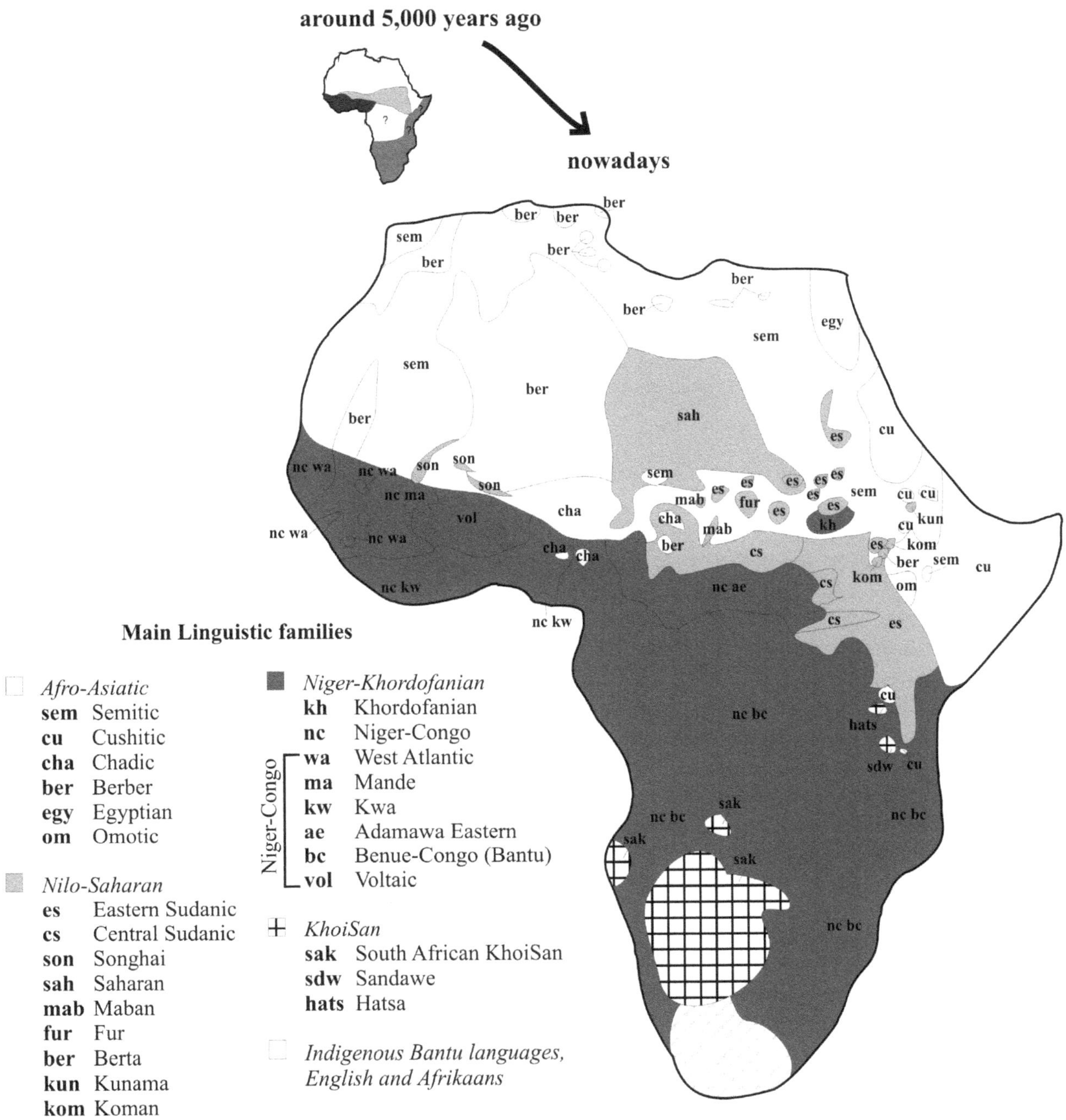

Figure 10. Distribution of major African linguistic families (line drawing after Greenberg, 1964; Murray, 1998, p. 26).

vocabulary analysis, can be used to show indices of early food production and to examine its complexity and diversity in Africa (Ehret, 1984, 1998, 2002a, 2002b). For example, Coupez (1989) observed a high abundance of agricultural vocabulary (e.g. yam, palm oil) not only in Bantu languages. Afro-Asiatic and Nilo-Saharan vocabulary also reflect a long acquaintance with food production (e.g. grain crops, herding).

The Bantu-speakers expansion appears not to be the earliest phenomenon related to food production. According to Vansina (1995), it must also have occurred after the introduction of pottery and farming, as the "proto"-Bantu linguistic reconstructions already identified the presence of vocabulary related to both pottery and cultivated root crops. These vocabulary observations could also date the first split of Bantu languages to *c.* 5,000 years ago.

De Maret and Nsuka (1977) also attempted to test

the question of whether there is any vocabulary evidence showing an association between Bantu-speakers diffusion and iron metallurgy. However, no convincing linguistic traces such as the presence of specific words ("metal" or "smelting") were found, and the diversity of terminology related to iron production was observed only in a few languages.

2.2.1.4. Expansion models

As the nature of the linguistic Bantu phenomenon is closely linked to the "wave model" (Vansina, 1995), it implies that changes did not occur within a stationary population.

Concerning both geographical origin and routes of the Bantu-speakers dispersal, there are two major linguistic models (Guthrie, 1962; Greenberg, 1964). These models do not differ much from each other, and from them several other, more detailed hypothetical scenarios have been derived (Bastin *et al.*, 1983; Heine, 1980) (Figure 11, p. 19).

The common ideas are, as follows: after a first differentiation and dispersal from a primary Bantu centre, there were several secondary centres of expansion from which western and eastern streams directed towards the south. Fissions then followed within each main of these main subdivisions.

First, Guthrie (1962) proposed that the spread of proto-Bantu languages started mainly in the south, between ancient Zaire and northern Zambia (Shaba region) (Figure 11a). He also envisaged a previous small migration of pre-Bantu-speakers from a region north of the forest close to Lake Chad (Central Cameroon and/or the Ubangi-Shari region). From the main "proto"-Bantu nucleus, radiating migrations would have occurred in all directions.

Second, Greenberg (1964), who partly disagreed with the previous model, localized the cradle of proto-Bantu languages close to the pre-Bantu centre of Guthrie between Nigeria and Cameroon (Figure 11b). This hypothesis is supported by other linguistic work developed in parallel as well as by archaeological finds, as will be discussed later.

In fact, Bastin *et al.* (1979, 1983) and Heine (1980) also determined a proto-Bantu nucleus, but more particularly in Cameroon (Grassfields), where the highest degree of "archaism" and diversity in Bantu languages is found. In agreement with Greenberg (1964), two lines of migration, a western and an eastern one, were also evident from the statistical trees (Holden, 2001; 2002; Rexova *et al.*, 2006). The western stream would have penetrated the forest and become highly diversified, probably in relation to the geographical conditions of forest life. Another stream would have expanded towards the east, following a route along the northern fringe of the forest.

The model of Bastin *et al.* (1983) viewed the two waves of migration as nearly simultaneous phenomena (Figure 11c). In contrast to the latter, Heine (1980) proposed a later eastern wave, originating from a secondary western Congolese centre (Figure 11d). This is supported by the fact that Eastern Bantu languages are less diverse than the West Central group, suggesting a more recent evolution (Heine, 1980; Ehret, 1997, 2002b; Holden, 2001; 2002).

Models of the exact route of the Bantu-speakers dispersal are not definitive, especially when the shape of linguistic trees change according to the number of included languages. However, most linguists, like archaeologists, agree that the Bantu-speakers expansion did not necessarily follow a regular pattern.

2.2.2. Archaeological approach and ethno-historical evidence

Although the archaeological approach has many methodological limitations, it can provide material and chronological evidence supporting the phenomenon of the Bantu-speakers expansion in terms of "culture" (Hodder, 1981). Other improved models of dispersal can be built, but now with the advantage of being confronted with both linguistic and paleoclimatological data.

The phenomenon of expanding "sedentism" is closely related to the Bantu-speakers dispersal (de Maret, 1980; Clist, 1991). However, it cannot be defined by a single aspect and a single geographical area in Africa, as food production is very diverse, as well as its emergence.

2.2.2.1. Advantages and limitations of archaeological approach

In contrast to Linguistics, Archaeology can provide an absolute chronological background: once mapped, radiocarbon dates can reflect past human occupation on both regional and continental scales (de Maret, 1985). However, the major obstacle

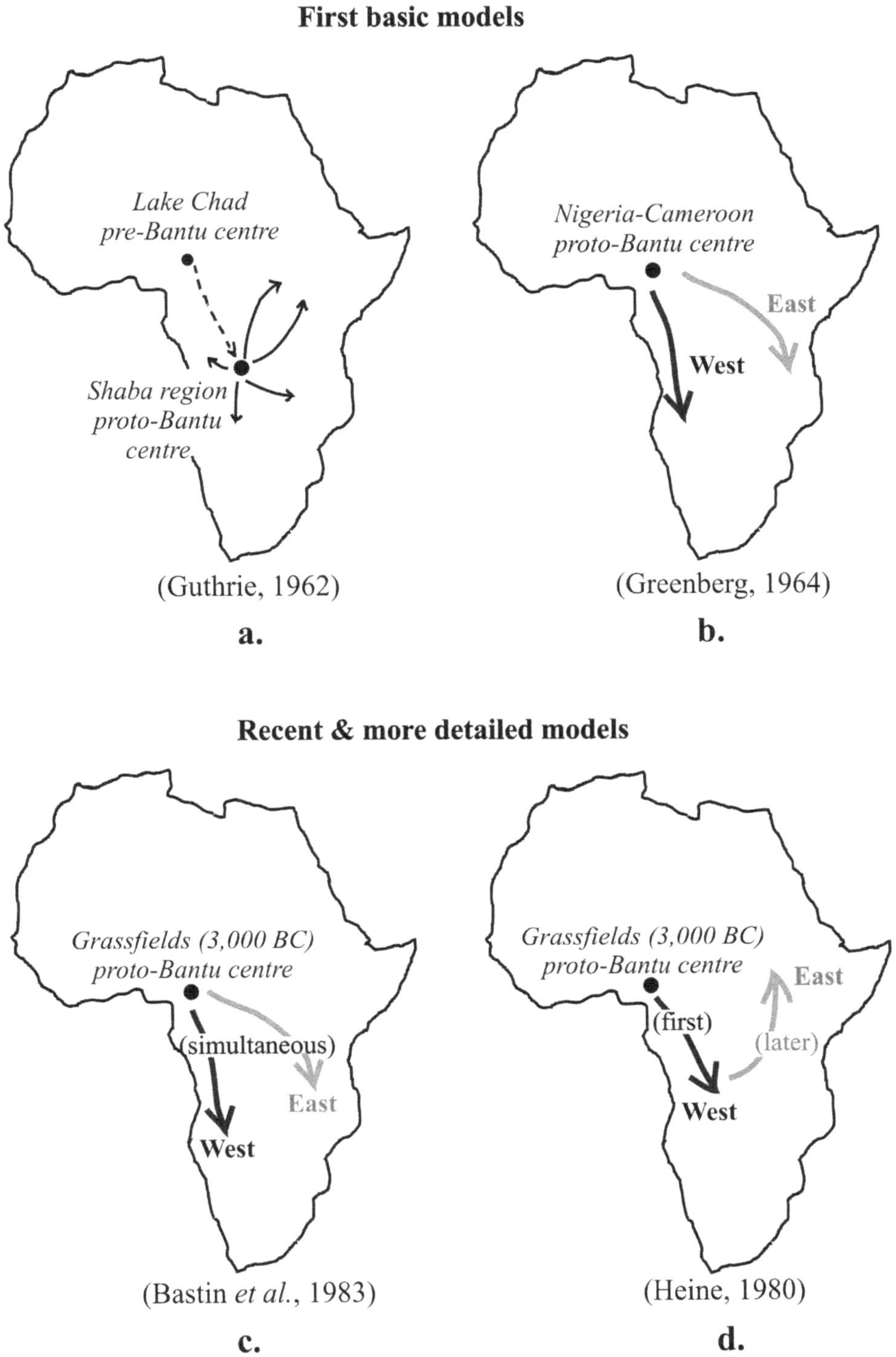

Figure 11. Major hypotheses about the Bantu-speakers expansion according to linguistic data.

concerning the Holocene history of West and Central Africa is related to the fact that the distribution of sites with human remains is quite biased (Figure 12, p. 20).

Organic materials in tropical areas do not preserve well, and so far there have been too few initiatives of extensive excavations in those areas (Clist, 1991). Therefore, in contrast to East Africa, the search for archaeological traces, especially in the equatorial forest, is relatively recent, although West Central Africa is a critical area for the origin of the Bantu-speakers (Cahen, 1978; Van Noten, 1982; Eggert, 1997; Ehret, 1972, 1973, 2001).

Furthermore, in Archaeology, the meaning of the word "bantu" is based on many assumptions (de Maret, 1989). For example, it is not possible to

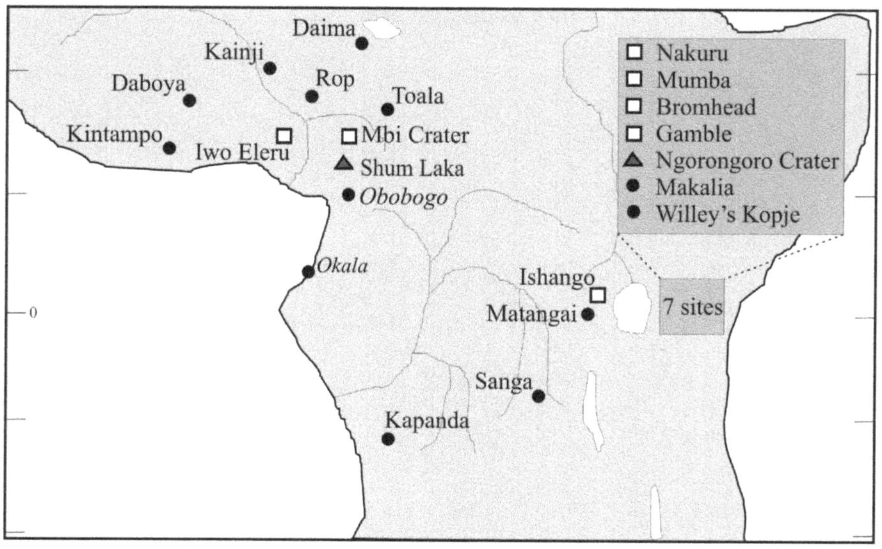

□ Later Stone Age (LSA)
▲ LSA-Iron Age
● Iron Age
Site with no human remains

West and Central African sites (from oldest to most recent ones)
Iwo Eleru (Brothwell & Shaw, 1971)
Mbi Crater (Asombang, 1988)
Ishango (Boaz *et al.*, 1990)
Shum Laka (de Maret *et al.*, 1993)
Kainji (Hartle, 1970)
Sanga (de Maret, 1985)
Toala (Menard, 1983)
Daima (Connah, 1976)
Kapanda (Gutierrez & Valentin, 1995)
Daboya (Shinnie & Kense, 1989)
Okala (de Maret, 1982)
Kintampo (Stahl, 1985)
Obobogo (de Maret, 1982, 1985)
Matangai (Mercader *et al.*, 2001)

East Africa (from oldest to most recent ones)
Gamble (Leakey, 1935)
Mumba (Brauer, 1980)
Bromhead (Protsch, 1978)
Nakuru (Rightmire, 1975)
Ngorongoro Crater (Magory & Day, 1994)
Makalia (Leakey, 1935)
Willey's Kopje (Leakey, 1935)

Figure 12. Later Stone Age and/or Iron Age sites in Equatorial Africa (line drawing after Orban *et al.*, 1996, p. 214).

relate an artefact to an ethnic affiliation or a linguistic group. However, according to archaeological evidence, the "cultural" meaning of the word "bantu" is closely related to the process of sedentism. Once settled for a long-term period, people left many artefacts that are related to the beginning of food production on a wider area (Andah and Bagodo, 1993).

2.2.2.2. Evidence of sedentism: dates, sites and artefacts

Since the transitional period of the Later Stone Age to Iron Age (*c.* 4,000-1,000 BC), an increasing number of sites with traces of sedentism (pottery, domesticated plants and animals, later iron production) have been observed in Africa (de Maret, 1989; Phillipson, 1994; Vansina, 1995). In fact, this phenomenon in sub-Saharan Africa has approximately the same time-depth as the linguistic Bantu expansion (the last 5,000 years). Therefore, archaeologists have inferred that the cultural meaning of the word "bantu" is closely related to the emergence and diversification of various kinds of food production (horticulture, agriculture and herding).

Archaeological evidence is not the same between West and East Africa, as they probably reflect different environmental conditions (Cornevin, 1993). In Eastern African savannas and North West Africa, grain crops and domesticated animals could have appeared earlier in time (*c.* 6,000 BC; Nabta Playa) and prior to the Bantu-speakers expansion (Wendorf *et al.*, 1984). In the forests of West Central Africa, artefacts, such as polished stone tools, grinding stones and palm oil nuts, appeared around late first millennium BC (Lavachery, 1996; Lavachery *et al.,* 1996) as well as banana cultivation and animal husbandry (Mbida *et al.*, 2000).

In Cameroon (Obobogo) and Gabon (Okala), traces of village compounds with refuse pits, pottery, extensive domestic grinding equipment, animal husbandry and evidence of horticulture have already been discovered *c.* 1,500 BC, (de Maret, 1982, 1985; Clist, 1987). These Later Stone Age – Iron Age sites are also very important for the archaeology of West Central Africa, as they can add supporting evidence for the hypothesis of a linguistic "proto-Bantu cradle" in the Grassfields (Bastin *et al.,* 1983).

Therefore, are Bantu-speaking populations synonymous to sedentism, as is suggested by archaeological evidence? It is probably easier to answer this question for early "proto"-Bantu-speaking people, prior to subsequent expansion and diversification.

2.2.2.2.1. Pottery: earliest indication of (semi-) sedentism

The presence of pottery is one of the main indices of the beginning of (semi-) sedentism, as it is related to the storage and control of food resources. In West Central Africa, the earliest evidence of pottery has been dated to later periods than in the Sahara (before 9,400 BC; Huysecom *et al.*, 2009): *c.* 4,100 BC in Ghana (Kintampo) (Stahl, 1985; 1993) and *c.* 1,000 BC in Gabon (Okala) (Eggert, 1993; Vogel, 1997). This implies that the Bantu-speakers dispersal probably occurred after the manufacture of pottery had developed indigenously.

In East Africa, the first wave of the earliest Iron Age Bantu-speakers has been correlated with the presence of the Urewe ware that appeared *c.* 500 BC (Phillipson, 1977b, 1985; Stahl, 1993). Archaeologists tried to understand the diversity in pottery styles in these regions, showing affinities to a common Urewe "culture". However, they were also aware that in this kind of typological study there is always the risk to both underestimate the regional variability of pottery styles and to result in an excess of simplification in the cultural meaning of the term "bantu" (Ogoti Kiriama, 1993).

2.2.2.2.2. Stone tools: a common range but differing proportions through Holocene

Stone tools have also to be analysed with caution, as there is not a single type associated with the "Bantu culture". A lithic industry cannot be directly associated with a linguistic group. The differences in lithic patterns reflect rather differences in patterns of subsistence. For example, the Tshitolian industry (12,000-2,000 BP), which is related to past hunter-gatherers is proof of long occupation in the forests of Central Africa (Eggert, 1993; Newman, 1995). In fact, it is composed of hoes and picks, which were, and remain to be, mainly used to collect roots from the ground (Cornelissen, 2002).

Through the Holocene in sub-Saharan Africa, various groups of "cultures" have been characterized by a common range of functional activities (hunting, fishing and/or collecting). Their lithic assemblages or "techno-complex" differed mainly in the proportions of artefacts, although the latter belonged to the same general families (microlithism, macrolithism) (Holl, 1997b).

According to Clist (1991), polished tools are an example of confusion with the word "bantu": their presence is not strictly related to a period of sedentism. Polished tools already appeared in the Later Stone Age and before the Bantu-speakers expansion, and secondly, in different contexts localized south and north of the forest.

In West Central Africa, the site of Shum Laka in the Grassfields (Cameroon) illustrates technological changes that occurred *c.* 3,000-1,000 BC (de Maret, 1990). The presence of a macrolithic industry of basalt (e.g. hoes, picks) increased at the expense of a microlithic industry of quartz (Cornelissen, 1997). This phenomenon may reflect a subsistence shift in West Central Africa, where horticultural practices using hoes and picks intensively developed in association with hunting.

2.2.2.2.3. Lack of botanical proof

According to the evidence from stone tools, the early proto-Bantu-speakers were probably horticulturists (yam growers), but afterwards food production diversified into a variety of subsistence strategies that do not allow reference to a common or single "Bantu" system of food production.

Direct evidence of food production is rare in West

Central Africa, as horticultural remains are highly biodegradable. There is at present no botanical proof for domesticated root crops, which represent an important supply of carbohydrates for forest fringe populations. Yam cultivation is only suggested by the presence of tools (e.g. macrolithic industry) that were used to collect them. However, evidence of intensive oil-palm exploitation was found at Shum Laka (Cameroon) *c.* 1,000 BC (de Maret, 1990).

In Africa, there is substantial evidence for a multiple origin of plant domestication, as it appeared in several areas according to the needs of each environment (Shaw, 1980; Harlan, 1993). For example, in a single country such as Nigeria, there were probably various times and places of food production origin, as suggested by two sites: in the south (Rop), polished tools and palm oil nuts were found, dating to *c.* 25 BC (Fagg, 1972), and in the north (Daima), cattle pastoralism was probably practised *c.* 600 BC (Connah, 1976). However, before the Bantu-speakers expansion, it is unknown how and to what extent these food production strategies interacted with each other.

2.2.2.2.4. Later appearance of iron

According to de Maret (1980), there is no reason to link the Iron Age and the Bantu-speakers expansion, as the appearance of iron production is archaeologically not represented by the expansion of (semi-) sedentary food producers. Traces of iron are present in Cameroon later (*c.* 8th century BC) than the first traces of sedentism (de Maret, 1985; Essomba, 1992). They are slightly older than those found in Nigeria (*c.* 5th century BC; Holl, 1993) but more or less contemporaneous to those found in East Africa (*c.* 7th century BC; Phillipson, 1994).

During the last 2,000 years, there are numerous sites with traces of both sedentism and iron throughout sub-Saharan Africa. They suggest that iron was a factor in speeding up the Bantu-speakers expansion, especially during the first millennium AD. For example, it is believed to have encouraged sedentary settlements in new empty areas, such as along the northern fringe of the equatorial forest (Toala, Central African Republic) (Vidal *et al.*, 1983).

2.2.2.3. Additional evidence from ethno-linguistic data: Murdock's approach

Murdock (1959), who analysed the present distribution of food production in Africa, also provided further information about the origin of the Bantu-speakers expansion. Although a diversity of subsistence strategies was revealed, the dispersal as a large-scale phenomenon could be deduced intuitively from the "apparent" homogeneity of Bantu-speaking agriculturists living mainly in sub-Saharan Africa.

The author subdivided the African continent into several main "cultural" provinces that group together peoples of essentially identical language and "culture" (Figure 13, p. 24). Murdock's classification (1959) therefore provided a summary of environmental and ethnographical data. Although Murdock also approached various other ethnographical aspects (kinship, housing, social and political organization), food-producing activities and linguistic affiliation were the key elements of his research.

According to the present distribution of indigenous cultivated plants, Murdock (1959) also suggested various hypotheses about the origin of different subsistence economies in Africa. For example, yam exploitation in West Central Africa and along the northern fringe of the forest seemed to be not only a long-term process, but also an independent one from cereal cultivation in the savanna of East Africa.

Among the seven "cultural" provinces of Murdock (1959, 1981), two of them (the "Yam Belt" and the "Bantu") are related to the Bantu-speakers expansion in both its initial and later phases respectively. They are also the most geographically unified provinces, like the "Western Sudan" and the "East African pastoralists" provinces.

Firstly, the "Yam Belt" province could have been initially a potential area for the proto-Bantu-speaking horticulturalists. Geographically, it extends from Liberia in the West up to the northern limit of the interlacustrine area in the East. On this long narrow strip of savanna, people presently cultivate bananas, yam, taros and a few other plants, such as palm oil and various crops.

Murdock (1959) considered this area as an "uninterrupted corridor of agricultural peoples", who adapted to a savanna environment at the edge of the forest. The province is subdivided into six groups, and it corresponds to peoples that speak Niger-Congo languages such as the Kwa in West Africa (Ashanti) or the Adamawa Eastern towards the East (Azande) (Olson, 1996; Yakan, 1999).

Secondly, the "Bantu Africa" province is the largest group, as it probably resulted from the later expansion and diversification of all Benue-Congo or Bantu-speaking populations in a large part of sub-Saharan Africa. At least 16 groups can be subdivided within this expansion. These groups include: the "Cameroon highlands" has the particularity in belonging also to the "Yam Belt" province, although most of their "ethnic" groups (Bamum, Ndu) speak Benue-Congo languages. Nowadays, various plants are cultivated in this province, as agriculturists expanded into different environments. In the north-west area (N'Komi), bananas, yam and taro are used for agriculture. In the central region along the Equator (Basoko), fishing and boat trade are also practised in addition to agriculture (Olson, 1996; Yakan, 1999).

Other "Central Bantu", as well as the "Mongo" and "Luba" groups (Basuku, Batetela), sometimes use crops such as millet, sorghum and fonio from the Sudan. In the "Kenya highlands" and "Interlacustrine" groups, Bantu-speakers (Wapare, Haya) cultivate cereal grains. In addition, amongst the southern Bantu-speaking groups, like the "Nguni" (Xhosa, Zulu), animal husbandry is as important as agriculture (Olson, 1996; Yakan, 1999).

Murdock's classification (1959) is far from perfect, as it has been neither updated by very recent historical developments, nor does it include all African populations. However, it is considered so far as one of the best tools for understanding present ethno-linguistic diversity in Africa.

In addition, when describing present and increasingly complex economies, he also took account of the impact of imported plants from both Asia (e.g. sweet potato, peanuts) and the Americas (e.g. maize, manioc).

2.2.3. Linking archaeology to other fields

This section attempts to provide a preliminary summary of all data regarding the Bantu-speakers expansion that has been obtained from Linguistics, Paleoclimatology, Archaeology and Ethnography. The comparison of these with biological data will be treated separately, as it involves a different methodological approach. Only certain demographic data and epidemiological observations that have been used by archaeologists, are briefly discussed here (Clark, 1980; Van Bakel, 1981).

2.2.3.1. Broad consensus on the Bantu-speakers expansion

Although both the geographical origin and routes of the Bantu-speakers dispersal are still debated, the common main lines of consensus are the following:

- after a first differentiation and dispersal from a "proto"-Bantu centre in the Grassfields (*c.* 3,000-1,000 BC); there were several secondary centres of expansion from which western and eastern streams moved south, followed by dividing within each main subdivision.
- according to the archaeological data above, these "proto"-Bantu-speakers who lived north of the forest and expanded southward via clearings and river valleys, were village-dwellers; they cultivated yams and vegetable crops, using both axe/hoes and pottery, and they probably also raised sheep and goats.
- the first trajectory must have occurred through the forest belt and/or along the latter; however, there were probably two streams (western, eastern) of dispersal reflecting two major ecological zones (forest, savanna), and different hypotheses about the earliest stream have often been proposed.

2.2.3.2. Interactions between culture and climate: ecological expansion models

In the light of paleoclimatological data, two ecological models for the Bantu-speakers expansion have been proposed. They allow us to specify the ecological implications of this dispersal.

According to the first model (Clark, 1980), the Bantu-speakers dispersal was initiated by a population that was not in equilibrium with its environment and had both food production shortages and an increasing population density. These conditions could have occurred during the dry phase *c.* 2,000-1,000 BC (Muzzolini, 1993). In order to set up a symbiotic relationship between people and environment without any competition, Bantu-speaking populations had to move away from "repulsive" areas.

However, the dispersal might have been confronted by numerous ecological barriers, such as the limit of summer rainfall for which cereal cultivation would not be possible in East Africa

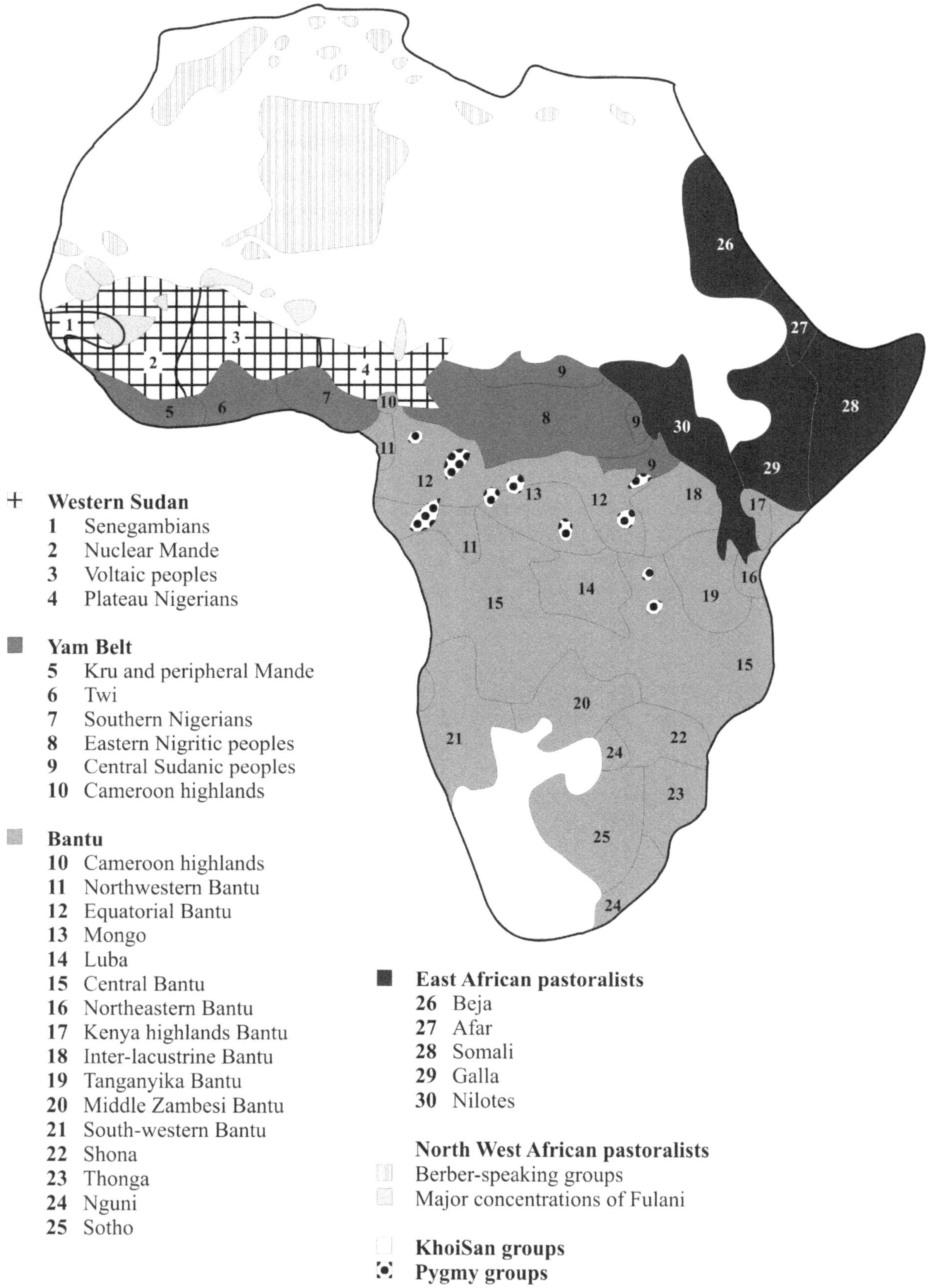

Figure 13. Subdivision of Africa into "cultural" provinces according to present food productions (line drawing after Murdock, 1959, p. 2-15).

(Phillipson, 1994). In addition, cattle rearing, which was adopted by Bantu-speaking agriculturists from pastoralists, could spread only into specific areas (free of the tsetse fly, far from lowlands during wet seasons, among populations without lactose intolerance).

The second model was proposed by David (1980, 1982), who summarised the dispersal according to the six following conditions:

i) a single source area,
ii) only one adaptive shift,
iii) no competition with other food producers,
iv) sufficient time,
iv) natural diffusion routes, and
vi) a successful technology among Bantu-speakers.

The author clarified even more than previous models, adding another fact that was supported by archaeological evidence: the zones at the edge of the forest (Sahelian and Sudanic areas) were probably the first ones to be exploited intensively by Bantu-speaking populations. They transformed their ecosystem, as the savanna areas increased due to intensive forest clearing and fires. Therefore, the Bantu-speakers expansion is considered not only as a movement of people looking for better "ecological" conditions, but also as a transformation of the environment occurring *in situ* wherever the dispersal was progressing.

However, can we also interpret these ecological models in terms of demography? According to Van Bakel (1981), a rapid expansion, whatever the size of the initial population, can be explained by a moderate growth rate of 1% per year. It did not reach over-population, as population growth rate was probably decreasing in later waves of demic expansion. Over-population was never reached, as when one area was exhausted beyond the level of carrying capacity, the groups segmented and colonized new areas (Vogel, 1997).

Therefore, the Bantu-speakers dispersal may have been a very slow continuous process in terms of distance and time. This fact is also suggested by Clist (1991), who estimated an annual speed *c.* 1.2 km per annum, as Eastern Bantu-speaking peoples reached South Africa by the 4th century AD. In addition, this fact is supported by Vansina's opinion (1995): the Bantu-speakers expansion was not a giant wave initially only triggered by over-population and climate change, but it was rather a dispersal, which proceeded irregularly, with reverses and advances, although the causal factors cited above must have been the same.

2.2.3.3. Increasing complexity in time and space: synthetic expansion models

In order to explore the limits of the possible interpretations about the Bantu-speakers dispersal, the major routes of the latter have also been debated by various researchers (Figure 14, p. 26).

Firstly, in agreement with Heine (1980), Phillipson (1977b, 1985) favoured an "anti-clockwise" model (Figure 14a): the western Bantu-speakers wave was most probably the earlier one, as there are too few archaeological traces along the northern fringe of the forest between the North-West region of Cameroon and the Eastern Lake Region. For example, along this area, which corresponds to a distance of approximately 3,000 km, only the megaliths of Bouar in the Central African Republic may correspond to the traces of proto-Bantu-speakers migrating eastward before later waves of Ubangi-speaking populations (Vidal, 1992). Other hypotheses could also favour the fact that the eastward spread of the latter may have occurred in parallel to the Bantu-speakers expansion (Ehret, 1997, 2001). Therefore, the hypothetical former traces of the eastern Bantu-speakers branch (between Nigeria and Great Lake regions) may have been obliterated by subsequent speakers belonging to the Adamawa Eastern branch of the Niger-Congo family (Heine, 1980).

Secondly, a "clockwise" model with a first expansion towards the east has also been proposed by some archaeologists (de Maret, 1980, 1989), as it does not imply a change in environment and therefore in subsistence pattern (Figure 14b). However, linguists such as Heine (1980) disagree with this hypothesis, which does not explain the absence of Bantu linguistic traces along the northern fringe of the forest.

Finally, in agreement with the model of Bastin *et al.* (1983), Vansina (1995) proposed another kind of ecological expansion model, mainly underlining the importance of riverine systems (Figure 14c).

The latter are ubiquitous in the forest and probably channelled the two streams of migration in a more or less simultaneous manner. Core areas were localized for the western stream in the northern Congo Republic, and for the eastern stream at the eastern confluence of Ubangi-Congo. After a first expansion of thinly scattered sites, a thickening of

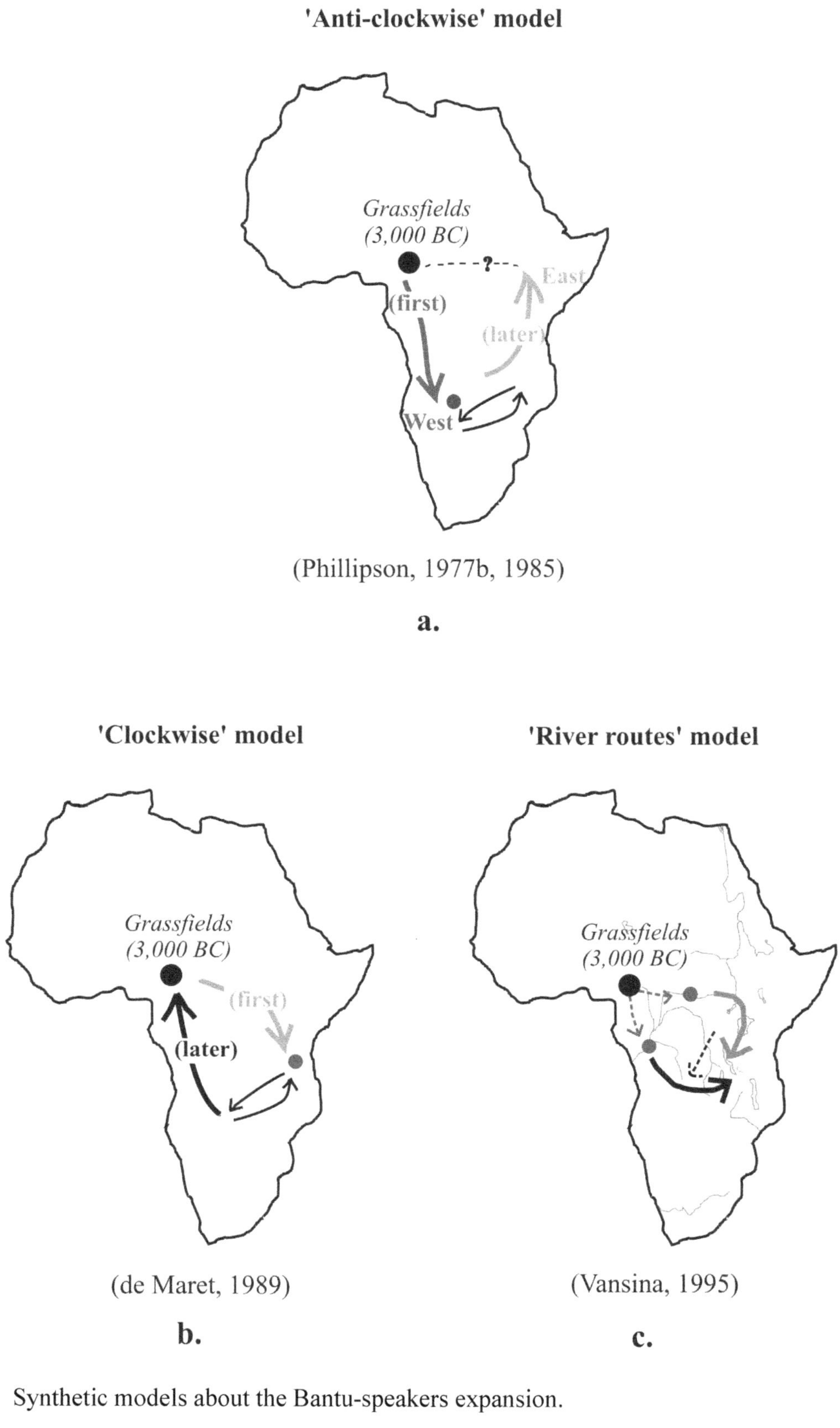

Figure 14. Synthetic models about the Bantu-speakers expansion.

settlements occurred in relation to the introduction of iron, new crops and cattle herding.

Human adaptation of Bantu-speakers in Central Africa seemed therefore to be archaic as their proto-Bantu vocabulary included words that relate to general transportation in a forest environment (e.g. "pirogue") (Guthrie, 1967-1971).

The expansion became even more complex during

the Iron Age. For example, the Bantu-speaking populations from the east filtered back to the west.

In a first phase of still ongoing debates, linguists and archaeologists focused mainly on the question of the exact trajectory of the Bantu-speakers dispersal. Models of expansion are never definitive, as they change or are clarified by further data. Although, there is a consensus nowadays about at least the geographic origin of the Bantu-speakers expansion, there are still some assumptions about the "cultural" meaning of the term "Bantu".

As suggested by Vansina (1995), whatever was the direction of the wave, "anti-clockwise" or "clockwise", we have also to focus on other related factors, such as the interactions between Bantu- and non-Bantu-speaking populations. Before the expansion, the indigenous populations of West Central and East Africa were probably even more diverse, both linguistically and culturally, than the present descendants, who later had been in part acculturated and/or replaced. After the expansion, all of sub-Saharan Africa became broadly united linguistically. However, we cannot view the Bantu-speakers dispersal as a simple cultural homogenization, as "culture" involves ecological parameters, which are extremely diverse in Africa.

2.3. Evidence from Physical Anthropology and Genetics

In order to understand the Holocene peopling of West and Central Africa, the present section focuses on the evidence obtained from both physical anthropology and genetics. These are logical following steps after having approached the "cultural" aspects of African populations. Biological diversity is the key element of our review and discussion, raising questions especially in relation to microevolutionary processes (Campbell and Tishkoff, 2010; Reed and Tishkoff, 2006). Therefore, the Bantu "problematic" is approached here in biological terms within a broader context of studies on population history.

2.3.1. Re-birth of Physical Anthropology

For a long time, anthropological conclusions especially in craniometric studies were masked by "racial" and typological interpretations, which were at a peak before the Second World War (Montandon, 1928; Kappers, 1936; Herskovits, 1937; Gebara, 1947). Biology and linguistics can too often be intermixed with "racial" interpretations as criticized by recent literature (MacEachern, 2000, 2006, 2008). Initially, physical anthropology fell into the typology of biological categories. However, later the Neo-Darwinian approach developed concepts about between- and within- population variability in relation to both genetic inheritance and phenetical plasticity (Mayr, 1974). Within this context of the re-birth of biological concepts, craniometric approaches, as well as all biological studies in general also took a new direction of research.

2.3.1.1. Updated concepts: population, ethnicity and complexity of migration

It is first necessary to discuss the definition of key concepts that will be used frequently during the present work, such as "population", "ethnic" group and "migration" event.

According to the biologist Mayr (1974), a local "population" is a community of individuals with potential panmixia or random interbreeding and exposed to selection in a particular place. The people can share the same gene pool, but the extent of sharing varies greatly sometimes, and it is difficult to quantify these differences. Therefore, intra-population variation due to individual variability is the key element of human clinal biological variation, as levels of inter-population variation are very low on a worldwide scale.

As also suggested by Mayr (1974), the definition of a "population" proposed by Hiernaux (1955a, 1974) is based mainly on the endogamy notion that is in turn related to various geographical levels (village, region). In sub-Saharan Africa, many "ethnic" groups have a high degree of endogamy, thus they should represent a suitable basic unit to use when analysing biological diversity. However, the delineation of a breeding population is often difficult, as it can also change through time. This is the case when some ethnic groups are conquered and absorbed by other peoples, such as the Kwanda by the Lozi in Zambia (Giles *et al.*, 1997a).

"Ethnic" identities are not related to nationhood, as the boundaries of countries in Africa have been defined very artificially during the colonial periods. For example, Nigeria and Cameroon share a high number of common ethnic groups (Ekoi, Boki, Anyang) in a region where the highest levels of ethno-linguistic diversity has been also observed (South Eastern Nigeria, South Western Cameroon; Figure 13, p. 24) (Murdock, 1959; Olson, 1996).

So far, "ethnic" units seem the best way of defining a "population" when describing the "cultural" diversity within Africans (Murdock, 1959).

However, when applying the notion of "population" to the past, a paleoanthropological approach has to make more assumptions, as the historical context is not so well known. Are the human skeletons excavated on a site the reflection of a group of people sharing similar features of not only behaviour but also biology? In addition, did their breeding vary from local to larger groups? For both past and present "populations", the choice of units of study will depend on the problem and geographical scale investigated (regional, continental or worldwide).

After having defined the primary units of analysis, it is important here to raise the following problem: what are the main factors in changing the structure of a population? In replying to this question, the concept of migration has often been one of the primary objects of research in microevolutionary studies (Johnston, 1913; Kappers, 1936). However, since a "retreat from migrationism" (Brothwell, 1963; Creighton, 1966), other phenomena appeared essential to understand biological diversity. The complexity of a migration event that is synonymous with gene flow, was also underlined, as it can be associated with several evolutionary mechanisms (founder effect, admixture, drift and/or selection) (Hiernaux, 1965a; Fix, 1999; Relethford, 2001a, 2001b; Campbell and Tishkoff, 2010).

Although subsequent levels of diversity can give an indication of the size of the founder population, several other aspects remain to be specified. What is the geographic origin of the immigrants? Did the latter invade a new territory either continuously or irregularly through space and time?

Recent methods for establishing proof of population expansion have been based on both "cultural" and biological processes and "causal" models have been built in a theoretical manner (Renfrew, 1992; Barbujani *et al.*, 1995). They often explain a population movement into a new location as resulting from an "economic" reason, such as in search of better land (Lasker and Mascie-Taylor, 1988; Weiss, 1988).

First, the "convergence" model of Renfrew (1992) for example, has been used to explain the pre-state agricultural societies on a worldwide scale. A low change in gene frequencies will result from a demic diffusion involving gene flow over a large area. In order to fulfil these conditions, the territory to be invaded must be nearly empty and previous inhabitants acculturated progressively by linguistic absorption and/or interbreeding.

A second example of Renfrew (1992), the "divergence" model explains a different situation: groups split into several units, leading towards a change in gene frequency (through genetic drift) and linguistic features. In addition, this can also be associated with a founder effect phenomenon.

2.3.1.2. Present morphological diversity in Africa

In the light of updated concepts and genetic data (Mayr, 1974), present morphological diversity in Africa has been re-evaluated by various anthropometrical studies (Auger, 1967; Gomila, 1969, 1971; Crognier, 1972; Huizinga, 1968a, 1974, 1977). First, Hiernaux's work (1953-1977) showed the complexity of microevolutionary processes. Then, various craniometric approaches (Howells, 1989; Froment, 1992a; Lahr, 1996) also provided a new interpretation of the morphological variation, all underlining the importance of geography as a key element in the differentiation of populations.

2.3.1.2.1. Hiernaux's approach: from several case studies to a final synthesis

Avoiding a terminology based on previous biological categories ("negroid", "caucasoid") (De Quatrefages and Hamy, 1882), the work of Hiernaux (1953-1977) was the first to throw some light on the inter- and intra-population variability in sub-Saharan Africa. The author provided both case studies and syntheses on the biological diversity of African living populations, exploring the impacts of geography, ecology and/or history.

Hiernaux (1953, 1964a, 1964b, 1965b) studied various ethnic groups in both Central and East Africa using biometrics. In addition to case studies, Hiernaux (1955b, 1965a, 1966a, 1966b, 1968, 1974, 1976a) provided global synthesis on the biological diversity observed in Equatorial Africa. His corpus of data represented the largest data set recorded on living populations (at least 267 ethnic groups). Hiernaux's research was also based on a wide range of characters in relation not only to anthropometry (stature, cephalic index) and genetics (dermatoglyphs, blood groups), but also to various social, cultural and environmental factors. Therefore, Hiernaux (1966b, 1976a) was able to test whether various kinds of data sets agree with each other or not.

The results obtained showed that the ethnic groups did not aggregate morphologically and genetically. A discrepancy was also found between metrical and blood polymorphism distances. As a consequence, the author put forward the explanation of a differential sensitivity to climate of the variables under study.

2.3.1.2.2. Importance of selective pressures

Microevolutionary mechanisms reflect various aspects of not only complex historical background, but also "plasticity" or adaptability to various environments. Two major questions therefore arise: What are the most important selective pressures for humans? What morphological features of the body are determined environmentally rather than genetically?

2.3.1.2.2.1. Phenotypic plasticity as an effect of adaptation

In a preliminary stage, a biological adaptation at the individual level corresponds to an "acclimatization" phenomenon: all organisms can respond to environmental changes by short-term physiological adjustments (Frisancho, 1996). A major problem arises from the fact that it is very difficult to sort out the effects due to "adaptation" from those due to "acclimatization".

At what limit or threshold does the environment involve plastic (phenetic) changes and have a subsequent genetic impact? In order to answer this question, physical anthropologists compared genetically related populations living in different environments because of recent migration. For the analysis of two Tutsi groups, Hiernaux (1965a, 1965b) could not distinguish between acclimatization and adaptation processes, as this kind of question needed a very long-term study. This fact was supported by another anthropometrical study of Froment and Hiernaux (1984): Dogon populations that migrated five centuries ago in the Middle Niger valley, do not yet present morphological affinities with Sahelian populations.

Therefore, morphological traits, thought to be controlled by several loci, appear to have a complex nature. Furthermore, so many variables are involved in shaping present diversity, it is very difficult to demonstrate the direct effects of natural selection apart from rare examples (haemoglobin genetic polymorphisms) (Hiernaux, 1976b).

2.3.1.2.2.2. Body morphology: variable sensitivity to environment

Another important question has to be raised especially for craniometric studies. Is the skull less affected by selective pressures than the postcranial skeleton? Or, in other words, does body morphology have a variable sensitivity to environment?

Climate (temperature, air moisture, UV radiation levels) has been recognized as a major environmental factor influencing morphology (e.g. body proportions, skin pigmentation), the effects of which, both on a worldwide scale and within Africa, can result in genetic adaptation (Hiernaux, 1965a; Jablonski and Chaplin, 2000; Katzmarzyk and Leonard, 1998; Ruff, 1994). However, the impact of non-climatic variables on morphological variation, such as energy efficiency, cannot be neglected (Kurki *et al.*, 2008). Generally, but not always, some principles can be applicable to morphological variation. Bergman's rule (relationship of body mass to surface ratio) and Allen's rule (shape of body and appendages) are supported by the example of the Tutsi, who are resistant to desert heat because of their morphology (high surface for weight unit, little fat) (Hiernaux, 1963, 1965a). In comparison to the Tutsi, the Peul and the Somali could therefore be considered as a climatic convergence as they do not share any genetic affinity. But, this question remains unresolved.

In comparison to limb proportions and stature, adaptation of head shape to climatic factors is less obvious and more difficult to demonstrate. As suggested by Hiernaux (1963), skull morphology does not follow very clearly thermoregulatory principles such as Allen's and Bergmann's rules. However, according to Howells (1973), prognathism, facial height and vault breadth are highly correlated with climate. On a worldwide scale, Beals (1974) also noted that in cold areas, the most advantageous head shape would be close to the spherical ideal, as broader and shorter cranial vault (e.g. brachycephalism) is more present in higher latitudes. However, this phenomenon of brachycephalism is in reality difficult to explain in "climatic" terms, especially within Africa where it may have occurred in several areas (Pales, 1953; Chabeuf, 1959).

Before Hiernaux's work (1974), no obvious relationship was found between craniometric variables and climatic parameters such as between nose shape and temperature (Steegman, 1975). In fact, Hiernaux and Froment (1976) observed a high correlation between nasal breadth and average rainfall. Narrow noses appear to work better as a humidifier for dry air, as it was already suggested by Davies (1932), Thomson and Buxton (1923) and Weiner (1954).

From various field data, Hiernaux (1974) also noted that, nasal breadth is part of a series of morphological features reflecting "climatic" adaptation to wet and dry areas along the Equator. For example, in wet tropical areas where nasal breadth tends to be larger, trunk length is also greater, limbs are shorter, and the skull tends to be low and broad. In contrast, in dry hot conditions when the nose tends to be narrower and higher, the trunk is shorter, the limbs are elongated and the skull also tends to be higher and narrower (Hiernaux 1974; Ruff, 1994).

Nevertheless, hypotheses of a "climatic" adaptation of the skull shape remain very speculative so far, as other factors (e.g. history, genetic background) may be responsible for morphological changes. In fact, recent studies (e.g. Relethford and Harpending, 1994) showed that multivariate cranial analyses provide the same population relations as genetics. However, caution is suggested when using skull morphology for phylogenetic studies, as some anatomical regions can preserve climatic signatures (Harvati and Weaver, 2006).

2.3.1.2.3. Origins of morphological diversity

What is the nature of the differences between modern sub-Saharan populations? Do most morphological and genetic criteria reflect climatic or historical events? How much of the variation is due to differences in intensity of specific selective forces, and/or length of time, since the populations shared common ancestors? In fact, all these questions raise the problem of the origin of morphological diversity. As short-term and long-term adaptations, as well as historical factors, remain difficult to separate, there are no simple explanations for the biological variation observed within modern populations.

2.3.1.2.3.1. Head shape: historical factors versus selective pressures

Although various degrees of correlation have been observed between morphological traits and ecological variables (Hiernaux and Froment, 1976), the relationship of morphology with climate does not always mean a direct causal association. However, can anthropometry accurately predict population affinities and reflect historical processes? Are the historical relationships blurred by the role of natural selection?

In fact, the genetic stability (balance between plasticity and heritability) of morphological traits is not completely elucidated. Nevertheless, as suggested previously by Hiernaux (1963), vault shape (cephalic index) seems to have a higher heritability than the rest of the body (stature). In this case, the differentiation between historical and environmental processes is perhaps less problematic. Therefore, craniometry should provide a much less detailed picture of the selective pressures. But, according to Varela and Cocilovo (1999), some facial features (minimal maxillary height) could be more influenced by environmental factors (biomechanics of the masticatory system) than the rest of the skull.

In order to test whether climatic adaptive traits of the skull mask the real differences and/or similarities due to historical factors between populations, Guglielmino-Matessi *et al.* (1979) analysed again the craniometric and genetic data sets of Hiernaux (1963, 1976a). For this purpose, the authors compared results both with and without corrections for the climatic effects, and they found that the latter can "destroy" similarities due to genetics on a worldwide scale, such as between Europeans and Africans, and Americans and Australians. However, all these studies cited above used classic genetic markers, which could be themselves under selective pressures. So, they probably reflect not only history but also adaptation to climate, environment and disease (Relethford, 1994).

In addition, more recent studies (Friess *et al.*, 2002; Lieberman *et al.*, 2002; Lieberman, 2008; Roseman and Weaver, 2004; Betti *et al.*, 2009) showed that, although modern cranial diversity reflects mainly geography, it also reflects selective pressures, especially in the nasal region. In fact, their study supports other work suggesting the importance of selection for cranial form (Hiernaux and Froment, 1976).

2.3.1.2.3.2. Increasing complexity: convergence, admixture and mating patterns

When attempting to understand the origin of morphological diversity, an increasing number of factors are involved (e.g. convergent evolution, population admixture, mating patterns).

According to Hiernaux (1966a, 1968, 1970), convergent morphological evolution in the equatorial forest could be the main cause of observed similarities of forest groups. The morphological similarities among Congolese pygmies, for example between the Twa and the Mbuti, were explained in terms of adaptive convergence rather than shared genes (Hiernaux *et al.*, 1976). The authors suggested that a kind of "tropical" adaptation (short stature, limb proportions and facial traits) could be a phenomenon that occurred several times on different gene pools, and in different places of the equatorial forest under similar selective pressures.

The anthropometrical distinction between Pygmies and Bantu-speaking populations of Central Africa also remains difficult not only because of long-term forest occupation, but also admixture with Pygmies (Froment, 1993). For example, according to Hiernaux (1966c), the Twa are a reduced form of "forest" Africans (Konda) rather than a Pygmy group. Therefore, debates on the origins of "tropical" adaptations are still ongoing (Shea and Bailey, 1996). The hypothesis that Pygmies could derive from a biological stock common to all Central African populations cannot be excluded. However, recent genetic studies also highlighted the diversity present among various Pygmy populations (Coia *et al.*, 2002).

In fact, genetic drift has been greatly over-estimated in the past, especially in relation to the origin of Pygmy diversity that was in turn under-estimated. The isolation of hunter-gatherers was over-stressed, and as a consequence the flexibility and variability of various social features (e.g. mating patterns) were both underestimated.

The mating pattern (endogamy, exogamy) adopted by a population can also play a role on the resulting biological diversity. For example, the high rates of endogamy observed in some Senegalese (Bedik) could be responsible for a general high degree of homogeneity coupled with a reduction in size (Gomila, 1969, 1971). The work of Crognier (1972) also found homogeneity, especially for vault indices, in another endogamous African group (the Sara from Chad).

Furthermore, in endogamous populations, the degree of intra-sex variability is similar between males and females (very low sexual dimorphism), in contrast to the opposite situation of exogamy. This fact has also been noted when analysing non-metrical morphological variability (Corruccini, 1974, 1976; Prowse and Lovell, 1996). Higher frequencies of very rare traits (such as metopism) that have a high degree of heritability could support the hypothesis of a high endogamous rate and/or the presence of "family" units in a group.

In these last few examples, the level of analysis, which is the ethnic group, is quite important. A broad regional and ecological study would not have reached the same kind of accuracy, especially when looking at specific socioecological features such as mating patterns.

2.3.1.2.3.3. Resulting broad geographical differentiation

Whatever the causes of the biological diversity, morphological differences often follow the geographical distribution of the populations under study. Geographical groupings tend to reflect predominant gene flow, depending however on the level of analysis chosen (regions, countries, continents). This phenomenon has been observed through the analysis of various morphological features related to living populations on a broad geographical scale (Hiernaux, 1974) and sometimes even intra-regionally (Auger, 1967). In the latter study, Auger's work found a high correlation between morphology and geography (r=0.58) among Southern Cameroonian populations (Banen, Basa, Eton, Ewondo, Duala and Bamileke). This fact was interpreted as probably resulting from a recent homogenization (intense gene flow).

A study on dermatoglyphs showed continuous variability, precluding classification of the African populations (Hiernaux, 1964a). Only the position of the Efe Pygmies could suggest a certain degree of isolation. However, a geographical tendency appeared, showing along a north-south axis an increase in loops and a decrease in whirlpools. This observation could suggest a primary direction of continental gene flow.

Although Vecchi *et al.* (1983, 1987) analysed again cephalic indices with univariate analyses, they had the advantage of using a mapping method (trend-surface analysis) for illustrating the distribution of the morphological features (cranial indices, stature, trunk length). Their sample composed of sub-Saharan African males, was based on the earlier survey of Hiernaux (1974). The authors could observe that, on the whole, geographical variability for cephalic index was relatively small within Africa. More differences were observed for facial traits (nasal index, facial index), especially in a west-east direction, between forest and savanna areas. Their groupings in fact supported previous hypotheses of Hiernaux (1974) on "ecological" morphological adaptations.

The work of Howells (1973, 1989, 1995) used a long series of metric traits related to not only the cranial vault but also both the base and the face. A more representative picture of the variability was then provided especially in comparison to earlier work (Kappers, 1936; Gebara, 1947). Using multivariate biodistance statistics, marked differences were observed among recent populations of different geographic areas, despite the presence of a general morphological homogeneity. Most of the variation in cranial shape was located in the upper face (upper nose, border of orbits). Furthermore, the discriminant analyses of Howells (1989) showed that geographical groupings (Europe, Asia, Americas, Australia, Polynesia and Africa) were very effective (total correct rate of classification: 97.5%).

Combining the use of both metric and non-metric traits of the skull, Lahr (1992, 1994, 1995, 1996) also observed a relative closeness in skull morphology between all modern populations. However, a different pattern of regional differences was discovered especially in relation to cranial robusticity: most gracile skulls such as sub-Saharan Africans and South (East) Asians differed markedly from most robust ones such as Melanesians and Fueguians-Patagonians. In addition, as suggested previously by Howells (1989), sexual dimorphism appeared not to mask the regional differences, because the intra-population variation due to sex differences was less important than the general features of the population as a whole. This fact has also been recently supported by the work of Pietrusewsky (1990, 2008), who analysed craniometric variation among various populations (e.g. South East Asians, Australians).

Finally, the various craniometric works (e.g. Howells, 1989; Froment, 1989, 1992a, 1993, 1998; Hanihara, 1996; Lahr, 1996; Bruner and Manzi, 2004; Roseman and Weaver, 2004; Pietrusewsky, 2008) that provided synthetic views on morphological variation, are in agreement on the following three aspects:

i) the continuous variation of cranial morphologies (especially the face) in relation to geography;
ii) the presence of confounding factors (e.g. climate, history) when attempting to understand diversity; and
iii) the probable low influence of sexual dimorphism on the final results.

2.3.2. Biological effects of a historical phenomenon within sub-Saharan Africa: the Bantu-speakers expansion

The previous section underlined the importance of various biological processes that are involved in the differentiation of African populations. It is also essential to link these processes to complex historical events, such as the Bantu-speakers expansion that might also have shaped present biological diversity of sub-Saharan Africa.

2.3.2.1. No biological entity for the term "bantu": geography and mating patterns

First, what does the term "bantu" mean in a biological perspective? Intuitively, it is not possible to find a simple answer, as there is no equivalent biological category. Hiernaux's studies (1953-1977) confirmed the fact that there is no biological entity for Bantu-speaking populations. Combining both somatic and genetic data, Hiernaux (1968, 1976a) found by multivariate analyses that there is a kind of quasi-continuum variation. As we saw previously, some anthropometrical variables (e.g. nasal width, stature) correlated very well with a broad geographical differentiation that reflected indirectly the effects of selective pressures on morphology.

However, the relationship between linguistic and somatic data did not appear significant, especially within Bantu-speaking populations (Hiernaux and Gauthier, 1977). Nevertheless, the correlation between the two data sets became much higher on a broader linguistic level, when Froment (1995) included in the analysis not only the Bantu- or Benue-Congo-speaking group, but also several other linguistic branches of the Niger-Congo family (Kwa, Adamawa).

In fact, the gross similarity among Bantu-speakers could reflect many aspects such as not only the similarity in environment, but also the hybridization between groups on both inter- and intra-ethnic levels (e.g. endogamy, exogamy). According to Hiernaux (1974), these phenomena probably had an effect on the shape of the clinal variation within Bantu-speaking populations.

These observations suggest that the low levels of heterogeneity among Bantu-speaking populations must have been maintained by intense gene flow, or be the result of a progressive replacement process. The phenomenon probably acted in a variable manner within each region but mainly on an intra-ethnic level, although interbreeding between different ethnic groups could have occurred.

2.3.2.2. Evidence of one centre and two main streams?

Anthropologists have been facing the problem of analysing both present and past sub-Saharan Africans (Hiernaux, 1974; Brauer, 1980; Boaz *et al.*, 1990). It is difficult to analyse the biological affinities of only a few archaeological specimens, as their variation is probably under-estimated in comparison to modern populations.

According to Hiernaux (1974, 1976a), the modern variation, which is of a continuous or clinal nature, suggests that Bantu-speaking groups may descend from an old common African stock, which is also relatively recent. Using dental morphology, Irish (1995, 1997, 1998) also reached the same conclusions but in broader temporal and spatial terms, as dental evolution perhaps involves more conservative and slower evolutionary processes than bones.

In addition to this "apparent" morphological homogeneity, is it possible to observe inter-regional differences reflecting gene flow corresponding to different streams of the Bantu-speakers expansion? Using both somatic and genetic features, Hiernaux (1974, 1976a) could not answer this question, as probably the Bantu-speaking populations both expanded and intermingled very often. However, in his multivariate analyses, several ethnic groups from Cameroon (Basa, Ewondo and Fang) seemed to be in a central position, having the lowest mean distances in comparison to all other groups. Therefore, they could be considered most probably the least modified descendants of a "proto"-Bantu-speaking population originating from the edge of the forest. The closest groups to the Ewondo correspond to populations originating from not only West Central (Luba-Katanga) but also Eastern regions (Luo, Sukuma). These observations could support indirectly the location of a primary "proto"-Bantu centre in West Central Africa proposed by linguists (Bastin *et al.*, 1983), archaeologists and historians (de Maret, 1989; Vansina, 1995). On the biological level, Guthrie (1962) and Vansina (1995) suggested one hypothesis involving several secondary centres for the Bantu-speakers expansion in both West Central Africa and East Africa.

To summarise, according to Hiernaux (1968, 1976a), three conditions seemed to have been necessary for the presence and conservation of low biological distances between Bantu-speaking populations: (i) a relatively recent common origin, (ii) little differences between habitats, and (iii) an absence of important admixture with genetically very different groups (e.g. progressive replacement). Those parameters, therefore, produced an apparent homogenization in sub-Saharan Africa, with low inter-population differences in a broad southern direction.

2.3.2.3. Agreements and disagreements between various data sets

Other studies, such as those of Rightmire (1972, 1976a), also used the Bantu-speakers as a test framework. Rightmire (1972) analysed various cranial skeletal features on a sample of interlacustrine East Africans and South African Bantu-speakers of known ethnic affiliation. The author aimed at testing whether metric and non-metric traits agree or disagree with the question of the Bantu-speakers expansion as a continuous and homogenizing process.

His distance statistics showed that metric traits were in close agreement with the ethnic affiliations of the populations under study. A high proximity was noted between several Bantu-speaking populations that were geographically and historically related, like the Interlacustrine East Africans (Rwanda, Rundi). The Venda appeared to be in an intermediate position between the latter and various other South Africans (Zulu, Xhosa and Sotho). Historical data support this proximity of the Rwanda and Rundi, as the Venda represents one of the latest waves of Bantu-speaking populations originating from an eastern stream (Lake Nyasa). The exact date for the arrival of the

Venda in northern South Africa is uncertain (*c.* 15th century AD), but according to various ethnographic and linguistic data, it occurred after the settlements of the Sotho and Nguni (Zulu, Xhosa) (Wilson, 1969; Olson, 1996; Yakan, 1999).

However, the statistical results obtained from the non-metric traits did not support the previous picture, as the Venda appeared to be close to the Xhosa. According to Rightmire (1972, 1977), this disagreement could reflect different facts: the complex expression of morphology in relation to plasticity, the specificity of the mechanisms related to the underlying nature of discrete variants, and the relatively small size of the sample under study (N=150). Consequently, as Rightmire was confronted with these methodological problems, he could not reach any conclusion on the nature of the phenomenon itself (Bantu-speakers expansion), in contrast to Hiernaux (1974).

Previous observations showing relatively low levels of metrical variation within Bantu-speakers (Hiernaux, 1974, 1976a; Rightmire, 1976a, 1976b) have been supported by Froment's work (1992a, 1992b, 1993, 1995, 1998). Froment noticed no marked differences within the Bantu-speaking populations, and between the latter and the Nilo-Saharan group. However, there were marked differences between each major linguistic group (Niger-Congo, Afro-Asiatic and KhoiSan) and the Pygmies.

Froment (1995) also suggested that the Bantu-speakers dispersal was probably not a continuous phenomenon, involving several population contractions and expansions in an irregular way, as people had to adapt to new environments. However, random genetic drift might have also played a minor role in the slight morphological differentiation of Bantu-speakers, especially when the founder group was of small size.

2.3.3. Skull morphology through time

In all craniometric studies, the main question that is raised is whether it is possible to detect subtle decreases and increases in gene flow through levels of diversity in skull morphology? According to the results already obtained from modern populations (Howells, 1989; Froment, 1992a; Lahr, 1996), the geographical variation suggests that, despite an "apparent" homogeneity on a worldwide scale, sub-Saharan Africans cannot be described as sharing a single morphology.

2.3.3.1. Focusing on population history: morphological diversity

The present morphological polymorphism observed in Africa (Hiernaux, 1953-1977) is far beyond "racial" categories proposed in the past (Leakey, 1935; Coon, 1971).

For example, Froment (1998) re-evaluated some of the metric traits used by Chabeuf (1959) to define sub-Saharan populations (e.g. dolichocephaly, platyrrhinia, low orbits, alveolar prognathism). His study showed that only half of the crania studied (42% for East Africa and 8% for West Africa) had a long, high and narrow vault (dolichocephaly) and a large nose (platyrrhinia). Even fewer Europeans (about 10 to 30%) appeared to be characterized by those traits. Some of these features (prognathism, large nose) related to this sub-Saharan morphological gradient were also observed in both Asia and Australia, although in the Australian group the prognathism appeared to be more marked on a facial level (Pietrusewsky, 1990; 2008).

The antiquity of these traits that could be considered as apomorphies or adaptations to tropical climate (Hiernaux, 1974) is unknown, but according to Protsch (1975) probably not as old as the appearance of the KhoiSan morphological pattern. This question also remains unresolved because of the scarcity of human remains in Africa during the past 100,000 years.

2.3.3.2. Affinities of a few post-Pleistocene African populations

The question of the origin and evolution of different groups on the African continent has been explored very irregularly (Figure 15, p. 35). Due to the scarcity of paleoanthropological evidence, and research interests focused elsewhere, Central Africa in particular, has not been explored extensively. However, in the last thirty years, an increasing number of multivariate studies were carried out, in order to evaluate the affinities of various post-Pleistocene African populations with modern samples (e.g. Rightmire, 1976b; Dutour, 1989, 1992; Froment, 1998).

2.3.3.2.1. West Africa and North Africa

For the Early Holocene (10,000-6,000 BC), West Africa provided skeletal remains, mostly in Mali (Asselar, Hassi-el-Abiod, Kobadi, Karkarichinkat), showing morphological affinities with some of the

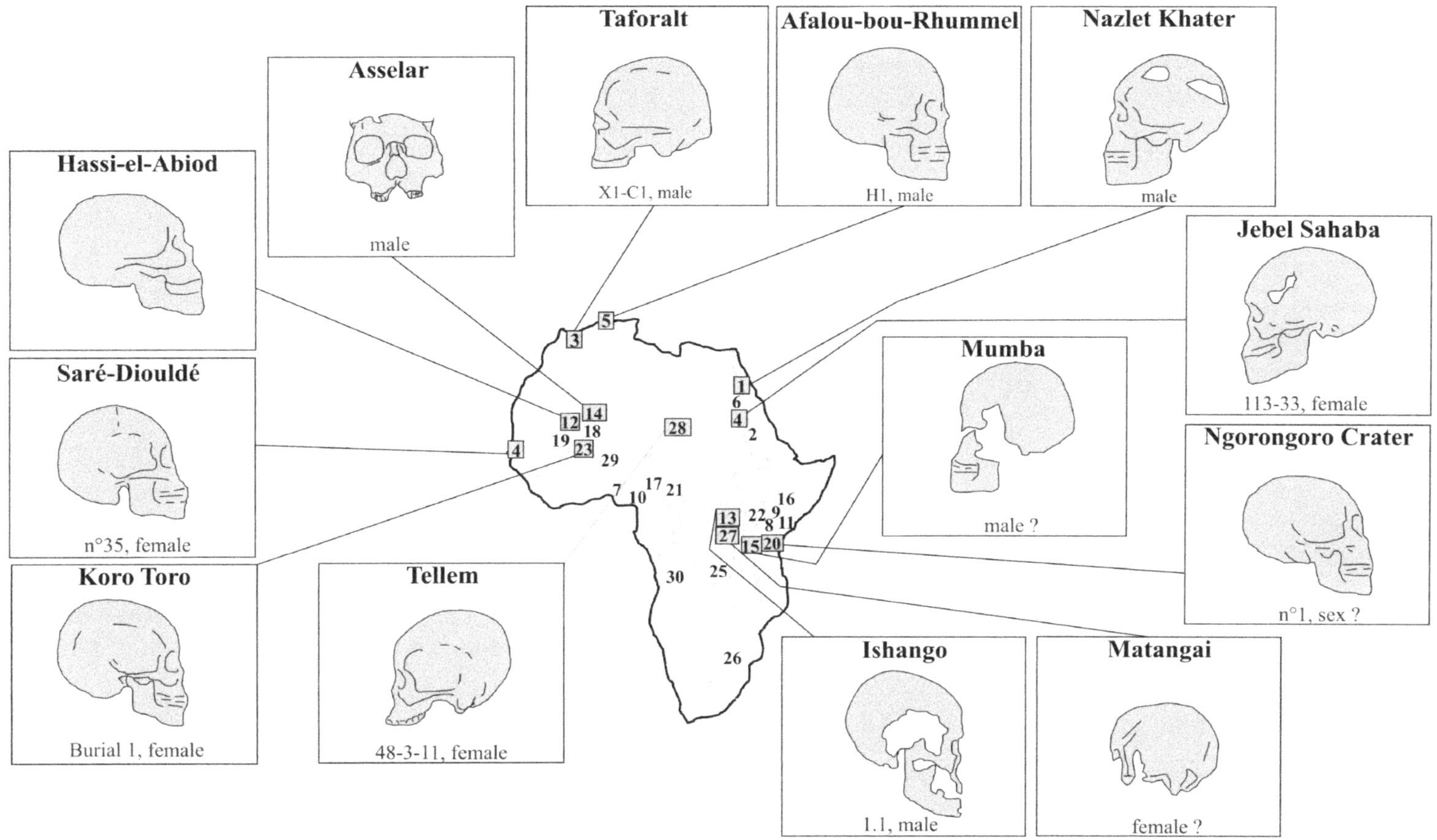

Before Holocene
1 Nazlet Khater (33,000-28,000 BC) (Vermeersch *et al.*, 1984)
2 Singa (uncertain date: 17,300±2,000 BP) (Stringer, 1979)
3 Taforalt (13,750 BC) (Chamla, 1978)
4 Jebel Sahaba (14,000-12,000 BP) (Anderson, 1968)

EarlyHolocene (10,000-6,000 BC)
5 Afalou-bou-Rhummel (10,500-8,500 BC) (Chamla, 1978)
6 Wadi Halfa (12,500-6,400 BP) (Hewes *et al.*, 1964)
7 Iwo Eleru (11,200±200 BP) (Brothwell & Shaw, 1971)
8 Gamble Cave (8,210±260 BP) (Protsch, 1978)

Mid-Holocene (6,000-2,000 BC)
10 Mbi Crater (7,790±80 BP) (Asombang, 1988)
11 Bromhead (7,410±160 BP) (Protsch, 1978)
12 Hassi-el-Abiod (7,000 BP) (Dutour, 1989)
13 Ishango (6,890±75 BP) (Boaz *et al.*, 1990)
14 Asselar (6,390 BP) (Oakley & Campbell, 1967)
15 Mumba (3,670 BC) (Brauer, 1980)
16 Lowasera (3,500-2,500 BC) (Schepartz, 1987)
17 Shum Laka (3,000-1,000 BC) (Orban *et al.*, 1996)
18 Karkarichinkat (4,000-3,600 BP) (Smith, 1974)
19 Kobadi (2,415±120 BP) (Georgeon *et al.*, 1992)
20 Ngorongoro Crater (2,260±180 BP) (Magory & Day, 1994)

Late Holocene (after 2,000 BC)
21 Toala (17th C. AD) (Menard, 1983)
22 Makalia and Willey's Kopje (2,000 BP) (Leakey, 1935)
23 Koro Toro (before 5th C. AD) (Treinen, 1975)
24 Saré-Diouldé (6th-16th C. AD) (Thilmans *et al.*, 1980)
25 Katoto and Sanga (8th-15th C. AD) (de Maret, 1985)
26 Mapungubwe (10th-13th C. AD) (Henneberg & Steyn, 1994)
27 Matangai (1,235 AD) (Mercader *et al.*, 2001)
28 Tellem (11th-13th C. AD) (Huizinga *et al.*, 1979)
29 Kainji (after 16th C. AD) (Wade, 1971)
30 Kapanda (16th-18th C. AD) (Gutierrez & Valentin, 1995)

Figure 15. Geographical distribution of some African Late Pleistocene and Holocene human remains having provided some information on biological affinities.

Late Pleistocene Cro-Magnoid populations of North Africa (Taforalt, Afalou) (Boule and Vallois, 1932; Chamla, 1978).

Although the Asselar specimen was thought to be closer to KhoiSan because of its cheek and orbit morphology, it then appeared to be much closer to contemporaneous (Hassi-el-Abiod) or earlier (Taforalt, Afalou) neighbouring populations (Dutour, 1989; Brauer and Rimbach, 1990). According to vault and facial features (broad and high vault, large nose), its cranial morphology was also intermediate between the so-called robust "Mechtoids" from North Africa and the much more gracile "Neolithic" populations from the Sahara. Furthermore, Asselar's morphology did not show any affinities at all with modern Malian populations (Dogon), who are characterized by a long narrow vault and narrow nose (Dutour, 1992).

These similarities between North African and Malian archaeological specimens support the hypothesis of migration and/or hybridization of populations within the Western and Southern Sahara from the Late Pleistocene up to the Late Holocene (Petit-Maire and Dutour, 1987). These phenomena probably occurred irregularly due to climatic fluctuations (Muzzolini, 1993) and genetic drift. Early West African populations (Kobadi) with affinities with the Asselar and Hassi-el-Abiod specimens could have persisted up to the first millennium BC (Georgeon *et al.*, 1992).

Early North Africans might also be the result of complex and long term gene flow with both sub-Saharan Africans and Mediterraneans (Rightmire, 1990; Chamla, 1978; Keita, 1988, 1990, 2007; Brauer and Rimbach, 1990). Although the morphological pattern of early Maghrebian crania is unique, Dutour (1992) underlined the fact that Mechta-el-Arbi specimens combined features, which were encountered in both sub-Saharan Africans and Europeans.

Brauer and Rimbach (1990) also showed similarities between an Upper Paleolithic Egyptian (Nazlet Khater) individual and various Africans and Europeans dated between 200,000 BP and 5,000 BP. The Nazlet Khater specimen is characterized by unique features (alveolar prognathism, very robust mandible) that are similar to those observed on Late Pleistocene Nubians (Jebel Sahaba, Wadi Halfa) (Anderson, 1968; Thoma, 1984). Recently, the cranium was re-analysed (Pinhasi and Semal, 2000; Crevecoeur, 2008; Crevecoeur and Trinkaus, 2004; Crevecoeur *et al.*, 2009) and exhibited very high levels of diversity for Late Pleistocene human skulls.

In West Africa, since the Early Holocene, geographical key areas along the savanna belt have probably played an important role, acting as population corridors. During the last millennium, successive micro-scale events have also been identified in the Middle Niger Valley. Due to the presence of fertile soil and springs at the base of escarpments (Bandiagara cliffs), it attracted many populations for various ecological reasons (Huizinga *et al.*, 1979).

For example, during four centuries (11th-15th centuries AD), the Tellem populations lived close to the Bandiagara cliffs probably because of political reasons, as according to oral traditions, they were trying to escape from the control of the Malian empire.

Once corrections between measurements taken on living populations and archaeological remains had been done, craniometric studies revealed two aspects: (i) the morphological uniqueness of the Tellem populations; and (ii) the importance of gene flow originating from sub-Saharan Africa (Huizinga, 1968b; Huizinga *et al.*, 1967).

The Tellem people represent a unique population with homogeneous features (brachycephaly, wide nose). They appear to be close to Central Africans rather than modern neighbouring Dogon and Kurumba (dolichocephaly, narrow nose).

These metrical differences have been supported not only by "cultural" data but also by additional analyses obtained from non-metric variation (Knip, 1971). Before the complete extinction of the Tellem populations, the hypothesis of a period of cohabitation with the Dogon was not excluded, as the latter, originating from south west Mali arrived in the region *c.* 15th century AD (Huizinga, 1968b). However, the hybridization of remaining Tellem with neighbouring populations has not been demonstrated anthropometrically so far.

2.3.3.2.2. (West-) Central Africa

Further south, just below the Western Sahara, the archaeological specimens dated to the Early Holocene are less numerous, especially within the equatorial forest. However, the few individuals that have been discovered in those areas (Iwo Eleru, Shum Laka, Mbi, and Ishango) seem to fit morphologically within the variation of the Early

Holocene Malian populations rather than modern Africans (Ribot *et al.,* 2001).

For example, although Iwo Eleru (*c.* 9,000 BC) is characterized by features often encountered in modern Africans (flat nasal bridge, slightly protruding chin, wide palate), it also reflects a unique morphological pattern. It has both a large size (long low vault) and high degree of robusticity (wide mandibular ascending ramus) (Brothwell and Shaw, 1971; De Villiers and Fatti, 1982). This last mandibular feature in particular has been found in other specimens dated to later periods (*c.* 7,000-1,000 BC) such as Mbi Crater (Asombang, 1988), Shum Laka (Orban *et al.,* 1996) and Ishango (Orban *et al.,* 2001). Recent morphometric studies suggest that the Iwo Eleru specimen might be more archaic (Stringer *et al.*, 2010). Only further dating will help to confirm whether this specimen dates to the Late Pleistocene instead of the Early Holocene.

A relatively robust morphology was also observed on specimens discovered at Ishango (*c.* 7,000-5,000 BC) through both odontometrical (Twiesselman, 1990) and craniometric (Boaz *et al.*, 1990; Orban *et al.*, 2001) analyses. The latter study showed that the Ishango cranium (*c.* 5,000 BC) is very close to the morphological variation of both Asselar and modern Central Africans. However, Crevecoeur *et al.* (2010) re-analysed some of the Ishango specimens in closer relationship with their stratigraphic context. They discovered more morphometric affinities with Middle Paleolithic modern humans than recent populations.

In the Late Holocene, only few morphological differences have been noticed between present West and Central Africans and Iron Age specimens (Toala, Koro Toro, Saré-Diouldé, Kainji and Kapanda). Hiernaux *et al.* (1992) could observe a morphological continuity between Iron Age populations (Sanga, Katoto) and present Luba groups in south-east D.R.C. (Shaba). Both Sanga and Katoto samples appeared to be similar on a broad geographical level to other modern populations (Cameroonian) and Late Iron Age groups (Tellem, Mali). However, differences were also observed within the populations of the Shaba region. For example, the two archaeological populations tended to have a slightly narrower face (and shorter body proportions) than the present ones. This fact could be interpreted possibly as the result of different migration waves of Central African Bantu-speaking populations.

Two cautionary remarks should be made here in relation to the study of early West and Central African populations. Firstly, terms such as "proto-Negro" (Brothwell and Shaw, 1971) or "proto-Bantu" (Asombang, 1988) that have been used sometimes to describe archaeological specimens are anthropometrically not very reliable. In fact, so far, there is no correlation between morphology and biological categories, as well as linguistic affiliation (Hiernaux, 1974). Secondly, the identification of past Holocene "forest" populations through present Pygmies is also rather difficult, as no particular cranial morphology has been observed at Matangai (Mercader *et al.*, 2001).

For example, the use of short stature as an indicator of Pygmy affinities is very unreliable on small skeletal samples, as stature is quite variable among all Central Africans (Marquer, 1972; Orban *et al.,* 1996).

2.3.3.2.3. East Africa and South Africa

The peopling of East Africa is also the result of complex microevolutionary processes. The biological input from sub-Saharan Africa was much greater than previously thought (Rightmire, 1975). For example, the old "Hamitic" concept (Leakey, 1935, Coon, 1971) has been completely rejected by Hiernaux (1968), as population differences appeared to be the result of convergent adaptation to the environment. However, some later population admixture with non-sub-Saharan African populations also has to be taken into account. In fact, during the Late Iron Age period, many Afro-Asiatic-speaking populations from the Middle East arrived in East Africa (Oliver, 1991).

In particular, Rightmire (1975) who re-evaluated craniometrically the affinities of a few Mid-Holocene (6,000-1,000 BC) specimens from the Rift Valley (Gamble, Bromhead, Willey's Kopje, Makalia, Nakuru), showed that the "Nilotic Negro" concept could not be applied to early Kenyan populations. It was found that many of these skulls were excluded from probable affiliation from Egyptians, and that they had much firmer links with other modern and archaeological African groups of sub-Saharan regions.

A common morphological pattern was also identified among Later Stone Age – Iron Age East Africans. Protsch (1978) showed the morphological affinities between the Bromhead and Gamble's Cave specimens. The latter most probably lived in an environment where gene flow was not stopped by any major environmental barrier, and where the eastern wave of the Bantu-speakers dispersal flowed southward. The effects of the latter could be illustrated by numerous sites with skeletal remains (Morris, 1992). For example, Steyn and Henneberg (1997) showed that the Iron Age Mapungubwe individuals had odontometric and craniometric affinities with Southern African Bantu-speaking populations rather than KhoiSan.

As the origin of the KhoiSan populations has not been elucidated among most of the Early and Mid-Holocene populations of East Africa, it still remains a controversial subject (Brothwell, 1963). KhoiSan skulls have been described previously by Tobias (1955) as pedomorphic with a short vault, frontal bossings, reduced triangular face, little prognathism, protruding cheeks and wide nose. This unique morphological pattern, which is very different from all other African populations, suggests that it is probably the result of very ancient microevolutionary differentiations.

De Villiers and Fatti (1982) raised the question of the Singa specimen as a "proto-KhoiSan" candidate, because of its quite large and short vault and large interorbital breadth. But, in fact, this specimen is currently considered as an archaic *Homo sapiens* (Stringer, 1979, 1989).

The origins of the KhoiSan have often been sought in East Africa, as KhoiSan languages could have had in the past a much wider distribution according to linguistic affinities with East African groups (Sandawe, Hadza) (Greenberg, 1964). Then, later events (Bantu-speakers expansion, European occupation) could have been responsible for the contraction of KhoiSan languages distribution. However, this hypothesis has been weakened by recent linguistic and paleoanthropological studies.

Firstly, Blench (1993) argued that the Sandawe and Hadza are true linguistic isolates, as the latter share only phonological features rather than an evident common lexicon with KhoiSan languages. Secondly, the non-affiliation of the KhoiSan with East Africans dated to the Mid- and Late Holocene (Mumba, Lowasera, Ngorongoro Crater) was clearly revealed by several craniometric studies (Brauer, 1980; Rightmire, 1977; Schepartz, 1987; Magory and Day, 1994). Nevertheless, this debate is still in progress, as genetic data on the Y-chromosome showing ancestral affinity between Ethiopians and KhoiSan (Semino *et al.*, 2002), also contradicted previous results.

However, recent craniometric and genetic work supports a local origin (in South Africa) for the KhoiSan as being the result of a very ancient differentiation through isolation in southern Africa (Morris, 2002; Knight *et al.,* 2003; Morris and Ribot, 2006; Stynder, 2009; Stynder *et al.*, 2007a, 2007b).

Although craniometry showed that several morphological trends developed on the whole African continent, it provides insufficient evidence on the question of population origin. This fact is mainly due to the lack of human remains dated to the Upper Pleistocene in Africa.

2.3.4. Various issues on the genetic origin and differentiation of African populations

Through present and past morphology, previous sections underlined the various biological processes that have been involved in the differentiation of West Central African populations. However, it is also essential to link these phenomena of population assimilation and/or replacement with genetic data.

Interpretations of genetic data are more reliable when they are based on results obtained from different genetic systems, providing therefore a more representative picture of the total genome (Destro-Bisol, 1993; Foley, 1998; Tishkoff *et al.*, 1998; Kittles and Keita, 1999; Weber *et al.*, 2000; Beleza *et al.,* 2005; Berniell-Lee *et al.*, 2009; Tishkoff *et al.*, 2009; Bryc *et al.*, 2010; Schuster *et al.*, 2010). In particular, the non-recombining mtDNA and Y-chromosome loci that are uni-parentally inherited are ideal to assess the origins of modern diversity, as they are fast evolving and very sensitive to microevolutionary processes.

African genetic diversity appears to be the result of very ancient demographic events, especially in relation to founder effect phenomena (Anderson *et al.,* 1981; Cavalli-Sforza *et al.*, 1988; Underhill *et al.*, 2001; Ramachandran *et al.*, 2005). The "out of Africa" model for the origins of modern humans is supported by various studies, showing that the diversity observed in non-Africans was mainly a subset of African variation (Cavalli-Sforza *et al.,* 1994; Salas *et al.*, 2002; Tishkoff and Williams, 2002; Takasaka *et al.*, 2006).

Furthermore, the calculation of "coalescence" dates, although very variable (300,000-60,000 years ago), supported the antiquity of these dispersals as well as the presence of populations within sub-Saharan Africa that probably underwent successive population expansions and contractions in relation to climatic events (Cann *et al.*, 1987; Vigilant *et al.*, 1991; Watson *et al.*, 1996, 1997a, 1997b; Foley and Lahr, 1998; Ramachandran *et al.*, 2005; Takasaka *et al.*, 2006).

2.3.4.1. Correlation with geography, linguistics and/or phenetics

Global surveys on African genetic polymorphisms also provided information on microevolutionary processes that occurred more recently during the last 10,000 years (Excoffier *et al.*, 1987; Sanchez-Mazas *et al.*, 1992; Cavalli-Sforza *et al.*, 1994; Watson *et al.*, 1996; Hassan *et al.*, 2008; Castri *et al.*, 2009). Isolation by geographical distance and divergence from a shared population, appear to be again the key-elements for explaining the continuous genetic variation on a worldwide scale (Eller, 1999).

Genetic differences seem to correlate broadly with both geography and linguistics. According to the variability of different systems (Gm, Rhesus, HLA), Excoffier *et al.* (1987) observed several geographical clusters (West, Central, East) that could also overlap broadly the linguistic differences. Although a relative intra-group genetic homogeneity was observed at a continental level, the genetic patterns corresponding to the differentiation of several linguistic groups (Niger-Congo, KhoiSan, Nilo-Saharan and Afro-Asiatic) were still visible (e.g. Hassan *et al.*, 2008).

However, other studies showed that geographical distances reflect the processes of population differentiation better than linguistic affiliation, and this fact appears to be in agreement with craniometric data (e.g. Hiernaux, 1974; Relethford and Harpending, 1994; Froment, 1995; Relethford, 2002). For example, Sanchez-Mazas *et al.* (1992) using the same genetic systems analysed by Excoffier *et al.* (1987), found that the differentiation of African populations is mainly under the control of geography. Three major areas (West, East and South) were determined as probably resulting from geographical distance and predominant gene flow, and East Africans are the most heterogeneous group.

A few gene frequencies in some populations remain however difficult to explain in terms of geography and/or linguistic affiliation (Excoffier *et al.*, 1987). For example, why are the Bamileke from Cameroon closer to Senegalese than Bantu-speaking South Africans as observed by Scozzari *et al.* (1994)? These gene frequencies could reflect later population admixture that obscured the initial similarities, but they could also be explained in terms of selective pressures.

Are genotypes really acting in a different manner than phenotypes? Do they provide a different picture from phenetics? In answering these questions, Hitzeroth (1986) and Relethford and Harpending (1994) observed like Hiernaux (1976a), that there is less detailed correspondence between genetics and body morphology than between the latter and geography or climate. However, genetics appears to agree with craniometrics on the low levels of variation in humans (Relethford, 1994). Furthermore, the single origin model for modern *Homo sapiens* has also been supported by various researchers (Howells, 1989; Lahr, 1996; Relethford and Jorde, 1999; Mellars, 2006; Goebel, 2007; Campbell and Tishkoff, 2010).

2.3.4.2. Various questions related to the origin of African populations

In order to explain in evolutionary terms the African biological diversity, morphological and genetic studies cited above attempted to propose models derived from various data (Kittles and Keita, 1999; Tishkoff and Williams, 2002). In particular, recent genetic studies raised various questions concerning the origin of African populations, as well as the differentiation of the latter from Pleistocene to Holocene periods.

Firstly, do Africans correspond to the potential ancestors of all modern groups within and outside Africa? Did they differentiate in situ since the Pleistocene?

In fact, various molecular approaches accumulated consistent information supporting the hypothesis of a recent common African ancestry of modern humans (Rogers and Harpending, 1992; Jorde *et al.*, 1997; Shriver *et al.*, 1997; Fleagle and Gilbert, 2008). The later dispersals occurring from Africa into Europe and Asia have also been suggested by genetic signatures left behind as population expansions and bottlenecks (Harpending *et al.*, 1998; Reich and Goldstein, 1998; Kimmel *et al.*, 1998; Ramachandran *et al.*, 2005). For example, several markers identified on the Y-chromosome (e.g. M168 mutation) could represent traces of these recent migrations of modern humans (Underhill *et al.*, 2001).

During and/or after these migrations out of Africa, the remaining Mid-Pleistocene Africans have probably been the primary ancestors of all present Africans, as suggested by the data obtained from nuclear DNA (e.g. Alu deletion at CD4 locus on chromosome 12) (Tishkoff *et al.*, 1996). In fact, Tishkoff *et al.* (1996) noted that most of the African variation (associated with a wide range of STRP alleles) probably reflected an *in situ* differentiation since the Mid-Pleistocene.

Other genetic data obtained from both mtDNA and Y-chromosome supported again the hypothesis of ancestral African populations that were already highly differentiated (Watson *et al.*, 1997a, 1997b; Passarino *et al.*, 1998; Malaspina *et al.*, 1998).

In addition, demographic models that combined both genetic (mtDNA) and craniometric data, also agreed with this high degree of "polymorphism" (Relethford and Harpending, 1994; Relethford, 1998; Relethford and Jorde, 1999; Eller, 2001). They all showed that sub-Saharan Africa always presented the largest effective population size and the greatest heterozygosity through time. This high degree of "polymorphism" that could have been even more variable than among modern populations, is also indirectly supported by craniometric studies on Late Pleistocene – Early Holocene populations (Chamla, 1978; Dutour, 1992; Lahr, 1996; Froment, 1998).

Secondly, which modern African population is the closest to hypothetical ancestral groups? How did the differentiation between Africans occur during the Holocene?

Genetic studies have often focused on hunter-gatherer populations (Pygmy, KhoiSan), as the latter could represent the earliest inhabitants of sub-Saharan Africa before the expansion of agriculturists during the Holocene (Vigilant *et al.*, 1991; Rapacz *et al.*, 1991). However, the results did not always show that Pygmies and KhoiSan are genetically very different from other African populations (Rouger *et al.*, 1983; Destro-Bisol *et al.*, 2000). According to Cavalli-Sforza (1986), these differences are not sufficient enough to exclude the hypothesis of a common origin for all sub-Saharan African groups. In fact, recent genetic research (Destro-Bisol *et al.*, 2004a; Quintana-Murci *et al.*, 2008; Verdu *et al.*, 2009) supports that divergence between ancestral pygmies and non-pygmy populations occurred between *c.* 54,000-90,000 years BP, while the divergence between pygmy populations occurred during the early- to mid-Holocene. Both were well before the Bantu-speakers expansion.

In fact, genetic data on red cell polymorphisms seemed also to confirm the relative similarity among sub-Saharan Africans even geographically further apart, such as between Nigerians (Yoruba) and Angolans (Nijinga) (Ojikutu *et al.*, 1977; Nurse *et al.*, 1979). Additional studies on mtDNA also agreed with this fact, as some genetic features (e.g. L3b) have been found quite widespread in both West Africa (Senegalese) and South Africa (KhoiSan) (Scozzari *et al.*, 1994; Chen *et al.*, 1995; Soodyall *et al.*, 1996).

However, although a common origin is likely for all modern sub-Saharan Africans including Pygmies and KhoiSan, the ancestral genetic (and morphological) pattern has probably been modified by later microevolutionary processes (isolation, admixture). For example, features of microsatellite variation specific to Pygmy groups could be due to historical isolation and smaller effective population size that in turn could indirectly result from hunter-gatherer culture (Destro-Bisol *et al.*, 2000; Verdu *et al.*, 2009).

Furthermore, some unusual features of the genetic make-up of the KhoiSan, who present high frequencies of "Asian" alleles, could also reflect some degree of admixture with non-Africans (Excoffier *et al.*, 1987; Cavalli-Sforza *et al.*, 1994). In fact, population admixture on a large-scale often occurred in a multi-directional manner through successive migrations, for example during the Late Pleistocene and Mid-Holocene, between sub-Saharan Africa and North Africa (Krings *et al.*, 1999; Keita, 2005).

In addition, since the Mid-Holocene, various migratory events, such as the Bantu-speakers (Berniell-Lee *et al.*, 2009) and pastoralists (Henn *et al.*, 2008) expansions have left genetic traces in sub-Saharan African populations.

In summary, although Africans share a common ancestral gene pool, they highly differentiated through complex microevolutionary processes.

2.3.4.3. Homogeneity within Bantu-speaking populations

For the Mid-Holocene period, large-scale population movements have been shown genetically not only on a worldwide scale, but also within sub-Saharan Africa (Semino *et al.*, 1996;

Beleza *et al.*, 2005; Berniell-Lee *et al.*, 2009). In particular, this phenomenon is well illustrated by the genetic homogeneity that has been frequently observed amongst Bantu-speaking populations (e.g. Excoffier *et al.*, 1987; Cavalli-Sforza *et al.*, 1994; Biondi *et al.*, 1996; Pereira *et al.*, 2001; Ramachandran *et al.*, 2005; Wood *et al.*, 2005;Castri *et al.*, 2009; Tishkoff *et al.*, 2009; Bryc *et al.*, 2010). Both synthetic and case studies including various genetic markers seem to agree broadly on the process of homogenization, which occurred within Bantu-speaking sub-Saharan Africans.

Firstly, according to a large survey of various genetic polymorphisms including mainly classic markers, Cavalli-Sforza *et al.* (1994) showed that Bantu-speaking populations have high intra-population variation and low inter-population variation. Through a founder effect phenomenon, the dispersal produced an overall homogeneity within sub-Saharan Africa. This fact supports the hypothesis that Bantu-speaking people could be the outcome of the recent expansion of a relatively small group (Vansina, 1995). Still analysing mainly classic genetic markers (ABO, JK, MNS, Rh, ACP1) but with a new sample, Biondi *et al.* (1996) also found a good correspondence between both geographical and linguistic differentiation of Bantu-speaking populations.

Secondly, other molecular approaches with new genetic markers reflecting this phenomenon have also been revealed, such as on the non-recombining portions of both mtDNA (Bandelt *et al.*, 1995; Chen *et al.*, 1995; Soodyall *et al.*, 1996; Pereira *et al.*, 2001) and Y-chromosome (Passarino *et al.*, 1998; Scozzari *et al.*, 1994; Underhill *et al.*, 2001; Castri *et al.*, 2009).

For example, concerning mtDNA, Bandelt *et al.* (1995), suggested that the high frequency of a subset of cluster L1a (that is defined by a 9-bp deletion at np 8272-8289) could correspond to the traces left by a large dispersal within the Bantu-speakers. Another possible mitochondrial marker for this expansion could have been found by Vigilant *et al.* (1991) and Soodyall (1993) in a sequence motif of subset of L3b (np 16124-16223-16278). The latter was in fact widespread in both West and South Africans, all belonging to the Niger-Khordofanian family.

Results on mtDNA that have been obtained by Pereira *et al.* (2001), are similar to all those just cited above (e.g. Castri *et al.*, 2009). The sequences of haplogroups L3b and L3e1a and subsets of L1a and L2a could represent again the traces left by the Bantu-speakers dispersal, by showing affinities of Mozambican with other northern and southern Bantu-speakers. Furthermore, the dating of the major Mozambican mtDNAs components (e.g. L2a) also provided an age range more or less compatible with linguistic data (Bastin *et al.*, 1983) on the Bantu-speakers expansion, although slightly older (*c.* 6,000 years ago).

Concerning the Y-chromosome, the widespread distribution of a haplotype (group III) in sub-Saharan Africa could also indicate the homogenizing effects of a founder effect phenomenon in relation to the Bantu-speakers dispersal (Underhill *et al.*, 2001; Ramachandran *et al.*, 2005). This fact was also supported previously by the high frequency of a related sub-clade (e.g. RFLP 49f Ht4) (Seielstad *et al.*, 1994; Hammer *et al.*, 1997). Underhill *et al.* (2001) also suggested that the pre-existing African diversity probably became overwhelmed, according to the presence (e.g. M33, M75) or lack (e.g. PN2) of mutations that are confined only to the continent.

The above data obtained from different genetic systems are being renewed constantly, but all of them seem to support this founder effect phenomenon causing a decrease in diversity within Bantu-speakers.

2.3.4.4. Effects of recent events on a micro-scale: case studies on modern Cameroonian populations

Recent historical events that occurred on a micro-scale might have been superimposed on the previous ones related to the earlier Bantu-speakers dispersal. For example, Excoffier *et al.* (1987) found some KhoiSan Gm specificities within Bantu-speaking populations (Sotho, Venda, Nguni). In this case, recent genetic admixture could have attenuated the homogeneity produced by earlier founder effects phenomena.

Numerous genetic studies focused on these microevolutionary processes, especially within Cameroon, as it provides an extreme diversity of populations (Rossi *et al.*, 1991; Destro-Bisol and Spedini, 1991; Destro-Bisol *et al.*, 1994; Spedini *et al.*, 1999; Destro-Bisol *et al.*, 2000; Destro-Bisol *et al.*, 2004a). In order to explore inter-regional diversity, Spedini *et al.* (1999) analysed ten protein genetic polymorphisms on a high number of both

Northern and Southern Cameroonians. As a substantial genetic similarity was found between various linguistic groups, they suggested that recent genetic admixture could have played an important role in homogenizing present levels of diversity. Despite this "apparent" homogeneity, the South Western Cameroonians of the Niger-Khordofanian group (Bakaka, Bamileke, Bassa and Ewondo) appeared to be more clustered than the other linguistic groups.

Microsatellite variation also revealed an intra-regional homogeneity, especially within the Bamileke, who are the result of several Sudanic migrations after the Bantu-speakers expansion (17th-18th centuries AD) (Spedini and Bailly, 1987; Destro-Bisol *et al.*, 2000). However, this genetic phenomenon could be explained by a high number of processes like larger population size, more rigid respect of clanic exogamy, and higher matrimonial mobility. This last example in particular shows that the levels of genetic diversity can also be modulated by numerous social factors.

2.3.4.5. Levels of diversity related to mating patterns

Recently, geneticists showed that gender-based mobility related to mating patterns is most possibly responsible for differentiating and/or homogenizing inter- and intra-population variation (Melnick and Hoelzer, 1993; Rogers and Kidd, 1993; Mitchell, 1996; Seielstad *et al.*, 1998; Jorde *et al.*, 2000; Wood *et al.*, 2005; Destro-Bisol *et al.*, 2004b; Verdu *et al.*, 2009).

Amongst non-human primate species, differences in relation to exogamy and/or endogamy have been observed, resulting in either male or female sedentism (Rhine, 1986). In macaques, according to mtDNA analyses, as matrilocality decreases female mobility, it increases indirectly genetic homogeneity within a female population (Melnick and Hoelzer, 1993; Morales and Melnick, 1998). Therefore, if there is differential mobility pattern between males and females, Y-chromosome data provide an opposite picture to the previous one.

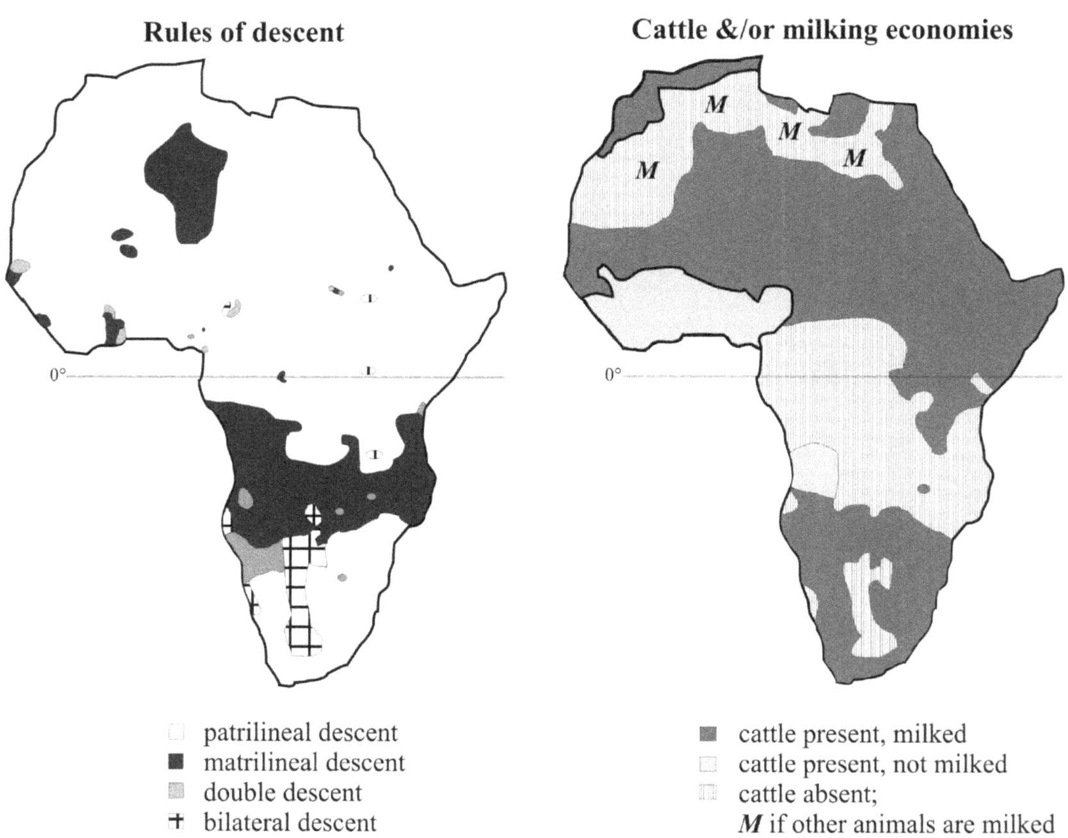

Figure 16. Comparison of the distribution of various rules of descent and cattle and/or milk-based economies within Africa (line drawing after Murdock, 1959, p. 20 & 28).

genetic system	between continents	within continents	within populations[1]
autosomes (♀, ♂)	↓[2]	↓	↑
mtDNA (♀)	↓	↓	↑
Y chromosome (♂)	↑	↑	↓

[1] A population is defined here as a group of pooled ethnic groups that correspond to a broad geographical region.
[2] Genetic meaning and interpretation of the arrows: ↓ = homogeneity as an indicator of sedentism; ↑ = heterogeneity as an indicator of mobility.

Table 1. Trends in levels of diversity (single nucleotide polymorphisms) for three genetic systems (data from Seielstad *et al.*, 1998).

In humans, all these phenomena are even more subtle, as mating patterns follow complex socio-cultural rules (e.g. marriage, kinship, rules of inheritance). Exogamy or endogamy is often related to the system of inheritance (patrilineal, matrilineal) practised by a society (Harris, 1997; Haviland, 2000). In Africa, marriages follow in general ethnic boundaries (Hiernaux, 1974), but it is not always the case, as it has been suggested by Spedini *et al.* (1983). According to erythrocyte and sero-genetic markers, the latter found a very low level of genetic differentiation in several Central African groups, reflecting probably a high degree of exogamy. However, to see whether males and/or females are responsible for this phenomenon, the ideal is to have parallel data from different genetic systems (e.g. autosomes, mtDNA, Y-chromosome) (Caldararo and Guthrie, 1998).

Therefore, several studies have explored these different levels of diversity (Melnick and Hoelzer, 1993; Seielstad *et al.*, 1998; Jorde *et al.*, 2000). In contrast to macaques, most humans provide a situation where mtDNA variation is mainly explained intra-regionally (Melnick and Hoelzer, 1993; Seielstad *et al.*, 1998). This fact suggests for the females a greater mobility on a regional level, while Y-chromosome regional diversity is low (Mitchell, 1996; Seielstad *et al.*, 1998).

All these genetic trends have been interpreted as an indicator of features related to patrilineal societies. They support the fact that on a worldwide scale, most populations (70%) adopted a patrilineal system where wives tend to move to the husbands home (Murdock, 1967; Burton *et al.*, 1996). However, these studies could have underestimated the impact of matrilineal societies that are also present in Africa (Figure 16, p. 42).

For example, Seielstad *et al.* (1998) analysed populations originating only from the savanna belt, where the patrilineal system has been mainly adopted. In fact, numerous African ethnic groups are also matrilineal, for example in Central Africa (Basuku and N'Komi), in East Africa (Dschagga), in West Africa (Ashanti) and in North Africa (Berbers) (Murdock, 1959). Especially in sub-Saharan Africa, they often correspond to agricultural people that did not adopt cattle in their economy, as pastoralism is often related to patrilineal societies.

However, according to mtDNA variation, hunter-gatherer populations of bilateral descent (e.g. KhoiSan) also suggest that females are less mobile because of matrilocality (e.g. Melnick and Hoelzer, 1993).

As gene flow due to mating patterns can have effects on both macro- and micro-scales, the level of study is very important for localizing differences in diversity. Seielstad *et al.* (1998) analysed therefore levels of polymorphisms found in three genetic systems, comparing the diversity between or within continents, or within populations. As they could better localize geographically the Y chromosome than mtDNA and autosomes, they also explained this phenomenon in relation to patrilocality (Table 1, p. 43).

Within a population, genetic variants of the paternal line are less frequent than those of the maternal line, as males tend to be sedentary and females more mobile. As a consequence, within a region, levels of diversity are low amongst males because of low male gene flow, in contrast to the heterogeneity observed amongst females. However, when comparisons are made on a macro-scale, the regional specificities of each male population are overwhelmed, and the levels of diversity of the latter appear much higher than those of the females (Wood *et al.*, 2005; Destro-

Bisol *et al.*, 2004b; Verdu *et al.*, 2009).

It is questioned therefore, whether gene flow is systematically exaggerated by the different mate choice systems. For example, in the case of matrilineal societies, are the genetic differences between groups on a regional level also modified because of low female gene flow?

This question has not been explored yet both genetically and morphologically, and it raises an important issue – genetic affinities between populations can be obscured by various factors – not only history, but also social systems (e.g. mate choice), as well as ecology (e.g. epidemiology). However, when research is undertaken on different levels of analysis, it could be attempted to distinguish between various types of gene flow (e.g. various migrations in relation to history, gender-based mobility within a group).

2.3.4.6. Co-evolution between genes, diseases and cultures

As epidemiological factors are often related to population density, they suggest that, selective pressures can also interact with history. Diseases probably played an important role in the geography of peopling, especially in West Central Africa (Froment, 1998; Tishkoff and Williams, 2002; Campbell and Tishkoff, 2008).

The selective value of a number of genetic defects illustrates the phenomenon of co-evolution between genes and cultures. For example, the distribution of low rates of lactose intolerance, which is a kind of nutritional adaptation, is very informative: it is highly correlated with the presence of pastoralist cultures subsisting on milk products, as well as the distribution of infested areas by tsetse flies (Figure 16, p. 42).

Some parasitic diseases coupled with a selective response, such as the interaction of malaria-resistance genes (e.g. Hb beta*S allele), could also be the result of increasing sedentism (Destro-Bisol *et al.*, 1996; Destro-Bisol *et al.*, 1999).

Erythrocyte polymorphisms observed within Africa support this fact: the frequency of Hb beta*S allele is very high amongst sedentary agricultural people living in regions with high levels of malaria endemicity, such as among the Southern Cameroonians (Bakakas) (Destro-Bisol and Spedini, 1989; Destro-Bisol *et al.*, 1992; Martorana *et al.*, 1996).

Therefore, the expansion of Bantu-speaking agriculturists within sub-Saharan Africa might have been very dependent upon the success of this genetic adaptation (Hiernaux, 1955b, 1974). As the latter was probably a very long-term process according to archaeological and ethno-historical data (Vansina, 1990, 1995), it also supports the irregular nature of the dispersal.

2.4. Complexity of a biological model for the peopling of West and Central Africa

As we saw previously, the present diversity of African populations is probably the result of complex microevolutionary processes that occurred during the Holocene, superimposed on much longer term processes. Different fields of research, such as linguistics, history, archaeology and biology suggest that the peopling of West and Central Africa was achieved through various demographic events of both expansions and contractions of human groups.

Large-scale population movements, such as the Bantu-speakers expansion, had a major impact on shaping both cultural and biological differences, especially amongst sub-Saharan Africans during the last 5,000 years. However, other events that occurred before, during and/or after this phenomenon also had additional effects.

Firstly, hypotheses on historical linguistics help us explore the complexity of population differentiation before the Bantu-speakers expansion. The latter is in fact a less well-documented period in both archaeological and biological terms. However, on its own, historical linguistics does not possess sufficient arguments; it will be integrated with data obtained from paleoclimatology, paleoanthropology, as well as genetics.

Secondly, for the periods during and after the Bantu-speakers expansion, as greater amounts of data are available from different fields of research, several biological models are proposed, focusing on the Bantu-speakers dispersal.

2.4.1. Before the Bantu-speakers expansion: greater regional diversity?

Since the Early Holocene, the expansion of both Niger-Congo and Afro-Asiatic families could be related to several agricultural dispersals within Africa (Diamond and Bellwood, 2003) (Figure 17a, p. 49). According to Blench (1993), who used cross-comparison of words of cultural significance, all African linguistic families (except for Afro-Asiatic and KhoiSan) originated along the northern fringe of the forest. Therefore, their present distribution reflects indirectly the intermingled expansion and fragmentation of groups associated with different food productions.

Linguistically, the KhoiSan populations seem to correspond to the earliest strata of peopling within Africa (Blench, 1993). Their hypothetical distribution all over the northern third of Africa is also still debated by both linguists (Blench, 1993) and paleoanthropologists (Schepartz, 1987). However, according to genetic findings, the KhoiSan speakers could belong like East Africans (although being also different from the latter) to the deepest clades of human phylogeny (Soodyall *et al.*, 1996; Tishkoff *et al.*, 1998; Semino *et al.*, 2002; Knight *et al.*, 2003; Reed and Tishkoff, 2006).

The Nilo-Saharan languages may be the oldest phylum after the KhoiSan, as their extreme diversity reflects a great antiquity (Blench, 1993). Prior to intensive desiccation in the Sahara, their distribution was much wider than present, probably ranging continuously from East Africa to West Africa. This fact coincides with the archaeological phenomenon of the "aquatic" life style expansion that was evidenced in the Sahara (MacDonald, 1999).

Does the fishing-based culture that persisted up to the first millennium BC along the Niger valley (e.g. Kobadi) correspond to the descendants of expanding Nilo-Saharan groups? The Gamble Cave specimens dated to the 7th millennium BC, could they be considered as the ancestral groups? These questions have not been resolved, but previous craniometric studies suggest that regional diversification occurred through time in a different manner within both East Africa and West Africa (Rightmire, 1975; Dutour, 1992). This phenomenon could also result from the fragmentation of the Nilo-Saharan group that occurred both before and contemporaneously to the Bantu-speakers expansion.

In addition to the two previous linguistic groups, is the origin of the Niger-Khordofanian family also related to a pre-agricultural phase? In approaching this problem, Blench (1999) proposed a macrophylum model where both Niger-Congo and Central Sudanic languages have common linguistic roots with the ancestral Nilo-Saharan family. However, one hypothesis involving separate origins with very ancient dates cannot be excluded.

Genetic data on mtDNA (e.g. L1b marker) could perhaps support indirectly the hypothetical antiquity of the Niger-Khordofanian phylum, as Watson *et al.* (1997b) identified a pre-Holocene population expansion restricted to West Africa with coalescence time of *c.* 19,000 years. However, in order to further support this hypothesis, studies based on a greater amount of gene markers will be necessary.

Furthermore, is this period related to the regional diversification of robust populations, in various parts of Africa, such as in Southern Africa (Hofmeyr), in the Maghreb (Taforalt, Afalou) and in West Africa (Asselar)? This Late Pleistocene morphological diversity seems then to have reduced drastically during the Holocene (Grine *et al.*, 2007; Crevecoeur *et al.*, 2009), supporting a series of bottleneck and founder effect phenomena (Ramachandran *et al.*, 2005).

During the Late Pleistocene-Early Holocene period, the diversification of the Niger-Khordofanian and/or Nilo-Saharan languages could have coincided with a parallel biological phenomenon. In fact, craniometry showed not only marked differences between early North Africans and Nubians, but also a common morphology for early West Africans that are intermediate between the last two groups (Dutour, 1992; Groves and Thorne, 1999). However, the question of the expansion of proto-Niger-Khordofanian speakers in West Africa remains very speculative in the detail. For example, it is unknown whether the latter was related to both Early Holocene (Iwo Eleru) and Mid-Holocene (Kobadi) populations of sub-Saharan Africa.

The arrival of a third linguistic group in the Sahara, the Afro-Asiatic speakers, has also to be considered. According to Blench (1993), these languages originated in the Levant (*c.* 8,000-6,000 BC). However, this hypothesis has been recently challenged by Ehret (2002a, b), who supports an earlier African origin. In the Maghreb, Afro-Asiatic speakers could correspond to the Capsian populations that appeared morphologically different from earlier Iberomaurusians (Chamla, 1978). In West Africa, it is unknown to what extent the cultural and/or biological effects of these proto-Mediterranean populations are superimposed on earlier and other contemporaneous regional diversifications. However recently, genetic data

(Cruciani *et al.*, 2010) suggests that proto-Chadic Afro-Asiatic speakers migrated through the Central Sahara into the Lake Chad Basin during the Mid-Holocene. This supports the fact that giant paleo-lakes could have provided a corridor for human migrations across the Sahara.

2.4.2. During the Bantu-speakers expansion: homogenization mainly due to a founder effect phenomenon?

During the Mid-Holocene (*c.* 1,000 BC), effects of various other historical phenomena have probably been added to previous diversity (Figure 17b, p. 49). Population movements that were triggered by a major arid phase within the Sahara occurred both on a micro-scale (within West Africa) and a macro-scale (in all sub-Saharan Africa). The latter is related to the successful expansion of the Bantu-speaking peoples belonging to the Niger-Congo group of languages (Benue-Congo). Its success is also in part related to the fact that it encouraged the progressive emptying of the Sahara, as well as an acculturation process in different areas. In sub-Saharan Africa, most of the indigenous languages, apart from few cases (KhoiSan isolate), have been dominated and replaced by Bantu languages.

When attempting to reconstruct the patterns of morphological diversification such as a "migration" for the Bantu-speakers, data from various disciplines are needed to support the hypotheses. First, linguistics showed that through language clustering the Bantu-speakers expansion operated as a "wave model", changing from a mother language and evolving through the movement of populations into two major streams, a western and an eastern one (Bastin *et al.*, 1983; Rexova *et al.*, 2006).

These two migratory trends reflect the geographical fragmentation of the Bantu-speakers expansion, and it has been suggested archaeologically through the diversification of food productions towards the south (Vansina, 1995). This geographical differentiation that has also been shown genetically, seems to be superimposed on a phenomenon of homogenization (due to a founder effect) (Cavalli *et al.*, 1994; Biondi *et al.*, 1996; Beleza *et al.*, 2005; Ramachandran *et al.*, 2005; Berniell-Lee *et al.*, 2009; Castri *et al.*, 2009). In terms of cranial morphology, the same observations have been done, although a geographical differentiation within Bantu-speaking populations is not always visible (Hiernaux, 1974; Vecchi *et al.*, 1987; Froment, 1998; Hájek *et al.*, 2008).

Furthermore, as paleoanthropological evidence is too scarce in West Central Africa, morphological variation has not been identified for the period of the early Bantu-speakers dispersal. Only very hypothetically, the Shum Laka specimens could be interpreted as belonging to this key-phase of proto-Bantu-speakers (Orban *et al.*, 1996; Ribot *et al.*, 2001).

Therefore, keeping in mind all these facts, is it possible to reconstruct a biological model for the Bantu-speakers expansion? Or was the gene flow too complex, especially in later phases? What are the major biological consequences on the diversity of sub-Saharan Africans? Are there any confounding factors (sexual dimorphism, selection, culture) that will obscure the picture?

In order to begin with some hypotheses and review all the factors involved in the phenomenon, a range of models (1 to 5) from simplest to more complex are proposed (Figure 17b). For all of them, the common factors that are the most frequently recognized are: (i) a main initial impulse of population movements from West Central Africa (Nigeria-Cameroon); and (ii) a principle of movement in a southern direction. The latter will be discussed mainly in terms of biological mechanisms rather than in relation to the exact trajectory of "cultural" dispersals. In addition, for each model that is increasingly complex, an important confounding factor (geography, ecology, culture) is superimposed on the effects of the Bantu-speakers expansion, in order to provide a more realistic picture.

The ***model 1*** presents a single and punctual migration expanding into an elliptical area and encountering empty niches. Within sub-Saharan Africa, it will result in high levels of biological homogeneity, showing clearly a common origin for the source of diversity. Historical factors are more important than environmental ones, as the selective pressures have not yet affected the biology. This model, reflecting a very short-term process, is of course completely unrealistic for the Bantu-speakers expansion. However, it provides a sort of null-hypothesis where the amount of factors involved in the phenomenon is minimal.

The ***model 2*** adds a "time" effect, as the migration is considered as a process subdivided into several phases: migrations occurred in successive layers. This phenomenon is a continuous and regular wave (like a "bow wave" pattern), with population size increasing progressively (Bastin *et al.*, 1983; Vansina, 1995). The biological consequences of such a model are also homogenization, but coupled with a slight gradient of diversification reflecting the effects of natural geographical barriers on an inter-regional level. These differences related to predominant gene flow also reflect broadly the two main streams (west-central, eastern) of the dispersal. This model agrees with both genetic results (Cavalli-Sforza *et al.*, 1988, 1994) and linguistic data (Holden, 2001; Rexova *et al.*, 2006).

The ***model 3*** is more complex because of the expansion trajectory. The inter-regional differentiation is even more visible, especially through both west-central and eastern dispersals. However, the successive layers of migrations related to this dichotomous pattern are slightly irregular going not only southwards but also backwards. As suggested by archaeological and ethno-historical data, the eastern and western migratory trends had some irregularities, and they intermingled together in a few places (e.g. southern D.R.C., Zambia) (Vansina, 1990). The biological consequence is not only homogenization, but also a marked gradient of diversification reflecting the effects of drift and/or selection, as the migration is here a more long-term process. This model agrees with the opinion of Hiernaux (1974), who supports the fact that the main differences left between Bantu-speakers populations are due to selective pressures (e.g. climate) operating in a relatively short-term period. It also fits better the "reality" of the Bantu-speakers spread, which did not correspond to sudden and punctual event(s), in contrast to the peopling of Americas (Lahr, 1995).

The ***model 4*** is similar to the previous one, but it adds the important aspect of population admixture. For example, indigenous populations have a biological impact on expanding Bantu-speaking agriculturists. Acculturations and/or an increasing network of symbiotic relationships between these various groups encourage population admixture in different ecological areas (savanna, forest). Although admixture with Pygmy hunter-gatherers is already present in Central Africa prior to the Bantu-speakers expansion, it will probably intensify later, especially in an uni-directional manner (e.g. Pygmy females marrying Bantu-speaking males) (Hiernaux, 1974; Cavalli-Sforza, 1986; Beleza *et al.*, 2005; Verdu *et al.*, 2009).

In addition to contact with hunter-gatherers (Pygmy, Hadza, KhoiSan), "proto-Bantu" horticulturalists also admix with previous East African inhabitants that already adopted agriculture and/or herding (Vansina, 1990). The biological consequences are broadly visible on two different geographical scales: inter-regional diversity increases between West-Central Africa, East Africa and South Africa, and this fact is coupled with a decrease in diversity within each region. However, the different types of homogenization effects (founder effect, population admixture) are intra-regionally difficult to distinguish between each other. As suggested by Hammer and Zegura (1996), a founder effect phenomenon suggesting a single origin can be obscured by recent intensive gene flow that in turn, could mimic multiple origins.

The ***model 5*** is one hypothesis related to genetic data that focuses only on the possible effects of some social features that could be superimposed on previous hypotheses. The degree of genetic differentiation and homogenization varies in intensity, according to not only geography, ecology and history (e.g. Bantu-speakers expansion), but also sex (e.g. differential mobility due to mating patterns). In fact, (West-) Central Africa and East Africa can be subdivided along the equator into two common areas (where societies with a specific type of rule of descent are predominant), such as a northern belt corresponding to patrilineal societies, and a southern one corresponding to matrilineal societies (Murdock, 1959).

According to Seielstad *et al.* (1998), the following trends could be hypothesized within sub-Saharan Africa:

i) in both patrilineal and matrilineal societies, the populations of the "dominant" sex present a higher diversity on an inter-regional level and a lower diversity on an intra-regional level; an opposite picture is provided for the "non-dominant" sex;
ii) subsequently, inter-regional differences are increased in two cases, among the male populations of patrilineal societies, and possibly the female populations of matrilineal societies; but, for the intra-regional differences, an opposite picture is also provided (Table 1, p. 43).

These five models in relation to the Bantu-speakers expansion are of course very schematical. They do not take account of all the factors and changes that might have occurred in detail through time. Nevertheless, they could provide working hypotheses when interpreting morphological diversity on different levels (geography, ecology, history, social features).

2.4.3. After the Bantu-speakers expansion: additional homogenization due to increasing population admixture?

It is important to remember that there is an endless list of historical events, other than the Bantu-speakers expansion. During the last five centuries, African history becomes increasingly complex as new ecological and economic factors were also involved. For example, the new trade systems and products (e.g. slaves, foreign cultigens), the increasing desertification and urbanization, are a series of processes which populated or de-populated to various degrees West and Central Africa. Furthermore, European colonization also altered the pattern of inter-ethnic gene flow.

Since the 15th century AD, history shows that there is an exponential increase of population movements on a micro-scale because of recurrent droughts and/or political reasons (resistance to slave trade, influential empires) (Figure 17c, p. 49). As we saw previously (section 2.3.3.2.2, p. 36), the Middle Niger Valley has been the scene of successive layers of peopling such as the Tellem and Dogon (Huizinga *et al.*, 1967). According to Spedini and Bailly (1987), Cameroon has also been populated recently by a wide array of populations that were originally non-Bantu-speaking (Bamileke).

As suggested by Papastergiadis (2000), the complexity of forms of migrations had reached unprecedented levels, and it is impossible to map movements with a series of arrows.

All these recent events probably shaped the present linguistic, "cultural" and biological differences between Africans. However, the Late Pleistocene should also be explored, as human variation will be dramatically decreased in the Holocene (Ramachandran *et al.*, 2005; Crevecoeur *et al.*, 2009). As a consequence, in order to explore Holocene diversity within West and Central Africa, it was important here, through both ancient and recent periods, to review briefly the major aspects responsible for the variation.

Next (Chapters **4 and 5**), the microevolutionary models presented in this chapter will be tested systematically through a "multi-layered" approach.

Each potential source of diversity (geography, ecology, ethno-history and sexual dimorphism) will be analysed in relative isolation: it will be kept in mind that the effects of various factors are difficult to separate, especially when trying to understand the biological effects of the Bantu-speakers expansion.

In **Chapter 4**, ***model 2*** that underlines geography as the key-element of population differentiation will be tested on a broad inter-regional level. The latter will be also approached in relation to sexual dimorphism and selection (e.g. climate).

In **Chapter 5**, ***models 3*** and ***4*** that added a historical dimension will be tested through various ethno-linguistic data.

In fact, two models (***1*** and ***5***) will be used for only secondary purposes. ***Model 1*** is considered as the initial null-hypothesis where the effects of the Bantu-speakers expansion are not confused with other sources of diversity. In addition, ***model 5*** is presented as one hypothesis related to genetic data that may also explain morphology, although in a very speculative manner.

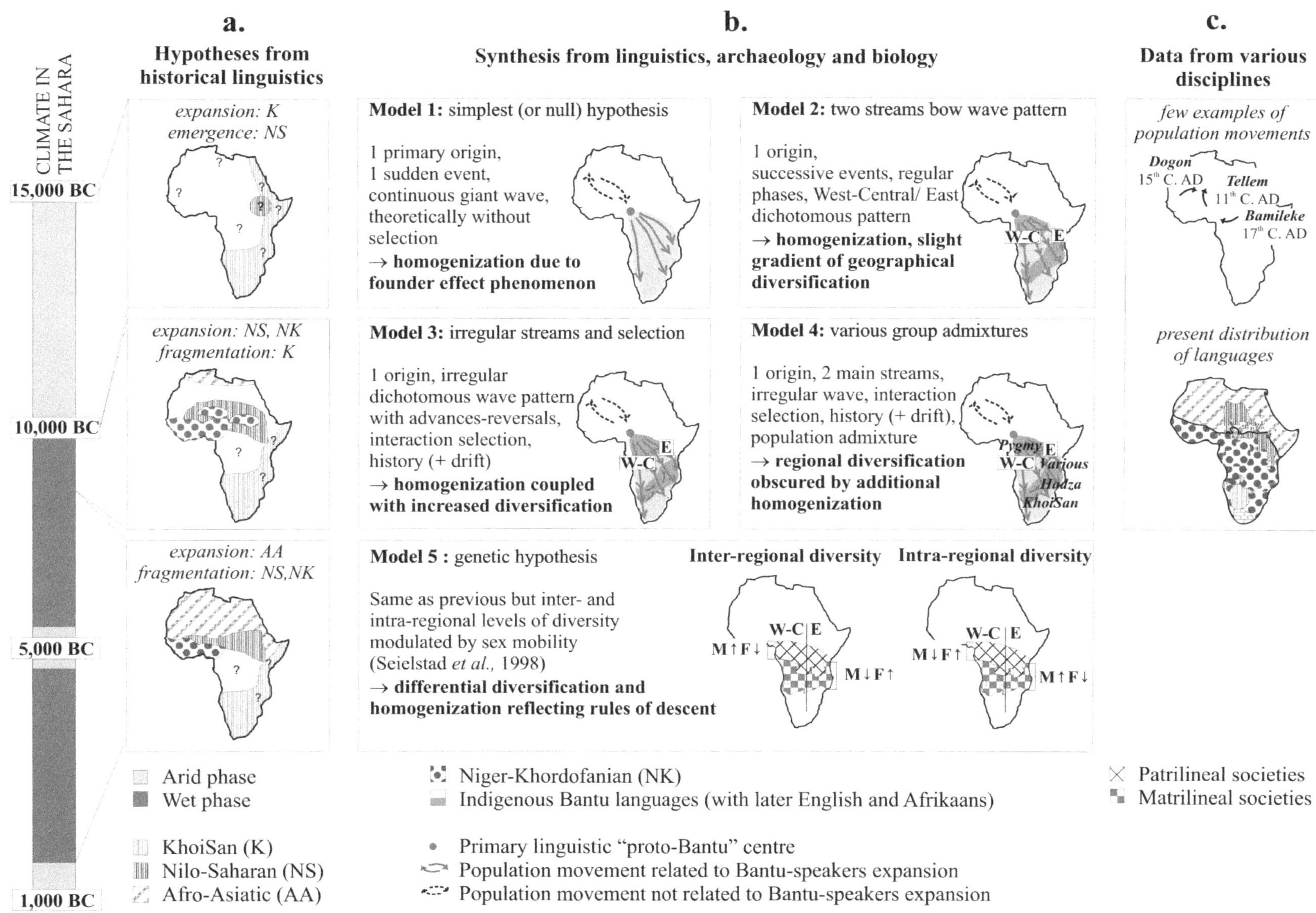

Figure 17. Major events having affected population differentiation and biological diversity within Africa during the Holocene. **a.** Before the Bantu-speakers expansion (15,000-1,000 BC); **b.** During the Bantu-speakers expansion (1,000-15th C. AD); **c.** After the Bantu-speakers expansion (since 15th C. AD).

CHAPTER 3

POPULATIONS AND VARIABLES UNDER STUDY

Prior to analysis, the nature of the sample and the selected methods of our present research are described. These preliminary aspects will be the key elements of our subsequent analysis and interpretations (Appendices I-V, p. 187-198).

3.1. The entire sample

First, the spatial and chronological distribution of the entire sample is presented on both continental and regional levels.

3.1.1. Chronological and geographical classifications

The entire sample was subdivided into two broad chronological groups:

i) the modern sample, dated to the last two centuries, coincides with both colonial and post-colonial periods;
ii) the archaeological sample, dated before the 19th century AD corresponds to pre-colonial times, from the Later Stone Age up to the Late Iron Age (last 16,000 years).

In order to classify the African sample into broad geographical parameters, the following regions were used for this purpose: West Africa (including West-Central region such as Cameroon and Nigeria), Central Africa (including Pygmy), East Africa (including the Horn), Southern Africa (KhoiSan and Bantu-speakers) and North Africa (Figure 18, p. 51).

These groupings allow us to perceive the broad geographical origin of each sub-sample under study. Although they are approximate, they are based on the existence of physical geographical features or natural barriers observed all over the African continent (Figure 3, p. 8). In fact, this classification has been already used by other authors for ethnographical purposes (Giles *et al.*, 1997a-e).

In order to test hypotheses not related to a purely geographical context, Pygmy and KhoiSan populations will also be treated separately as much as possible, and a few other ethnic groupings will be used (Murdock, 1959).

3.1.2. Summary

A total sample of 2966 individuals originating mostly from Africa (84%; N=2501) formed the empirical basis of this study. Table 2 (p. 51) summarises the material, according to skeletal origin of data (compilation or personal work), geography and broad chronological period. The various skeletal series originated from at least 20 known places of curation such as museums, universities and research laboratories (Appendices I & III, p. 187-191 & 195).

The sample consists of data, which have been both compiled from the literature (69%) and collected by the author (31%). The compiled data set originated from twenty-seven papers, and most of them (51%) corresponded to modern material.

On a broad geographical level, the majority of data originated from West Africa (32%), Central Africa (18%) and North Africa (10%), with smaller samples from KhoiSan South Africa (4%) and Pygmy Central Africa (1%). The West African sample accounts for 90% of the mandibular data, as well as 37% of the cranial data and 40% of the combined data (where a mandible is associated with a cranium).

Most of the sample (80%) dates within the last two centuries and the remainder comes from various Later Stone Age and Late Iron Age sites.

Most of the data (53%) is related to the cranium only; with the remainder corresponding to both the associated cranium/mandible and the mandible only. Eighty-seven percent of the mandibular data were compiled, as well as 60% of the cranial data, and 68% of the associated cranial/mandibular data.

Forty-seven percent of the total compiled data set corresponds to crania only. This is due to the fact that crania are often better preserved than mandibles (e.g. funerary rituals, interests of past collections).

Due to the better preservation of cranial samples, the majority (40%) of personal data includes crania without mandibles. Nevertheless 25% of the personal data includes mandibles with or without matching crania.

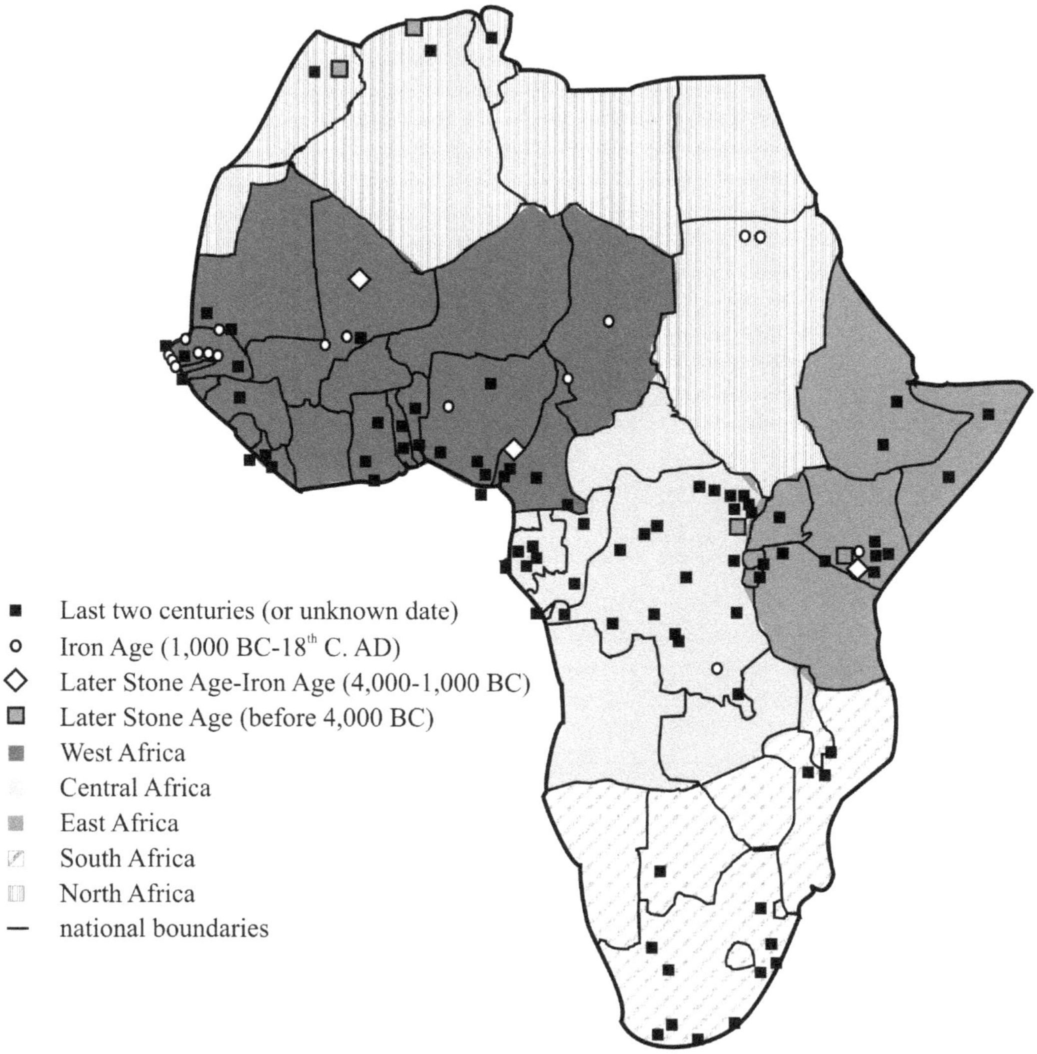

Figure 18. Chronological and geographical distribution of the sample under study.

	Modern data	Archaeological data[1]	Total sample
Origin of data			
Publications compilation	1527	522	2049
Personal data collection	854	66	920
Geography			
West Africa	531	416	947
Central Africa	447	15	462
Pygmy (Central & East Africa)	28	-	28
East Africa	577	24	601
Bantu-speaking Southern Africa	125	-	125
KhoiSan Southern Africa	116	-	116
North Africa	92	130	222
Worldwide	465	-	465
Total sample	2381	585	**2966**

[1] Before 19th century AD.

Table 2. Summary of all materials available: origin of data, geography and chronology.

3.2. Modern populations

The modern crania (N=2381) are described here in detail in Appendix I (p. 187-191), according to their geographical and ethnic origin, as well as various other aspects explored during the present study (e.g. geographical origin, environmental background, ethno-linguistic affiliation, mode of subsistence, historical data).

3.2.1. Compiled and personal data

Twenty-one papers have been used for data collection of modern material (N=1527). The author that provided the largest amount of data is Górnay (1957) on the Ugandese (N=176).

The latter represent 20% of the total compiled modern sample. Kitson (1931) and Benington (1912) also contributed to the data set with large samples (N>100) (18%).

A North African sample was obtained from Storto (*pers.comm.*) (N=81). It included data from four countries: Algeria (44%), Spain, Tenerife Islands (11%), Morocco (20%) and Tunisia (25%).

A sub-sample of the cranial database of Howells (1973, 1989, 1996) was also for comparative purposes on a worldwide scale. It consisted of 475 individuals, who originated from six broad geographical areas: Europe (25%), Australia including Melanesia (14%), Polynesia and West Pacific (24%), Americas (including Eskimos and Arctic) (28%), Asia (including India, Far East and South East Asia (6%) and North Africa (3%).

Data on both crania and mandibles (N=854) were collected during seven sessions of fieldwork between June 1998 and October 2000. Fieldwork focused on the skeletal collections of the Institut Royal des Sciences Naturelles of Belgium (IRScNB) at Brussels (N=328), as they had never been studied before, and are the largest ones for Central and East Africa. In fact, these African collections provided 45% of the total sample of the modern data that I collected personally.

3.2.2. Grouping variables

In order to test morphological patterns at different levels, the modern sample is also subdivided into different units using 10 grouping variables in relation to geography (N=3), ecology (N=5), and culture (N=2) (Table 3) (Appendix II, p. 192-194).

Geographical coordinates that were obtained by reading maps (Gazetters, 1954-1982; Times Atlas, 1999) are used as both continuous and categorical variables. Unfortunately, a large sample compiled from Górnay's data (1957) (N=176) could not be localized accurately within Uganda because of the lack of information. The same problem was encountered with other individuals originating

Variable	Type		References	Method of recording
	Continuous	Categorical		
GEOGRAPHICAL VARIABLES				
Broad regional groupings	-	X (6 entries)	Times Atlas (1999); Gazetters (1954-1982)	map data (scale: 1/5,000,000)
Latitude	X (in degrees)	X (9 entries)		
Longitude	X (in degrees)	X (7 entries)		
ECOLOGICAL VARIABLES				
Distance from equator	X	X (6 entries)	Times Atlas (1999)	(scale: 1/5,000,000)
Vegetation	-	X (7 entries)	Philip (1998)	map data: 1/8,000,000
Humidity in January	-	X (9 entries)	Jackson (1961)	map data (1/50,000)
Humidity in July	-	X (9 entries)		
Temperature	X	-	Idrisi (Eastman, 1997)	Collard (*pers.comm.*)
Rainfall	X	-		
CULTURAL VARIABLES				
Linguistic affiliation	-	X (8 entries)	Greenberg (1964); Holden (2002)	compiled literature
Ethno-linguistic units	-	X(15 entries)	Murdock (1959)	

Table 3. List of the geographical, ecological and cultural variables under study.

from North Africa (N=81; Storto, *pers.comm.*).

A range of ecological data was also assigned to each individual when the geographical origin of the latter was known. The first ecological variable, distance from equator, is categorical, and it was obtained from the previous geographical data based on latitude.

Two variables that were obtained from both the programme Idrisi (Eastman, 1997) and the personal data of I. Collard (*pers.comm.*) are continuous (annual mean temperature, total annual mean rainfall). Three other variables are categorical and they were also obtained from various atlases: mean humidity for both January and July (Jackson, 1961) and vegetation (Philip, 1998; Times Atlas, 1999).

Linguistic affiliation based on Greenberg's classification (1964) corresponded to the main subdivisions within the Niger-Congo family. The Benue-Congo languages, which correspond mainly to Bantu languages, are subdivided into three main groups according to the recent work of Holden (2001, 2002, *pers.comm.*).

Murdock's (1959) ethno-linguistic units are used here, as they combine various (geographical, ecological or linguistic) aspects. They correspond to nine Bantu-speaking units and six non-Bantu-speaking ones. However, at least 15% of individuals could not be assigned to any ethno-linguistic classification due to the lack of information again.

3.2.3. Large known ethnic groups

The modern sample under study corresponds to only a very small fraction (<5%) of Africa's many ethnic groups (around 1,800 communities; Olson, 1996). To a certain extent, it was an arbitrary choice, as it depended upon the availability of the written records necessary for checking the geographical and ethnic origin in the archives.

In total, 85 ethnic groups were identified (Figure 19, p. 54). However, only 10 of them had at least a sample size of 30 individuals (N=986; 33% of entire modern sample). These were: for East Africa, the Teita, the Bahutu, the Haya and the Dschagga; for Central Africa, the Basuku, the N'Komi and the Batetela; and for West Africa, the Ashanti, the Ibo and the Griots Lebu (Table 4, p 55).

Large samples that have been localized in Cameroon, such as the Grassfields (N=61) and the South-West region (N=32), represent at least 15 different ethnic groups pooled together. A number of individuals although geographically localized (N=415) remained of completely unknown ethnic origin, in particular the sample originating from Uganda (17%).

3.2.4. Ethno-historical information

Ethno-historical information presented below belonged mainly to four ethnic groups: the Bahutu, Basuku, Ashanti and Griots.

Bahutu (N=103)
Two samples of Bahutu (or Hutu) originating from Rwanda were available at the Institut Royal des Sciences Naturelles, Brussels. The first corresponds to a small sample (N=15) collected by R. Van Saceghem between 1922 and 1935 in the Bugoye region (Kissenyi territory). The Bahutu crania were found on small islands in Lake Kivu that were used as cemeteries for women, who had died during pregnancy. The second corresponds to a much larger series of crania (N=88) collected by N. Nenquin in 1960 during his stay in Rwanda. According to his notes in the archives of the Royal Museum of Central Africa (Tervuren, Belgium) that were sent to Brabant (1963), he found the crania in a volcanic cave close to the city of Ruhengeri. These crania most probably belonged to a Bahutu group, who had been hiding in the surroundings of the cave in order to escape the Germans during the First World War (1914-1918). They therefore date within the last century.

The Bahutu are mainly concentrated in Rwanda and speak an eastern Bantu language (Rwanda). They also resulted from one of a series of waves of the Bantu-speakers expanding southwards. At *c.* 2nd century AD, they probably migrated from the Chad-Niger region to the area of Burundi-Rwanda forcing the previous inhabitants (Twa pygmies) to retreat (Yakan, 1999). Between the 14th and 18th centuries AD, they were subjugated by the Tutsi, who arrived from the Nile Valley or Ethiopia and set up a system of feudal classes. Although they adopted cattle from the Tutsi, the Bahutu are still mainly agriculturists. Their society is also based on patrilineal clans ruled by kings (Murdock, 1959).

Basuku (N=168)
A large Congolese sample was also available (Institut Royal des Sciences Naturelles, Brussels).

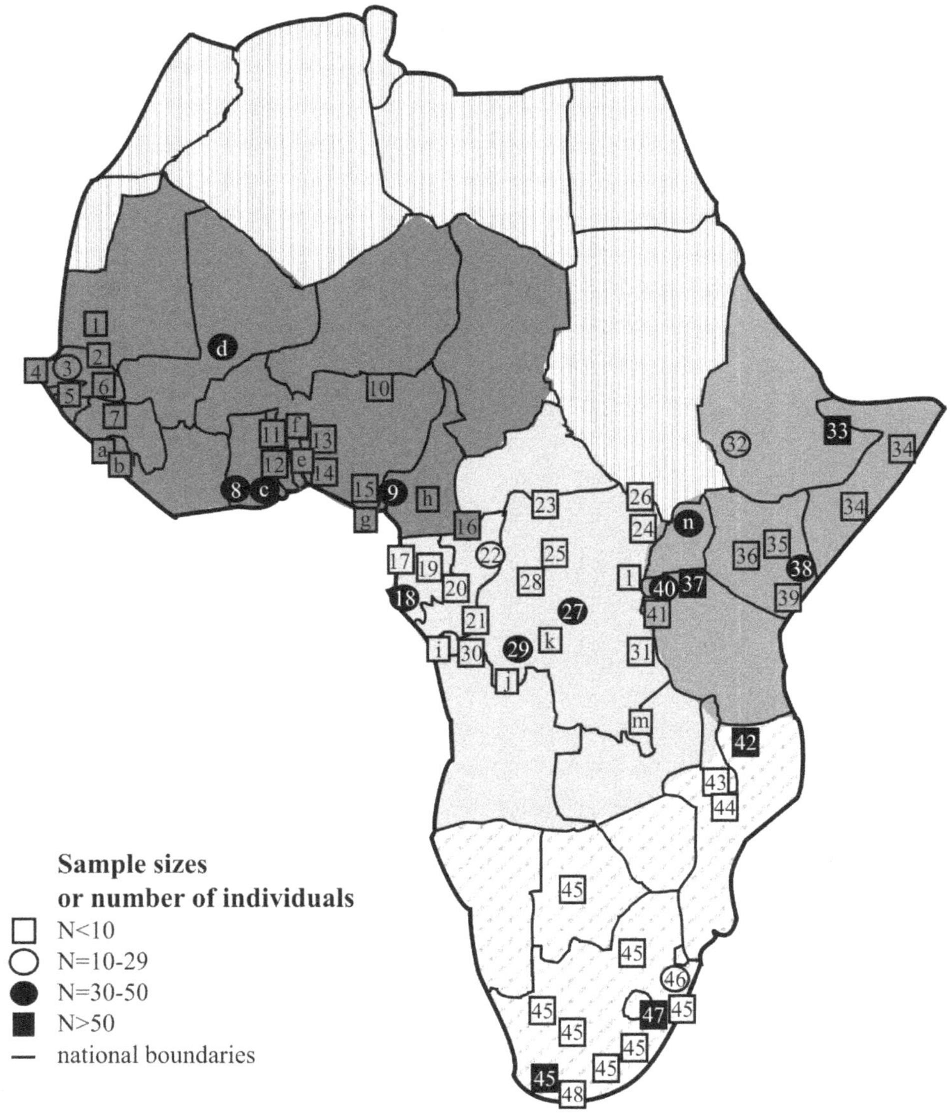

Ethnic groups

1 Maure (?)
2 Toucouleur
3 griots Serer
4 griots Lebu
5 Diola
6 Malinke (?)
7 Mandingo
8 Ashanti
9 various (Bana, Bali, Ndu, Bafulero, Basu, Bamum, Babungo, Mbo, Boki, Anyang, Banjangi, Anjang, Keaka, etc...)
10 Hausa
11 Yendi
12 Ewe
13 Yoruba
14 Ibo
15 Ekoi
16 Pygmy South Cameroon
17 M'Pongwe
18 N'Komi
19 N'Javi, Adouma, Ashango
20 Pygmy Babongo
21 Teke
22 Pygmy Babinga
23 Azande Mamvu
24 Pygmy Irumu
25 Bassoko
26 Medje, Mayanga, Makere, Bafulero, etc...
27 Batetela
28 Mongo
29 Basuku
30 Basundi
31 Warua
32 various Galla
33 Somali, Galla, Beja
34 Somali
35 Wapare
36 Massai
37 Haya
38 Teita
39 Dschagga
40 Bahutu
41 various Tutsi
42 Wahiou
43 Wahenga
44 Manganja
45 KhoiSan
46 Zulu
47 Xhosa
48 Angoni

Geographical locations

a Monrovia
b Vai Camp
c Gold Coast
d Nokara
e Abome
f Whydah
g Calabar
h Yaounde
i Loango, Kabinda
j Low Congo
k Kassai
l Kivu
m Sakania
n various Uganda

Figure 19. Distribution of the modern sample of known ethnic affiliation within sub-Saharan Africa.

Ethnic groups	cranium only	mandible only	cranium & mandible	Total sample
East Africa				
Teita	129	1	9	139
Bahutu	88	-	-	88
Haya	38	-	-	38
Dschagga	28	1	1	30
Central Africa				
Basuku	168	-	-	168
N'Komi	-	-	133	133
Batetela	77	-	-	77
West Africa				
Ashanti	59	-	37	96
Ibo	35	-	-	35
Griots Lebu	19	10	4	33
				986

Table 4. Modern sample of known ethnic affiliation with at least a sample size of 30 individuals.

According to the archived notes of Bequaert, Curator at the Royal Museum of Central Africa (Tervuren, Belgium), a few isolated burials were found in the territory of Feshi (south west D.R.C.). Van de Ginste (1946) collected various crania from this area. He was in fact carrying out an anthropometrical study on living populations of the same area.

Most of the sample probably belongs to the Bantu-speaking Basuku (or Suku), who represent the dominant group in the Kwango district (longitude 17-18°E and latitude 5-6°S). However, the hypothesis that other ethnic groups (Batsamba, Bawongani, Bapende, Baluwa-Basonde) were present, cannot be completely excluded. According to Bequaert's notes, these groups show affinities with the Basuku in many aspects (e.g. habitat, language, origin, customs), but the Baluwa-Basonde may present some differences. According to Van de Ginste (1946), Basuku and Batsamba originated from the Congo Kingdom.

After the 16th-18th centuries AD, they migrated southward to the Kwango district, where they had contact with the Bawongani and Bapende, who had been displaced as well. By the end of the 19th century AD, the Basuku also came into contact with Portuguese traders and settlers. Recently they have been influenced by the (Ba-)Kongo people from the Kasai region.

The Basuku and their neighbouring ethnic groups are essentially agriculturists. In addition, the Batsamba are blacksmiths, but they do not own land apart from their house and fields, in contrast to the Basuku. Both Batsamba and Basuku are matrilineal societies, with the latter being governed by royal lineages (Yakan, 1999).

Ashanti (N=96)

Two samples of Ashanti (or Asante) crania from Ghana were available. The first series (N=58) studied previously by Shrubsall (1899) was collected during an expedition to Kumasi. The second is part of the German Van Luschen collection that was donated in 1924 to the American Museum of Natural History in New York, but its precise geographical origin remains unknown.

Ashanti people speak a Kwa language (Twi) of the Niger-Congo family (Murdock, 1959). Although they presently live mainly in Ghana, the hypothesis that a portion of the sample may also come from Ivory Coast cannot be excluded completely.

According to historical evidence, their origin remains unknown. However, they probably moved between the 11th and 13th centuries AD to their present location, where they unified in the 17th century AD as a nation, after being joined by various clans (Giles, 1997; Yakan, 1999). They are currently farmers, but in the past they produced and traded much gold and kola nuts to both northern and coastal areas. Ashanti societies consist of clans comprised of groups of families that are mainly matrilineal (Yakan, 1999).

Griots (N=45)
A sample of Griots crania from West Senegal were studied at the IFAN (Cheikh Anta Diop Institute, Dakar). It was collected by Dr. G. Thilmans between 1959 and 1965 in Dakar (Lebu) and Serer regions (Thilmans, 1968). The skeletons were originally buried within the trunk of a baobab, as the Griots had funerary practices specific to their group (Mauny, 1955). In fact, they were considered as a lower professional cast of bards or story tellers, living among the Serer and the Lebu groups. They were endogamous within their group, and their society was essentially egalitarian (Murdock, 1959; Giles, 1997e). During the last few generations, this cast has been assimilated and is difficult to isolate as a group.

3.3. Later Stone Age and Iron Age samples

Most of the archaeological samples (N=585) is dated to the (Late) Iron Age period (82%) with a small portion dated to the Later Stone Age (11%) and the transitional Later Stone Age-Iron Age period (7%) (Table 5, p. 57). Furthermore, 74% of the mandibular data corresponds to Iron Age samples.

Information on data compilation, personal work, geographical origin and dates is summarised for each sample in Appendix III (p. 195).

3.3.1. Compiled and personal data

Nine papers provided the compiled data set for the Later Stone Age and Iron Age samples (N=535; 89% of total archaeological sample) (Figure 20, p. 58). Most of the publications (seven) correspond to very small samples (N<10). The other two provided the most. Huizinga *et al.* (1967) included 337 Malian individuals (55% of total sample), followed by Knip (1970) with 71 Sudanese individuals (12% of total sample).

Archaeological collections studied by myself (N=66) were deposited at different institutions, but most of the personal data (74%) was collected at the Institut Cheikh Anta Diop in Dakar.

The majority of the specimens under study originated from West Africa (69%) and were dated mostly to the Iron Age period (82%; N=495). However, 40 individuals originating from various regions were dated to the Later Stone Age-Iron Age transition (e.g. Shum Laka, Hassi-el-Abiod).

Finally, for the Later Stone Age, the number of skulls available for sub-Saharan Africa was much smaller (N=9). A North African sample was added mainly from the personal data collected by Lahr (N=57).

3.3.2. Archaeological information

3.3.2.1. Main samples

The archaeological background related to eighteen sites, where most of human remains (86%; N=521) were discovered, is briefly described in Table 5 for each broad chronological period. However, additional information on the remaining archaeological sample originating from seven other sites (N=82*)* is also presented in Appendix III (p. 195) and in **Chapter 2** (p. 34-38).

3.3.2.2. The Iron Age Senegalese samples

The series of Senegalese Iron Age crania originate from four coastal settlements (Cap des Biches, Faboura, Dioronboumak and Cayar) and one inland site (Walalde) (Figure 21, p. 59). Three sites (Cap des Biches, Faboura and Dioronboumak) are dated to the first millennium AD and two others (Cayar, Walalde) to the first millennium BC.

Cap des Biches (N=12)
A series of skeletons close to Dakar and Rufisque in Western Senegal were found during the construction of an electrical power station in 1965 by Drs Hugot and Ziegler (Thilmans, *pers.comm.*). Only relative dating, based on pottery styles similar to those of Faboura, dated the remains to *c.* 1st-7th centuries AD.

Faboura (N=13)
Coastal shell middens found during road works in Senegal revealed human remains (Descamps and Thilmans, 1997). The site has been radiocarbon dated to a period between the 1st to 7th century AD (1,315±80 BP-1,940±80 BP).

Dioronboumak (N=18)
Two tumuli were excavated at Dioronboumak in the Saloum delta of the West-Central region of Senegal. They were built over shell middens during at least a 500 year period from the 4th century AD to the 8th century AD (1,580±80 BP-1,140±80 BP) (Thilmans and Descamps, 1982). From then until now, populations in this region have exploited marine resources, accumulating shells in deposits of various sizes.

SITE NAME	LOCATION	DATE	ARCHAEOLOGICAL CONTEXT	REFERENCES	N[1]
IRON AGE (1,000 BC – 18th century AD)					
Kainji	North West Nigeria	after 16th C. AD	Late Iron Age	Wade (1971)	10
Tellem	Niger valley, Central Mali	Caves B to Z: 11th-13th C. AD	caves containing disarticulated skeletons	Huizinga *et al.* (1967) Knip (1971)	266
Sanga	Upemba Valley, South East D.R.C.	8th-15th C. AD	Kamilambian & Kisalian pottery traditions	Nenquin (1963) Hiernaux *et al.* (1971) de Maret (1977,1985)	6
Duweishat	North Sudan, western bank of Nile	Christian period	Christian cemetery	Knip (1970)	37
Attiri	North Sudan, western bank of Nile	6th-14th C. AD	Christian cemetery	Knip (1970)	34
Cap des Biches	near Dakar & Rufisque, West. Senegal	1st-7th C. AD	individual burial with pottery style similar to the site of Faboura	Thilmans (*pers.comm.*)	12
Faboura	Saloum delta, West Central Senegal	1st-7th C. AD (1,315±80 BP-1,940±80 BP)	coastal shell middens	Descamps & Thilmans (1997)	13
Dioronboumak	Saloum delta, West Central Senegal	4th-8th C. AD	two tumuli built over coastal shell middens	Thilmans & Descamps (1982)	18
Koro Toro	Nemra region, Central Chad	Before 5th C. AD	burial mounds of Pre-Islamic period	Treinen (1975)	1
Tellem	Niger valley, Central Mali	Cave A: 4th C. BC	caves containing disarticulated skeletons	Huizinga *et al.* (1967) Knip (1971)	71
Cayar	north Dakar, North West Senegal	around first millennium BC	coastal shell middens	Thilmans (*pers.comm.*)	2
Walalde	Middle Senegal Valley, North East Senegal	not < 6th C. BC	(semi-) sedentary settlements	Abu Deme (*pers. comm.*), McIntosh & McIntosh (1992), Bocoum *et al.* (1999)	4
LATER STONE AGE &/or IRON AGE (before 1,000 BC)					
Shum Laka	Grassfields, North West Cameroon	1st millennium BC	transitional phase: rockshelter burials	de Maret *et al.* (1987, 1993)	1
Ishango	Upper Semliki Valley, North East D.R.C.	5,000 BC-13th C. AD	"Mesolithic" & more recent levels	De Heinzelin (1962) Boaz *et al.* (1990)	6
Hassi-el-Abiod	North West Mali	5th millennium BC	burials with or without tumuli	Dutour (1989)	27
Bromhead	West of Rift Valley, South Central Kenya	6th millennium BC	"Elmenteitan Neolithic"	Leakey (1931, 1935) Protsch (1978)	11
Gamble Cave II	West of Rift Valley, South Central Kenya	7th millennium BC	"Upper Capsian C"	Leakey (1931, 1935) Protsch (1978)	2
				Total =	**521**

[1] Number of individuals.

Table 5. Archaeological information for sites under study with human remains.

Cayar (N=2)
Coastal shell middens have been also excavated at Cayar in a region just north of Dakar in North West Senegal. The human remains have been dated to *c.* the first millennium BC (Thilmans, *pers.comm.*).

Walalde (N=4)
The site of Walalde in the Middle Senegal Valley of North East Senegal provided four skeletons that have not been radiocarbon dated yet (McIntosh and McIntosh, 1992; Bocoum *et al.*, 1999). According to Abu Deme (*pers.comm.*), the occupation probably belonged to a period not before the 6th century BC. A series of other contemporaneous sites (e.g. Culabel, Sioure, and Malam) in the same region could reflect some sort of (semi-) sedentary settlements.

This area is nowadays peopled essentially by Toucouleur populations. The question of a possible "cultural" and biological continuity between the latter and the Walalde specimens has not yet been tested.

3.4. Sex determination

As information on sex determination was not always provided in the publications, it was attempted to compensate for this lack of information during data collection.

Although the skull does not provide the most reliable features related to sexual dimorphism, sex

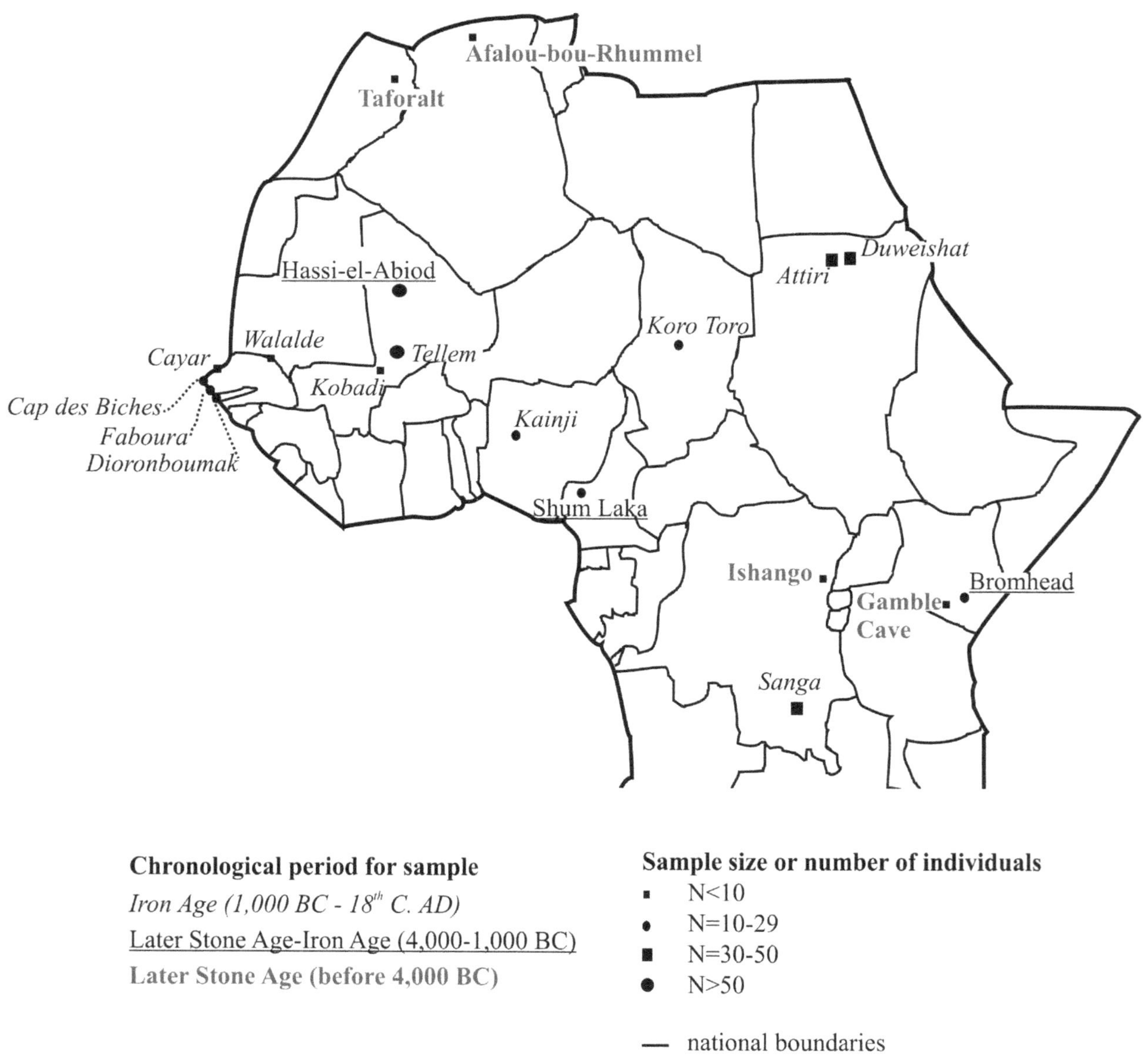

Figure 20. Geographical distribution of Later Stone Age and Iron Age African samples under study.

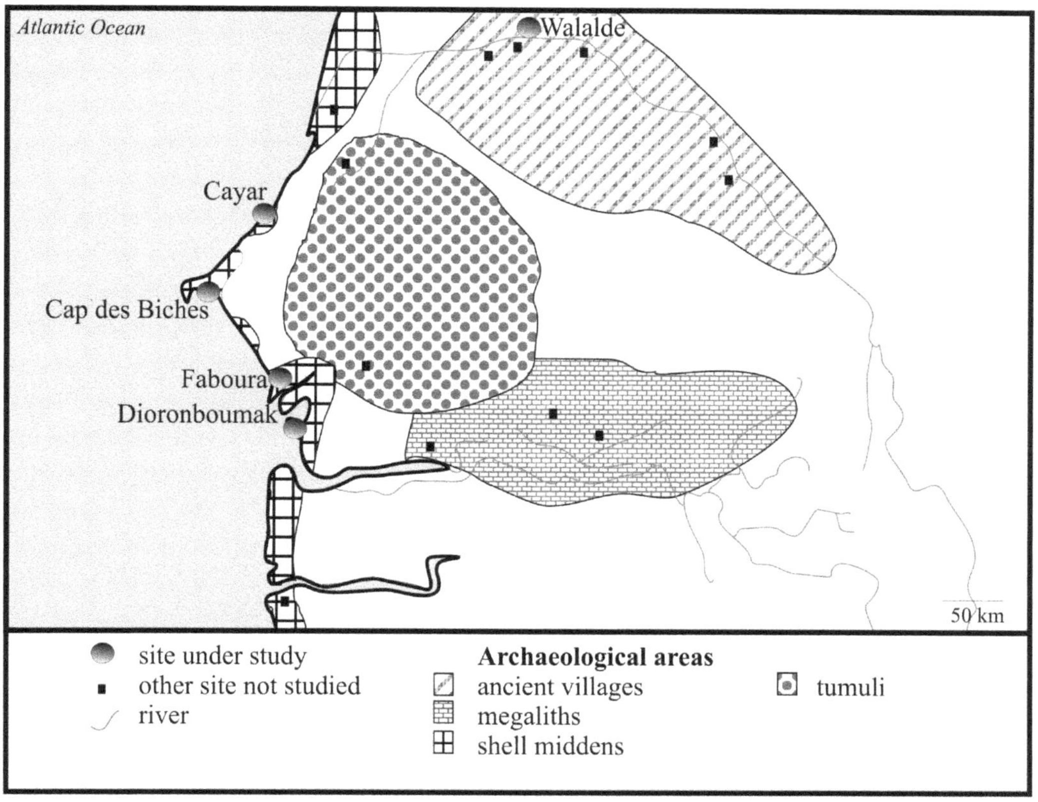

Figure 21. Map showing the detailed distribution of the Senegalese Iron Age samples.

was assessed morphologically according to the methods used by Ferembach *et al.* (1980).

However, as the sex-related morphological traits, dealing essentially with robustness are recommended especially for European populations, they had to be adapted to African populations. For example, the frontal inclination is not considered here, because both females and males amongst Africans generally tend to present a full rounded frontal contour (Krogman and Iscan, 1986). As Africans also often have a more U-shaped dental arch than Europeans, this last feature is also not taken into account for sex determination.

In total, most of the skulls were male (42%). Thirty percent of our entire sample could not be sexed properly, due to the lack of indicative features or information in the publications.

3.5. Metric traits

Sixty-one metric traits for the skull were recorded. Their definitions are presented in detail in the Appendices IV & V (p. 196-197 & 198), and their anatomical localization is also broadly represented on schematic drawings (Figures 22-24, p. 60-62). All measurements are taken to the nearest millimetre on both sides, using several instruments (e.g. sliding and spreading calipers, coordinate caliper, mandible board). However, the CC variable was not measured, as it was obtained only from the compiled data.

Agreements and disagreements for each measurement definition have been reviewed in various scientific reviews of anthropometry, which were in use before the Second World War in both England (e.g. *Journal of the Royal Anthropological Institute of Great Britain & Ireland, Biometrika*) and Germany (e.g. *Archives für Antropologie*), as well as in several reference manuals (e.g. Martin and Saller, 1959; Howells, 1989; Lahr, 1996). Eighty-four percent of the variables relates to the cranium (51 variables) and most of which relates to the face (41%; 25 variables). For the mandible, 10 metric traits were recorded.

Table 6 (p. 63) shows the number of variables that have been used previously by different authors. Twenty-nine variables under study (48%) correspond to those that were selected (and sometimes devised) by Howells (1989). However, 48 metric traits (79%) were also used by various authors, who published in *Biometrika* (e.g. Shrubsall, 1897; Benington, 1912). Although they

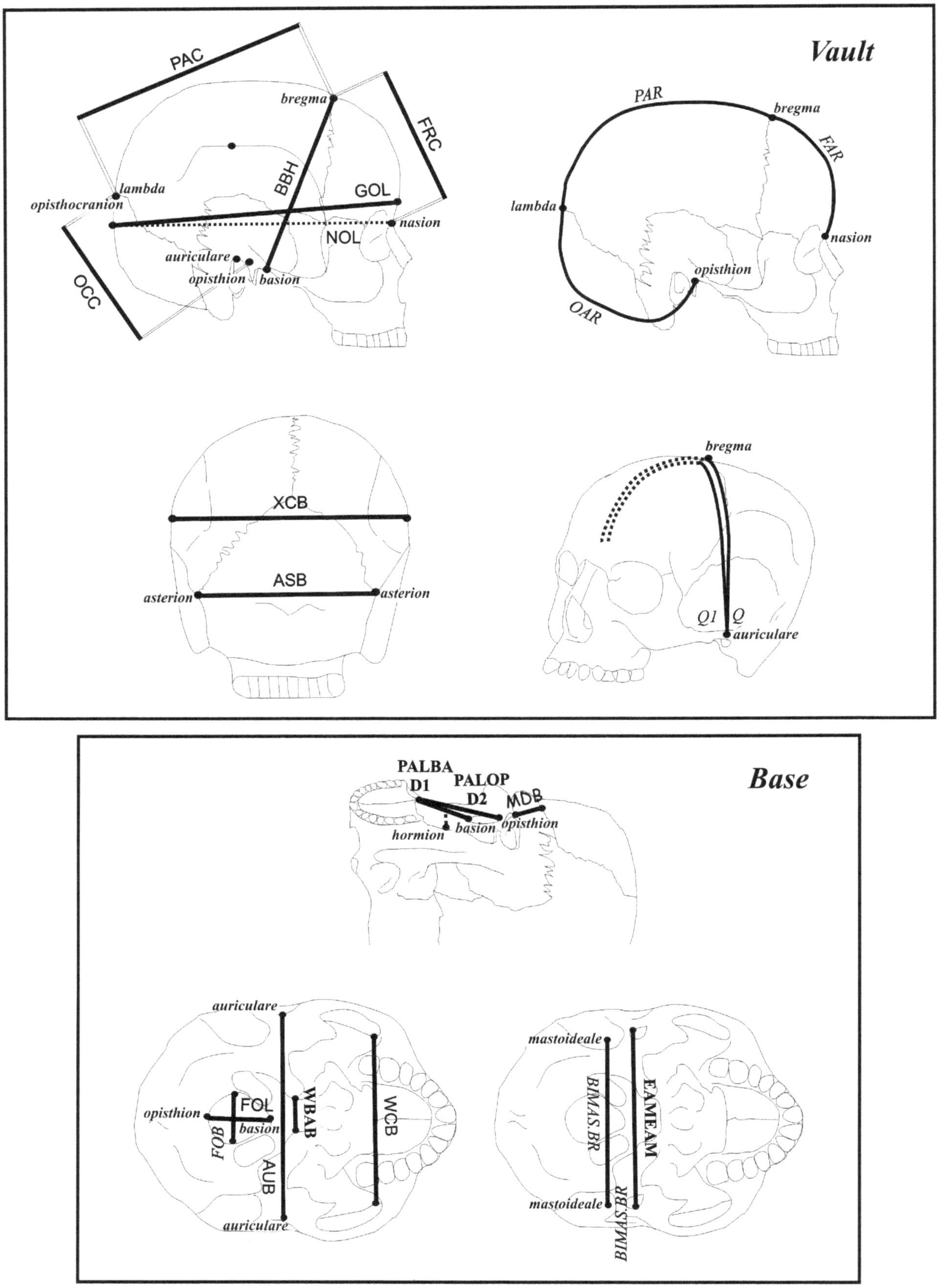

Biometrika &/or Martin & Saller, 1959
Howells, 1989
Lahr, 1996
Giles & Elliot, 1963

Figure 22. Schematic drawings of the skull showing the anatomical localization of metric traits related to vault and base.

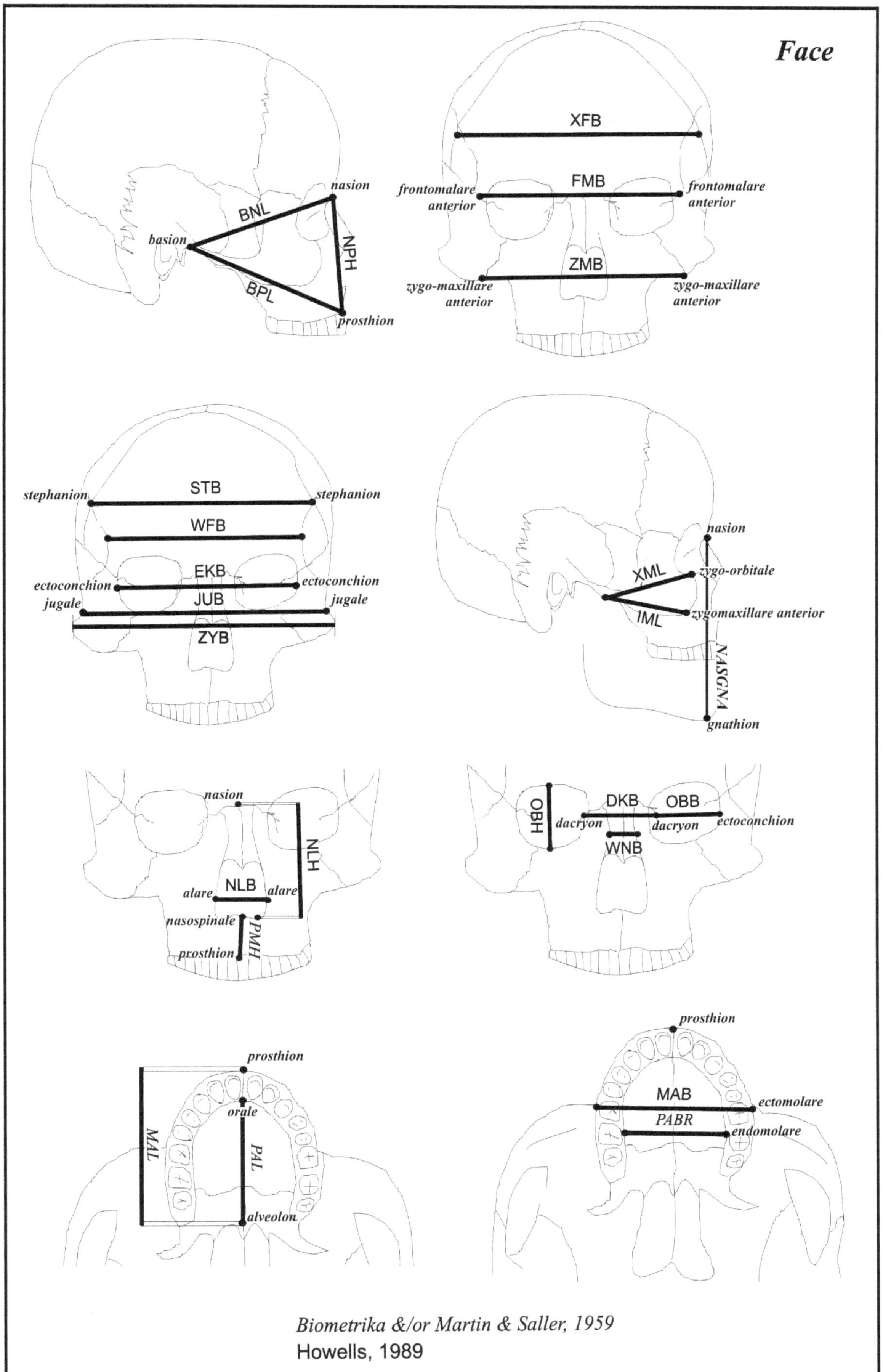

Figure 23. Schematic drawings of the skull showing the anatomical localization of metric traits related to face.

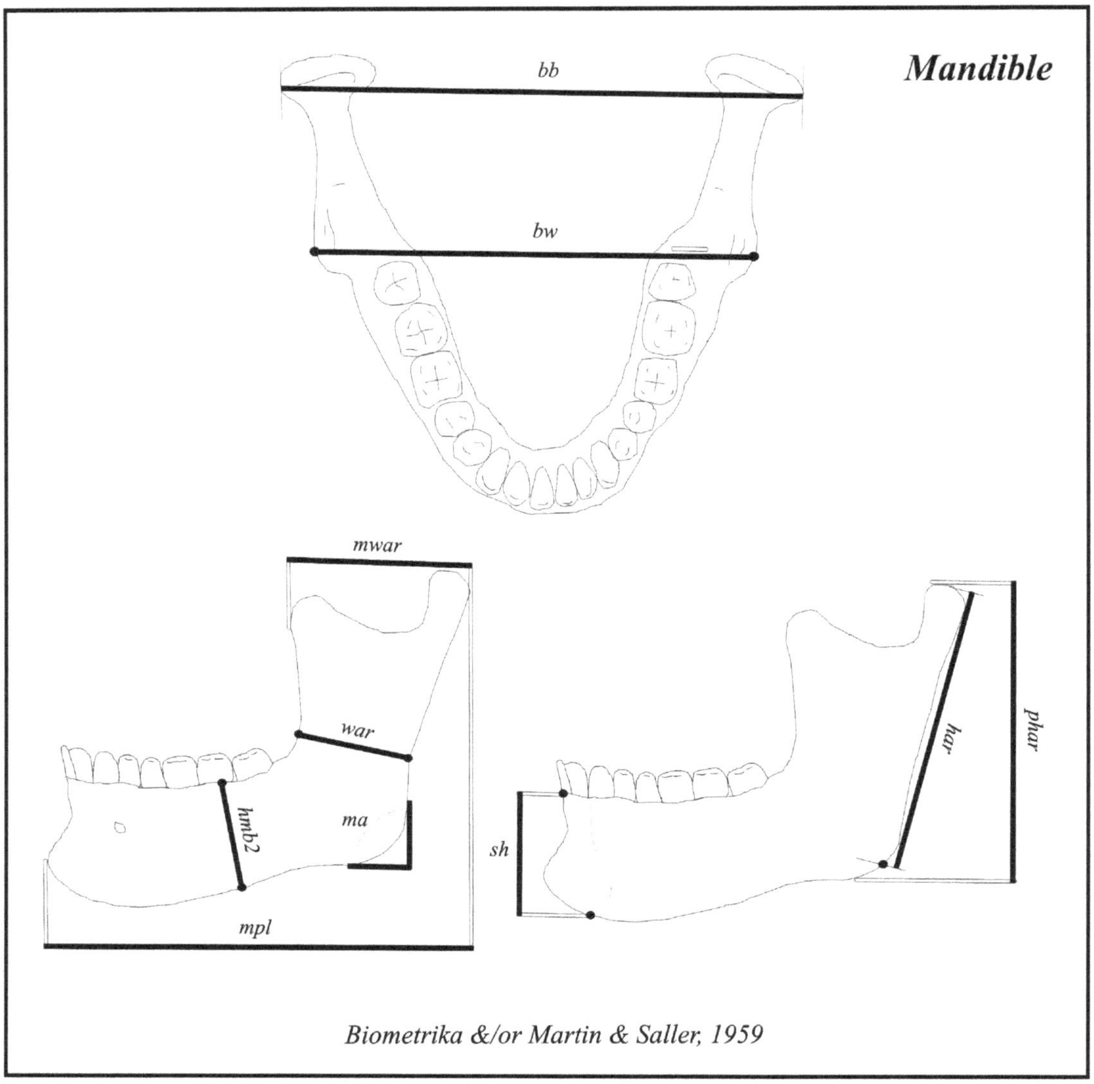

Figure 24. Schematic drawings of the mandible showing the anatomical localization of metric traits.

are listed in the reference manual of anthropometry (Martin and Saller 1957), some (19 variables) were not used by Howells (1989). In addition, 6 variables (10%) were devised by Lahr (1996) for the cranial base.

In contrast to the craniometric studies of Froment (1992a, 1992b, 1998), mandibular data were added. In fact, the latter have not been explored in detail for modern sub-Saharan populations since the study of Salmons (1925).

In order to test, from a methodological point of view, differences and/or similarities between variables, a series of very similar metric traits were selected in relation to the following aspects: cranial length (e.g. GOL, NOL), cranial breadth (e.g. Q1, Q), cranial base (e.g. BIMAS.BR, EAMEAM), palate (e.g. MAL, PAL, PABR) and ascending ramus of the mandible (e.g. har, phar; war, mwar).

3.6. Preliminary methodological aspects: intra- and inter-observer errors

Previous anthropometrical literature reveals that craniometric studies do not deal systematically with the problem of measurement errors. No tests about intra- and inter-observer errors were carried out in *Biometrika* papers, which were published at the beginning of the last century. Since the seventies, interest and questioning of the reliability of anthropometrical studies generally increased, and a few publications on various skeletal populations included tests for error measurements (Howells, 1973; Zegura, 1975). However, analyses focusing only on cranial variation in sub-Saharan Africa never raised such methodological issues, even though their results were based on data sets originating

	Biometrika	Martin & Saller, 1959	Howells, 1989	Lahr, 1996	Total of variables used
Vault	13	13	8	11	15
Base	2	4	2	9	11
Face	17	22	19	18	25
Mandible	8	9	-	6	10
Total	**40**	**48**	**28**	**44**	**61**

Table 6. Number of metric traits for different parts of the skull and according to previous literature.

from various publications (Hiernaux, 1974; Froment, 1992a, 1992b).

As our objectives are to re-evaluate craniometric variation in sub-Saharan Africa with both compiled and personal data, it was necessary first to validate the "heterogeneous" nature of the sample under study (involving at least twenty-nine different observers).

First, the term "measurement error" has to be defined, as it is related to various concepts, such as "true" values, reliability, random errors and systematic bias (Ulijaszek and Kerr, 1999).

An observation corresponds to both a "true" numerical value of a biological variable and an error or disturbance (Hunter, 1980). As "true" values are impossible to determine, a measurement that measures a metrical characteristic is "valid" or reliable, but never completely accurate. Therefore, a degree of indeterminacy should be accepted. The general assumption is that errors should be random rather than systematic, and therefore, the mean of each variable will be unaffected by the fact that several investigators independently took these measurements.

Although, a random error contributes to both intra- and inter-observer differences, it is not serious, as it does not introduce bias. Repeated measurements can vary, as they can be affected by random mechanical factors, such as when two individuals do similar measurements (e.g. reading errors, recorder error).

However, systematic errors significantly depart from the "true" values and introduce systematic bias. Measurements that are taken on different occasions by the same or different observers can then differ consistently due to measurement technique and/or instrument error due to equipment calibration.

Inter-observers errors are more likely related to differences in measurement techniques than intra-worker errors, although the latter can sometimes reflect a systematic tendency to drift, during a long work period, in relation to both small changes in measurement techniques and observer fatigue.

In summary, errors which are random are secondary for both intra- and inter-worker studies. However, the systematic ones in relation to differences in measurement techniques are the primary source of risk for studies based on several observers.

For studies on skeletal (and living) populations, the methods most commonly used for measurement error evaluation are the simple calculation of the mean differences (in mm) obtained from two compared data sets, the paired t-tests and secondarily the correlation coefficients (r) (Jamison and Zegura, 1974; Zegura, 1975; Stefan, 1999). For example, Zegura (1975) concluded that intra-observer errors were not affecting the results, as the mean differences between repeated measurements were inferior to 1mm. The variables that presented both more than 1mm difference (but still below 2mm) and low r coefficients, often corresponded to measurements without very well defined end-points, such as maximum cranial breadth, minimal frontal breadth and interorbital breadth (Page, 1976; Utermolhe and Zegura, 1982).

However, it is also important to determine the effects of the inter-observer errors through multivariate analyses, as the latter are very important for studies on population affinities. For example, are the effects of any bias detected in the univariate analysis then exaggerated in the distance statistics? In replying to this question, Jamison and Zegura (1974) determined significant effects in multiple discriminant analyses, especially when systematic bias identified through univariate analysis was not corrected afterwards. However, in the opinion of Page (1976), Principal Component analyses seem to be unaffected by measurement error: the variables with high inter-observer errors, in comparison to the unaffected ones, did not have significant differences in factor loadings.

Therefore, in order to assess measurement errors through both univariate and multivariate analyses, this preliminary study addresses two key questions:

i) Are there "unreliable" variables in our data set?
ii) Do intra- and inter-observer errors contribute to the distinction of populations?

The metrical data was tested in two steps:

i) For intra-observer error, between two sets of measurements of the same crania taken by myself (or duplicates).
ii) For inter-observer error, by comparing my measurements on certain crania to those of other authors (or repeats).

3.6.1. Metric traits

3.6.1.1. Data sets and variables tested

The sample under study (N=198) for methodological testing is described in Table 7 (p. 65). Fifty percent of the total sample was tested for intra-observer error and the other half for inter-observer error. For intra-observer error testing, 15 Kenyan crania were measured twice, a month apart, as well as 30 Rwandese crania, 30 Congolese and 25 Kenyan mandibles.

In particular, measurements for different data sets were repeated twice for two samples of crania and mandibles (with a six month interval), in order to ensure adequate training and quality control across time (from June 1998 to March 2000) for the same observer.

For inter-observer error testing, data sets from three different authors were also used (Benington, 1912; Kitson, 1931; Knip, 1970). As the Gabonese sample (N=35) was previously measured by Kitson (1931), it helped to check measurement definitions as provided in past publications (e.g. *Biometrika*).

All the metric traits under study (Appendices IV and V, p. 196-198) were analysed for intra-observer error, except for two variables (CC and NASGNA). For the inter-observer error, a shorter list was used (15 for the cranium and 6 for the mandible) to test previous data (Benington, 1912; Kitson, 1931; Knip, 1970).

3.6.1.2. Univariate analyses

All samples were checked for normality, using stem-and-leaf displays and tables for testing skewness (Zar, 1984; SPSS Base 9.0, 1999a, 1999b). In order to evaluate the differences between the various data sets obtained from both repeats and duplicates, the mean differences between each pair of samples was calculated and compared to the mean of each sample (Appendices VI & VII, p. 199-200 & 201).

Paired t-tests were also performed, as they provide a level of significance of the difference in mean between paired data sets. The Pearson's correlation coefficient also showed whether the two data sets covaried in a similar way for each variable.

3.6.1.2.1. Intra-observer error

The first set of intra-observer results obtained from the Kenyan sample (Table 7, **1a & 1b**) correspond to a first selection of 11 variables currently in use in the literature.

Mean differences in this case never reached 1mm. Only two t-tests are significant, and correlation coefficients between the two data sets are very high ($r \geq 0.89$).

The two other data sets obtained from the Rwandese (**2a & 2b**) and Congolese (**3a & 3b**) samples were used for a more detailed selection of 57 metric traits (48 for the cranium and 9 for the mandible). The mean differences were never above 2mm and most of them do not reach 1mm, except for 7 variables having mean differences that range between 1 and 2mm (e.g. Q1, MDB, WCB, XML, PMH, phar, ma). All the variables appear to present high correlation coefficient again ($r \geq 0.80$), and most of the t-tests (58/66) were not significant.

The comparison of the last two small data sets of 7 mandibular measurements obtained from a Kenyan sample (**4a & 4b**) also confirmed previous observations (e.g. mean difference below 1mm, $r > 0.93$, only two significant t-tests).

Across time (1998-2000), the mean differences are always below 0.8mm and 1.5mm for the cranium and the mandible respectively (Figure 25a & b, p. 67). They decreased slightly for 6 cranial

Sample	Origin	Place of deposit[1]	Data sets code	Name of observer	Dates	Number of variable tested cranium	mandible
tested for INTRA-OBSERVER ERROR							
15 crania	Kenya	Duckw	**1a** **1b**	Ribot	06.-07.1998	11	-
30 crania	Rwanda	IRScNB	**2a** **2b**	Ribot	07.-08.1999	48	9
30 mandibles	D.R.C.	IRScNB	**3a** **3b**	Ribot	07.-08.1999	-	9
22 mandibles	Kenya	Duckw	**4a** **4b**	Ribot	02.-03.2000	-	7
tested for INTER-OBSERVER ERROR							
35 crania	Gabonese	NHM	**5a** **5b**	Benington Ribot	1912 07.1998	10	-
38 crania	Kenya	Duckw	**6a** **6b**	Kitson Ribot	1931 03.2000	12	-
16 crania	Mali	Grav	**7a** **7b**	Knip Ribot	1970 06.2000	15	-
12 mandibles	Mali	Grav	**8a** **8b**	Knip Ribot	1970 06.2000	-	6

[1] **IRScNB** = Institut Royal des Sciences Naturelles de Belgique (Brussels); **Duckw** = Duckworth Laboratory (University of Cambridge); **NHM** = Natural History Museum, London; **Grav** = Gravenzande Museum, Leiden.

Table 7. Sample of modern skulls used for methodological purposes.

variables (e.g. GOL, XCB, BBH, BNL, ZYB, and OBB) and more sharply for three mandibular ones (e.g. phar, ma, and bb).

3.6.1.2.2. Inter-observer error

For testing inter-observer errors, 10 to 15 cranial measurements were selected, as they corresponded to those already measured by Kitson (1931), Benington (1912) and Knip (1970) (Appendix VII, p. 201).

For the Gabonese sample (**5a & 5b**), the mean differences for 10 cranial measurements were below 2mm, and 6 of them range between 1 and 2mm. Most of the t-tests (58/66) appear to be significant at varying levels. The correlation coefficients were also always above 0.85, except for the DKB variable.

For the Kenyan sample (**6a & 6b**), mean differences are below 1mm for 7 variables (e.g. XCB, BPL, NPH, NLB, DKB, OBB, WFB), between 1 and 2mm for 4 variables (e.g. GOL, BBH, BNL, ZYB) and above 2mm only for the PAL variable. Most of the t-tests (11/12) are significant, and the correlation coefficients are always very high ($r \geq 0.87$).

For the Malian sample of crania (**7a & 7b**) and mandibles (**8a & 8b**), all mean differences for 21 metric traits are below 2mm, except for MAB and mwar variables that are slightly above this value. Only 4 of them range between 1 and 2mm. All correlation coefficients are above 0.80, and most of the t-tests (15/21) are not significant.

In comparison to previous intra-observer analyses, the mean differences observed between the data sets of different observers are slightly higher, but most of them also tend to decrease across time, except for a few variables (e.g. NLB, XCB, BNL) (Figure 25c, p. 67).

In spite of slight differences across time and between different authors, these never reach the critical value of 2mm apart from a few variables such as PAL, MAB and mwar. The PAL variable, in particular, may have a problem of definition. Although it probably corresponds, in most cases, to the *orale-alveolon* distance (Martin and Saller, 1959: n° 62a), this measurement can be confused with a slightly different one - the *orale-staphylion* distance (Martin and Saller, 1959: n° 62). For example, in Shrubsall's (1897) publication, the definition of the variable is not very clear ("internal palate length"). Therefore, this variable was discarded from later analyses that involved data from several authors.

3.6.1.3. Multivariate analyses of error

In order to evaluate the effect of measurement errors when using multivariate statistics, several multiple discriminant analyses, as well as factor analyses, were performed (Tables 8-9, p. 68-69). Different combinations of data sets, measured at different times, were analysed.

For intra-observer error, a series of four stepwise discriminant analyses using Mahalanobis distance were performed with five cranial variables (Table 8, p. 68). The aim of this type of analysis is to produce a linear combination of independent variables serving as a basis for classifying cases into one of the groups (4 countries). Therefore, the best discriminating variables are selected in a stepwise manner, using the Mahalanobis distance method. Prior probabilities are also computed from group size.

In order to compare classification results and function coefficients deriving from similar samples (some as first measurements, some as second measurements), two pairs of separate analyses (I-II) were performed. In each case, two data sets were compared, and two of their sub-samples were changed (e.g. 1 and 2):

i) for the first pair of analyses (type I), GOL, XCB, BBH and NPH variables were entered, and two data sets (1a, 6b, 2a, 5b, 7b; and 1b, 6b, 2b, 5b, 7b) were compared, each including 119 crania;
ii) for the second pair of analyses (type II), GOL, XCB, BBH and OBB variables were entered, and two data sets (1a, 6b, 2a, 5b; and 1b, 6b, 2b, 5b) were compared again, each including 82 crania.

According to the results obtained for the two types of analyses (I & II), the total rates of correct classification vary slightly (<0.6%), the selected variables are always the same, and the function coefficients are nearly identical.

For testing measurement errors, a series of six factor analyses with varimax rotation were also performed for cranial and mandibular variables (Table 9, p. 69).

In order to compare eigenvalues, percentages of variance and factor loadings deriving from similar samples (some as first measurements, some as second measurements), three types of separate analyses were performed for intra-observer error (III, IV), and both intra- and inter-observer error (V):

i) for the first two pairs of analyses (types III & IV), only mandibular variables (e.g. bb, sh, war, ma) were entered. The first two data sets (3a, 4b; and 3a, 4a) consisted of 47 mandibles (3a, 4b). The last two (3b, 4b; 3b, 4a) consisted of 41 mandibles (3a, 4a);
ii) for the last pair of analyses (type V), only cranial variables (e.g. GOL, XCB, BBH, NPH, BPL, BNL) were entered, and data sets (1a, 2a, 5b, 6b, 7b; and 1b, 2a, 5a, 6a, 7a) were compared, each including 88 crania.

According to the results obtained (types III, IV, V), the number of factors and the selected variables are always the same. Furthermore, the eigenvalues, the percentages of variance and the factor loadings are nearly identical.

3.6.2. Conclusions

At the beginning of the section 3.6 (p. 62), two questions regarding the importance of systematic bias on the results were posed. According to both univariate and multivariate analyses obtained, these can now be answered:

1. One metrical variable (PAL) is "unreliable", as it presents the highest mean difference (2.48mm) between data sets of two observers, and therefore it was discarded.
2. The use of various sets of measurements according to different times of collection and/or authors does not give different results in the multivariate analyses. As it was suggested previously (Page, 1976), this supports the fact that the data sets can be pooled and used multivariately to address issues of population distances.

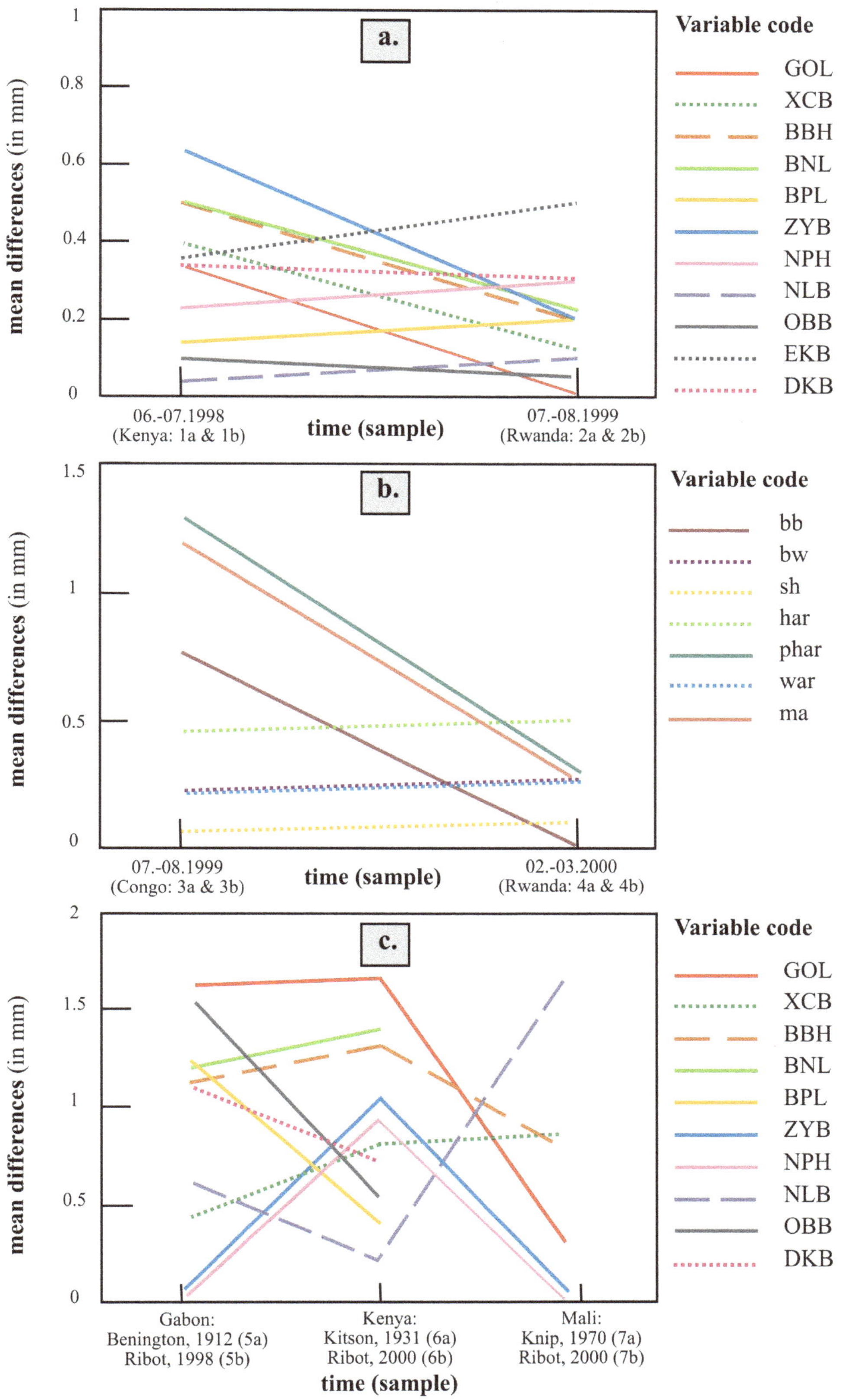

Figure 25. Graphs showing the mean differences for 18 measurements of the skull using various data sets through time. **a.** Intra-observer mean differences for 11 cranial measurements compared between Kenyan (06.-07.98) and Rwandese (07.-08.99) samples. **b.** Intra-observer mean differences for 7 mandibular measurements compared between Congolese (07.-08.99) and Kenyan (02.-03.00) samples. **c.** Inter-observer mean differences for 10 cranial measurements compared between our data and three authors (Benington, 1912; Kitson, 1931; Knip, 1970).

Type of analysis	Data set code	N	Variables entered	Country under study	Rates of correct classification in %		N° of functions	Selected variables	Standardized canonical discriminant function coefficients		
					by country	total			Function 1	Function 2	Function 3
I	**1a, 6b**	46	GOL, XCB, BBH, NPH	Kenya	69.6	69.7	3	GOL	-0.839	0.091	0.862
	2a	26		Rwanda	61.5			XCB	0.120	0.982	-0.412
	5b	32		Gabon	87.5			BBH	1.107	-0.076	0.335
	7b	15		Mali	46.7						
	1b, 6b	46	GOL, XCB, BBH, NPH	Kenya	67.4	69.6	3	GOL	-0.973	0.225	0.807
	2b	26		Rwanda	63.6			XCB	0.183	0.920	-0.530
	5b	32		Gabon	87.5			BBH	1.172	-0.078	0.367
	7b	15		Mali	46.7						
II								GOL	-0.786	-0.438	-
	1a, 6b	27	GOL, XCB, BBH, OBB	Kenya	70.4			XCB	0.213	0.804	-
	2a	23		Rwanda	78.3	81.7	2	BBH	0.996	-0.526	-
	5b	32		Gabon	93.8			OBB	0.399	0.651	-
								GOL	-0.854	-0.436	-
	1b, 6b	27	GOL, XCB, BBH, OBB	Kenya	74.1			XCB	0.250	0.743	-
	2b	23		Rwanda	70.0	81.0	2	BBH	1.007	-0.593	-
	5b	32		Gabon	93.8			OBB	0.469	0.606	-

Table 8. Summary for the multiple discriminant analyses performed on the various data sets used for intra-observer error.

Type of Analysis	Data sets	N	Variables entered	Factor 1 eigenvalue	Factor 1 (%)	Factor 2 eigenvalue	Factor 2 (%)	Cumulative % variance	N° of factors extracted	Selected variables	Factor loadings after varimax rotation: Factor 1	Factor loadings after varimax rotation: Factor 2
III	**3a, 4b**	47	bb, sh, war, ma	1.58	(39.58)	1.20	(30.08)	73.38	2	bb	- 0.106	0.767
										sh	0.132	0.815
										war	-0.877	0.221
										ma	0.878	0.254
	3a, 4a	47	bb, sh, war, ma	1.59	(39.75)	1.34	(33.42)	73.17	2	bb	-0.122	0.763
										sh	0.154	0.810
										war	-0.875	0.221
										ma	0.876	0.260
IV	**3b, 4b**	41	bb, sh, war, ma	1.70	(42.57)	1.20	(30.08)	72.65	2	bb	- 0.140	0.802
										sh	0.265	0.754
										war	-0.880	0.134
										ma	0.862	0.262
	3b, 4a	41	bb, sh, war, ma	1.71	(42.81)	1.19	(29.65)	72.47	2	bb	-0.154	0.806
										sh	0.297	0.738
										war	-0.874	0.136
										ma	0.860	0.265
V	**1b, 2b, 5b, 6b, 7b**	88	GOL, XCB, BBH, NPH, BPL & BNL	2.30	(38.30)	1.33	(22.18)	60.48	2	GOL	0.478	0.573
										XCB	-0.223	0.854
										BBH	0.333	0.508
										NPH	0.137	0.667
										BPL	0.784	0.193
										BNL	0.911	0.132
	1b, 2a, 5a, 6a, 7a	88	GOL, XCB, BBH, NPH, BPL & BNL	2.09	(34.81)	1.40	(23.41)	58.21	2	GOL	0.427	0.577
										XCB	-0.245	0.840
										BBH	0.368	0.407
										NPH	0.104	0.670
										BPL	0.773	0.289
										BNL	0.920	0.012

Table 9. Summary for the factor analyses performed on the various data sets used for both intra- and inter-observer errors.

CHAPTER 4

A RE-EVALUATION OF THE MODERN AFRICAN DIVERSITY

Modern African diversity is re-evaluated here through two main aspects: geography and ecology. This is analysed firstly before approaching the historical aspects in **Chapter 5**.

4.1. Geography (and sexual dimorphism) as a source of diversity?

Geography is a key-element in understanding population differentiation both on a worldwide scale and within a continent. Through skull morphology (using only linear measurements), it has been explored previously by three main recent studies (Howells, 1989; Lahr, 1996; Froment, 1998). They showed that not only vault but also various facial features are responsible for both inter- and intra-regional differences.

Firstly, using various samples (e.g. Teita, Dogons, Zulu, Bushman) (N=373), Howells (1989) noted that, on a worldwide scale, sub-Saharan African crania are characterized by the following morphological features and variables: narrow cranial base; very convex frontal; narrow and short face; large nasal breadth, eyes and face; and face bowed forward without nasal prominence.

Secondly, as already suggested by Howells (1989), Lahr (1996) also showed that, on a worldwide scale, sub-Saharan Africans have the most gracile skulls. For example, the latter (originating mainly from East Africa) show a low degree of robusticity.

Thirdly, in comparison to previous authors, Froment (1992a, 1992b, 1998) best evaluated the regional diversity present within sub-Saharan Africa, especially West Africa. In fact, he used a larger data set obtained from compiled publications (N=531) that were more representative of all regions. He observed marked differences in vault breadth (XCB) between the following regions: South Africans (e.g. Zulu) and West Africans (e.g. various Cameroonians) with broadest crania and East Africans (e.g. Teita) and other West Africans (e.g. Ashanti) with narrowest crania.

Finally, multivariate analyses by both Howells (1989) and Froment (1992a, 1992b) agreed on the following points:

i) Africans are set furthest away from other populations (especially American Indians and Far Easterners); they often appear close to both Australians and Melanesians. This observation is also supported by geometric morphometrics (Hennessy and Stringer, 2002);
ii) finally, these marked geographical differences and similarities can be demonstrated when using two cranial variables related to facial (ZYB) and nasal (NLB) breadth.

The question of the influence of sex on geographical differentiation of populations has not been approached systematically yet as an aspect of research on its own. In fact, craniometric variation due to sex differences especially in modern humans is very low in comparison to geographical differences (Howells, 1973, 1989, 1995; Lahr, 1996). Sexual dimorphism is in general not considered as a major obstacle for studying population differences on a worldwide scale. Howells (1989) who presented his distance statistics for each sex separately, found a marked agreement on population affinities between males and females.

Several authors focusing on African diversity (Rightmire, 1976a; Excoffier *et al.*, 1987; Vecchi *et al.*, 1987; Froment, 1992a, 1992b, 1998) used only male populations, as the latter often represent the largest sample available. However, if both female and male samples are pooled together, subtle differences between the sexes cannot be analysed. Therefore, using our large corpus of data on sub-Saharan Africans (N=1824), which is at least three times larger than previous studies, geographical differentiation can be re-evaluated through the analysis of both male and female populations.

Two main questions are addressed here:

1. Do results obtained from our large corpus of data agree with previous results on geographical differentiation of African and non-African populations? Do they provide additional information especially about sub-Saharan African diversity?
2. Do sexes when considered separately, provide a picture of population differentiation on a geographical level

similar to or different from when sexes are pooled together? In particular, are high levels of sexual dimorphism the result of some differential geographical pattern?

First, all samples within each region (both pooling and separating sexes) were checked for normality, using stem-and-leaf displays, Gaussian curves, Cox's test and tables for testing skewness (Zar, 1984). Then, various univariate and multivariate statistical tests were performed, in order to explore cranial morphological diversity in relation to geography (and sex).

4.1.1. Sub-Saharan Africa in a worldwide context

In order to compare African and non-African populations, the data were organized into 10 geographical units according to region of origin: Europe (1), Asia (2), Australia including Melanesia and Polynesia (3), Americas (4), West Africa (5), Central Africa including both Bantu-speaking and Pygmy populations (6 & 7), East Africa (8), and/or South Africa with only Bantu-speaking (9) and/or KhoiSan (10).

These ten geographical subdivisions were used here when dividing the data into sub-groups (n>20). Two sub-groups for both Central Africa (Bantu-speaking and Pygmy) and South Africa (Bantu-speaking and KhoiSan) were also considered separately as much as possible. All the following univariate and multivariate analyses were performed on 23 cranial variables (Appendix VIII, p. 202-203).

4.1.1.1. One-way analyses of variance

One-way analyses of variance were performed in order to compare both between- and within-population variation of sub-Saharan Africa on a worldwide scale (Appendix IX, p. 204).

When both sexes are pooled, results show that geographical differences are significant for all

Variables	Between non-Africans & Africans	within non-Africans	within Africans
GOL	X	-	X
NOL	X	-	X
XCB	X	X	X
PAC	X	-	X
OCC	X	-	-
BBH	X	X	X
WCB	X	X	X
BNL	X	X	X
BPL	X	X	X
NPH	X	-	X
ZYB	X	-	X
JUB	X	X	X
STB	X	X	-
XFB	X	X	X
FMB	X	-	X
ZMB	X	X	X
IML	X	X	-
XML	X	-	X
NLH	X	-	X
NLB	X	X	X
DKB	X	-	X
OBB	X	X	X
OBH	X	X	X

Table 10. Summary of differences localized between and within African and non-African populations according to post hoc multiple comparison tests.

metric traits at the highest level of probability (P<0.001). The most marked (with highest F-values: >30) are for the following variables: ZYB, DKB, BBH, XCB, WCB and NLH. These results agree with Howells (1989) and Froment (1992a, 1992b), who also showed marked geographical differentiation with the ZYB variable.

When both sexes are considered separately, similar results are obtained. However, the significance of geographical differences tends to decrease for most of the variables (except for GOL, OCC and FMB).

In order to localize the geographical differences between the various regions, post hoc multiple comparison tests were performed. When homogeneity of variance was significant using the Levene's test, Tamhane's test was used instead of the Scheffe's test.

Results are similar when both pooling and separating sexes. In summary (Table 10, p. 71), the geographical differences are localized especially between non-Africans (e.g. Europe, Asia, Australia, Americas) and Africans (e.g. West Africa, Central Africa, East Africa, South Africa) as well as within the latter. Geographical differences are more often absent within non-Africans (e.g. GOL, NOL, PAC, OCC, NPH, ZYB, FMB, XML, NLH and DKB) than within Africans (e.g. OCC, STB and IML).

These facts were already observed by previous craniometric analyses (Howells, 1989; Lahr, 1996) as well as genetic data (Cavalli-Sforza *et al.,* 1994) that showed a higher diversity within Africans.

On a worldwide scale, morphological differences are geographically the most widespread for the following variables: XCB, BBH, WCB, BNL, BPL, JUB, XFB, ZMB, NLB and OBB. Howells (1989) and Froment (1992a, 1992b) also suggested previously that both cranial breadth (XCB) and nasal breadth (NLB) reflect quite well the geographical differentiation of modern populations.

However, additional inter-regional differences and/or similarities have been here observed within African populations:

- for malar size (XML), between West Africans (long malars) and Bantu-speaking Central Africans (short malars); and
- for orbital height (OBH), between South African KhoiSan and Central African Pygmy (low orbits), and most of all the other groups (medium-size to high orbits).

4.1.1.2. Multiple discriminant analyses

In order to test the extent to which geographical differences on a worldwide scale are expressed through skull morphology, several multiple discriminant analyses were carried out with the three following series of variables (Tables 11-13, p. 73-75):

- for both shape and size of the vault: GOL, XCB, PAC, OCC, BBH, WCB and BNL;
- for the facial breadth: BPL, ZYB, JUB, STB, XFB, FMB and ZMB; and
- for various facial details: IML, XML, NLH, NLB, DKB, OBB and OBH).

Seven geographical groups were used, and samples were analysed systematically for pooled sexes and both males and females separately.

All analyses provided five to six functions and the standardized canonical discriminant function coefficients varied only slightly. Most of the variables were selected during the analysis, except for XML, JUB, STB and XFB.

According to highest standardized canonical coefficients, the variables that are most frequently effective in "classifying" groups on a geographical level are (on the first two functions):

- for the vault: XCB and BBH;
- for the face: FMB, XFB, ZYB and NPH; and
- for various facial traits: NLH, NLB and OBH.

Again, these results agree with the work of Howells (1989) and Froment (1992a, 1992b). The latter also found that breadth of both vault (XCB) and face (ZYB), nasal breadth (NLB) and orbital height (OBH) express quite well geographical differences on a worldwide scale.

Although sample size causes the rates to change, the total correct classification rates obtained are:

- for the whole cranium: between 46.3% and 60.5%;
- for the vault: between 46.3% and 59.1% (Table 11, p. 73);
- for the facial breadth: between 53.8% and 55.6% (Table 12, p. 74); and
- for various facial details: between 48.5% and 60.5% (Table 13, p. 75).

The accuracy of these results is much lower than

Groups	N	Classification results (%) 1.	2.	3.	4.	5.	6. & 7.	8.	Total correct rate	Selected variables	Standardized canonical discriminant function coefficients function 1	function 2	function 3	function 4	function 5	function 6
Pooled sexes (N=904)																
1. Europe	72	29 (**40.3**)	6 (8.3)	12 (16.7)	3 (4.2)	13 (18.1)	6 (8.3)	3 (4.2)								
2. Asia (various)	140	1 (0.7)	59 (**42.1**)	13 (9.3)	7 (5)	33 (23.6)	17 (12.1)	10 (7.1)		GOL	-0.269	-0.861	0.185	-1.044	0.904	-0.105
3. Australia (& Melanesia, Polynesia)	126	12 (9.5)	17 (13.5)	55 (**43.7**)	8 (6.3)	2 (15.9)	11 (8.7)	3 (2.4)		**XCB**	**0.693**	-0.368	0.092	-0.030	-0.334	**0.837**
4. Americas (including Eskimos)	100	6 (6)	6 (6)	11 (11)	39 (**39**)	17 (17)	16 (16)	5 (5)	**48.3%**	**PAC**	0.109	0.352	0.332	**1.232**	0.233	0.511
5. West Africa	198	0	21 (10.6)	15 (7.6)	12 (6.1)	81 (**40.9**)	55 (27.8)	14 (7.1)		**BBH**	-0.113	**1.032**	-0.335	-1.001	0.241	0.002
6. & 7. Central Africa (Bantu-speakers & Pygmy)	186	1 (0.5)	3 (1.6)	2 (1.1)	7 (3.8)	39 (21)	125 (**67.2**)	9 (4.8)		WCB	0.493	0.128	-0.539	0.359	0.377	-0.791
8. East Africa	82	0	2 (2.4)	2 (2.4)	6 (7.3)	12 (14.6)	11 (13.4)	49 (**59.8**)		**BNL**	0.246	0.041	**0.969**	0.659	**-0.924**	-0.381
Only males (N=508)																
1. Europe	43	12 (**27.9**)	3 (7)	11 (25.6)	5 (11.6)	9 (20.9)	1 (2.3)	2 (4.7)								
2. Asia (various)	102	0	52 (**51**)	11 (10.8)	11 (10.8)	16 (15.7)	8 (7.8)	4 (3.9)		GOL	-0.437	-0.819	0.332	-0.894	0.695	0.526
3. Australia (& Melanesia, Polynesia)	95	5 (5.3)	12 (12.6)	53 (**55.8**)	9 (9.5)	10 (10.5)	6 (6.3)	0		**XCB**	**0.609**	-0.484	0.147	0.027	0.192	-0.832
4. Americas (including Eskimos)	67	3 (4.5)	5 (7.5)	12 (17.9)	29 (**43.3**)	7 (10.4)	8 (11.9)	3 (4.5)	**46.3%**	**PAC**	0.208	0.300	0.118	**1.343**	0.274	-0.284
5. West Africa	101	0	25 (24.8)	11 (10.9)	9 (8.9)	35 (**34.7**)	17 (16.8)	4 (4)		**BBH**	0.087	**0.870**	-0.206	-0.904	0.470	0.144
6. & 7. Central Africa (Bantu-speakers & Pygmy)	66	1 (1.5)	6 (9.1)	1 (1.5)	5 (7.6)	14 (21.2)	38 (**57.6**)	1 (1.5)		**WCB**	0.565	0.062	-0.342	0.193	-0.189	**0.886**
8. East Africa	34	0	2 (5.9)	2 (5.9)	6 (17.6)	7 (20.6)	1 (2.9)	16 (**47.1**)		**BNL**	0.131	0.243	**0.860**	0.621	**-0.925**	-0.226
Only females (N=357)																
1. Europe	29	22 (**75.9**)	0	0	1 (3.4)	5 (17.2)	1 (3.4)	0								
2. Asia (various)	35	0	9 (**25.7**)	1 (2.9)	1 (2.9)	11 (31.4)	10 (28.6)	3 (8.6)		GOL	-0.130	0.803	0.108	0.209	-0.589	0.856
3. Australia (& Melanesia, Polynesia)	31	5 (16.1)	3 (9.7)	9 (**29**)	3 (9.7)	8 (25.8)	1 (3.2)	2 (6.5)		**XCB**	**0.714**	0.248	0.050	-0.399	0.663	0.406
4. Americas (including Eskimos)	33	1 (3)	1 (3)	3 (9.1)	17 (**51.5**)	5 (15.2)	5 (15.2)	1 (3)	**59.1%**	**OCC**	0.131	-0.419	-0.488	**0.882**	0.468	-0.252
5. West Africa	89	0	4 (4.5)	2 (2.2)	0	48 (**53.9**)	28 (31.5)	7 (7.9)		**BBH**	-0.475	**-0.646**	0.910	-0.043	-0.144	0.484
6. & 7. Central Africa (Bantu-speakers & Pygmy)	100	1 (1)	1 (1)	0	1 (1)	16 (16)	76 (**76**)	5 (5)		**WCB**	0.483	-0.445	**-0.097**	0.078	**-0.884**	-0.262
8. East Africa	40	0	0	0	3 (7.5)	0	7 (17.5)	30 (**75**)		**BNL**	0.359	0.472	0.226	0.120	0.516	**-1.114**

Table 11. Summary of multiple discriminant analyses: predicted geographical group membership on a worldwide scale with vault variables (GOL, XCB, PAC, OCC, BBH, WCB, BNL).

Groups	N	Classification results (%) 1.	2.	3.	4.	5.	6. & 7.	8.	Total correct rate	Selected variables	Standardized canonical discriminant function coefficients function 1	function 2	function 3	function 4	function 5	function 6
Pooled sexes (N=847)																
1. Europe	74	20 (**27**)	16 (21.6)	17 (23)	10 (13.5)	1 (1.4)	9 (12.2)	1 (1.4)		BPL	-0.276	0.708	0.096	-0.144	0.874	0.344
2. Asia (various)	138	7 (5.1)	79 (**57.2**)	21 (15.2)	10 (7.2)	10 (7.2)	10 (7.2)	1 (0.7)		NPH	0.160	-0.324	-0.718	0.625	-0.003	-0.398
3. Australia (& Melanesia, Polynesia)	126	5 (4)	14 (11.1)	79 (**62.7**)	16 (12.7)	2 (16)	10 (7.9)	0		**ZYB**	0.992	0.123	-0.722	**-0.688**	0.633	-0.868
4. Americas (including Eskimos)	99	15 (15.2)	11 (11.1)	14 (14.1)	49 (**49.5**)	2 (2)	5 (5.1)	3 (3)	**53.8%**	**JUB**	0.497	0.694	**0.865**	0.666	**-1.152**	1.067
5. West Africa	133	3 (2.3)	9 (6.8)	2 (1.5)	0	50 (**37.6**)	62 (46.6)	7 (5.3)		STB	-0.049	0.082	0.838	0.571	0.620	-0.997
6. & 7. Central Africa (Bantu-speakers & Pygmy)	222	2 (0.9)	11 (5)	7 (3.2)	2 (0.9)	23 (10.4)	172 (**77.5**)	5 (2.3)		**XFB**	0.088	**-0.728**	-0.460	-0.325	0.032	**1.655**
8. East Africa	55	2 (3.6)	5 (9.1)	1 (1.8)	4 (7.3)	16 (29.1)	20 (36.4)	7 (**12.7**)		**FMB**	**-1.364**	-0.089	-0.543	0.261	-0.312	-0.221
										ZMB	0.069	-0.247	0.739	0.260	0.045	-0.461
Only males (N=491)																
1. Europe	45	8 (**17.8**)	9 (20)	17 (37.8)	4 (8.9)	1 (2.2)	6 (13.3)	0		BPL	0.294	-0.604	0.258	0.018	0.451	-0.568
2. Asia (various)	100	2 (2)	64 (**64**)	16 (16)	7 (7)	8 (8)	3 (3)	0		NPH	-0.136	0.078	-0.662	0.175	0.582	0.140
3. Australia (& Melanesia, Polynesia)	95	4 (4.2)	10 (10.5)	63 (**66.3**)	11 (11.6)	1 (1.1)	6 (6.3)	0		**ZYB**	-0.520	**-0.607**	-0.689	**-0.576**	0.510	0.791
4. Americas (including Eskimos)	66	3 (4.5)	8 (12.1)	12 (18.2)	31 (**47**)	3 (4.5)	5 (7.6)	4 (6.1)	**55.2%**	**JUB**	-0.588	-0.049	0.940	0.815	**-0.632**	**-1.049**
5. West Africa	69	1 (1.4)	7 (10.1)	1 (1.4)	0	34 (**49.3**)	24 (34.8)	2 (2.9)		**STB**	0.158	0.167	**0.991**	-0.327	1.289	0.217
6. & 7. Central Africa (Bantu-speakers & Pygmy)	92	1 (1.1)	4 (4.3)	3 (3.3)	2 (2.2)	17 (18.5)	64 (**69.6**)	1 (1.1)		XFB	-0.253	0.606	-0.775	0.604	-0.920	-0.692
8. East Africa	24	1 (4.2)	3 (12.5)	0	3 (12.5)	8 (33.3)	2 (8.3)	7 (**29.2**)		**FMB**	**1.251**	0.052	-0.400	0.255	-0.182	0.540
										ZMB	-0.199	0.381	0.556	0.195	-0.055	0.738
Only females (N=360)																
1. Europe	29	11 (**37.9**)	5 (17.2)	3 (10.3)	5 (17.2)	1 (3.4)	3 (10.3)	1 (3.4)								
2. Asia (various)	38	5 (13.2)	22 (**57.9**)	3 (7.9)	0	2 (5.3)	6 (15.8)	0		**BPL**	0.256	-0.111	**1.044**	0.141	-0.381	-
3. Australia (& Melanesia, Polynesia)	31	1 (3.2)	6 (19.4)	15 (**48.4**)	5 (16.1)	1 (3.2)	3 (9.7)	0		**NPH**	-0.108	**0.914**	-0.465	-0.146	-0.520	-
4. Americas (including Eskimos)	33	2 (6.1)	2 (6.1)	5 (15.2)	23 (**69.7**)	0	1 (3)	0	**55.6%**	**ZYB**	**-1.276**	0.022	0.386	-0.103	0.547	-
5. West Africa	83	3 (3.6)	2 (2.4)	2 (2.4)	0	23 (**27.7**)	40 (48.2)	13 (15.7)		**FMB**	1.089	0.474	-0.091	-0.191	**0.671**	-
6. & 7. Central Africa (Bantu-speakers & Pygmy)	118	0	5 (4.2)	3 (2.5)	2 (1.7)	8 (6.8)	93 (**78.8**)	7 (5.9)		**ZMB**	0.067	-0.309	-0.381	**1.118**	-0.174	-
8. East Africa	28	0	0	0	2 (7.1)	7 (25)	6 (21.4)	13 (**46.4**)								

Table 12. Summary of multiple discriminant analyses: predicted geographical group membership on a worldwide scale with a first series of facial variables (BPL, ZYB, JUB, STB, XFB, FMB and ZMB).

Groups	N	Classification results (%) 1.	2.	3.	4.	5.	6. & 7.	8.	Total correct rate	Selected variables	Standardized canonical discriminant function coefficients function 1	function 2	function 3	function 4	function 5	function 6
Pooled sexes (N=864)																
1. Europe	74	27 (**36.5**)	16 (21.6)	17 (23)	10 (13.5)	2 (2.7)	2 (2.7)	0								
2. Asia (various)	140	5 (3.6)	71 (**50.7**)	15 (10.7)	18 (12.9)	20 (14.3)	9 (6.4)	2 (1.4)		**IML**	0.192	-0.158	**0.669**	0.535	-0.580	0.410
3. Australia (& Melanesia, Polynesia)	126	12 (9.5)	22 (17.5)	52 (**41.3**)	12 (9.5)	14 (11.1)	13 (10.3)	1 (0.8)		**NLH**	**-0.988**	0.181	-0.273	-0.131	0.242	**0.791**
4. Americas (including Eskimos)	96	4 (4.2)	16 (16.7)	8 (8.3)	52 (**54.2**)	9 (9.4)	7 (7.3)	0	**48.6%**	**NLB**	0.241	**0.783**	-0.216	0.712	0.245	-0.256
5. West Africa	199	3 (1.5)	18 (9)	12 (6)	5 (2.5)	109 (**54.8**)	45 (22.6)	7 (3.5)		DKB	0.670	-0.051	0.006	-0.628	0.248	0.534
6. & 7. Central Africa (Bantu-speakers & Pygmy)	172	1 (0.6)	7 (4.1)	6 (3.5)	7 (4.1)	47 (27.3)	99 (**57.6**)	5 (2.9)		**OBB**	-0.178	0.428	0.545	**-0.823**	0.001	-0.590
8. East Africa	57	0	4 (7)	1 (1.8)	3 (5.3)	19 (33.3)	20 (35.1)	10 (**17.5**)		**OBH**	0.407	-0.715	0.330	0.477	**0.673**	-0.334
Only males (N=499)																
1. Europe	45	20 (**44.4**)	9 (20)	12 (26.7)	4 (8.9)	0	0	0								
2. Asia (various)	102	3 (2.9)	59 (**57.8**)	12 (11.8)	13 (12.7)	10 (9.8)	4 (3.9)	1 (1)		**IML**	0.238	-0.161	**0.724**	0.224	**-0.599**	0.384
3. Australia (& Melanesia, Polynesia)	95	4 (4.2)	18 (18.9)	50 (**52.6**)	9 (9.5)	8 (8.4)	6 (6.3)	0		**NLH**	**-0.914**	0.337	0.108	0.176	0.306	**0.789**
4. Americas (including Eskimos)	65	2 (3.1)	15 (23.1)	9 (13.8)	32 (**49.2**)	5 (7.7)	2 (3.1)	0	**48.5%**	**NLB**	0.276	0.440	0.092	**0.884**	0.053	-0.354
5. West Africa	103	2 (1.9)	11 (10.7)	8 (7.8)	4 (3.9)	61 (**59.2**)	15 (14.6)	2 (1.9)		DKB	0.682	0.043	-0.100	-0.419	0.435	0.531
6. & 7. Central Africa (Bantu-speakers & Pygmy)	64	1 (1.6)	3 (4.7)	6 (9.4)	2 (3.1)	29 (45.3)	19 (**29.7**)	4 (6.3)		OBB	-0.101	0.552	-0.506	0.621	0.437	-0.505
8. East Africa	25	0	3 (12)	1 (4)	3 (12)	9 (36)	8 (32)	1 (**4**)		**OBH**	0.361	**-1.097**	0.118	0.320	0.303	-0.403
Only females (N=337)																
1. Europe	29	22 (**75.9**)	1 (3.4)	1 (3.4)	3 (10.3)	2 (6.9)	0	0								
2. Asia (various)	38	2 (5.3)	18 (**47.4**)	2 (5.3)	3 (7.9)	6 (15.8)	7 (18.4)	0		**IML**	0.151	-0.166	0.200	**0.835**	-0.619	0.174
3. Australia (& Melanesia, Polynesia)	31	1 (3.2)	5 (16.1)	10 (**32.3**)	4 (12.9)	6 (19.4)	4 (12.9)	1 (3.2)		**NLH**	**-0.748**	0.365	0.208	-0.602	-0.523	0.518
4. Americas (including Eskimos)	31	1 (3.2)	3 (9.7)	2 (6.5)	22 (**71**)	0	3 (9.7)	0	**60.5%**	**NLB**	0.281	**0.613**	**-0.818**	0.261	0.257	0.268
5. West Africa	88	1 (1.1)	7 (8)	3 (3.4)	3 (3.4)	58 (**51**)	19 (21.6)	4 (4.5)		DKB	0.684	0.051	0.552	-0.571	-0.232	0.060
6. & 7. Central Africa (Bantu-speakers & Pygmy)	94	1 (1.1)	4 (4.3)	3 (3.2)	2 (2.1)	10.6 (10)	70 (**74.5**)	4 (4.3)		**OBB**	-0.110	0.597	-0.553	-0.012	0.276	**-0.796**
8. East Africa	26	0	0	0	0	11.5 (3)	12 (46.2)	11 (**42.3**)		**OBH**	0.290	-0.477	0.223	0.436	**0.860**	0.527

Table 13. Summary of multiple discriminant analyses: predicted geographical group membership on a worldwide scale with a second series of facial variables (IML, XML, NLH, NLB, DKB, OBB and OBH).

those obtained by Howells (1989) (up to 98%). As his variables were more or less similar, it suggests that our sample contains much more variability. This may be the result of two factors:

i) the sample size for sub-Saharan Africans is at least three times larger than the African data set used by Howells (1989);
ii) the data set of non-Africans that was obtained from Howells (1989) did not correspond to his final published samples. Therefore, as it probably contained several geographical incorrect assignments, it also contributed to a lower accuracy in geographical classifications.

According to rates of classification, the female sample provides slightly more marked inter-regional differences on a worldwide scale (55.6%-60.5%) than when only using males (46.3%-55.2%). For example, in relation to height and breadth of nose, female crania express geographical differences better (60.5%) than male crania (48.5%) (Table 13, p. 75).

In addition, depending on the sex under study, West and Central Africans appear to be difficult to differentiate, especially in terms of facial features. For example, within sub-Saharan Africans, two marked misclassifications (>40%) were observed:

i) in females, for facial breadth, West Africans are misclassified with Central Africans; and
ii) in males, for nasal and orbital height, Central Africans are misclassified with West Africans.

4.1.1.3. Factor analyses

In order to assess morphological differences on a geographical level and to visualize the variation in a two-dimensional space, several factor analyses were performed (Table 14, p. 78-80).

The advantage of factor analysis is the identification of a relatively small number of factors that can be used to represent indirectly observable relationships among sets of variables (Eastman, 1997). In contrast to multiple discriminant analyses, the various types of measurements entered are not independent, but they constitute groups of inter-related variables or factors. According to Principal Component analysis, successive linear combinations of variables are extracted, progressively accounting for smaller portions of variance in the sample.

The rotation phase of factor analyses was also performed, in order to obtain a simple structure, or in other words, to transform complicated matrices into easier ones to interpret. In particular, the varimax rotation was used, as it minimizes the number of variables having high loadings on a factor, increasing therefore the interpretability of the factors.

In order to avoid the over-estimation of one type of variability (Leguebe, *pers.comm.*), a variable selection was first undertaken through correlation matrices. Therefore, five series of metric traits, presenting relatively low correlation ($r \leq 0.5$) between each other, were selected:

- for **analysis I**: GOL, XCB, ZYB, BBH, BPL, NPH, BNL and NLB;
- for **analysis II**: NOL, FMB, STB, IML, NLB, DKB and OBH;
- for **analysis III**: GOL, BPL, ZML, XML, NLH and OBH;
- for **analysis IV**: FRC, PAC, OCC, BBH, NPH, XFB, NLB and OBB; and
- for **analysis V**: WCB, BPL, JUB, NLH, NLB and OBH.

These series of metric traits were analysed systematically in three steps:

- with pooled sexes (analysis a);
- with only males (analysis b);
- with only females (analysis c).

In all analyses, at least two factors were obtained. As they presented eigenvalues (or total variance explained by each factor) greater than one, they were retained for the varimax rotation. The total cumulative percentage of variance obtained for the various factor models was relatively high, as it varied between 51.64% and 66.71%.

For both pooled and separated sexes, the variables that are present most frequently on the factors with the highest loading (≥ 0.7), are:

- on **factor 1**: BPL (e.g. analyses Ib, Ic, IIIa, IIIb, IIIc), NLH (e.g. analyses Va, Vb, Vc) and IML (e.g. analyses IIa, IIb);
- on **factor 2**: OBH (e.g. analyses IIc, IIIa, IIIb, IIIc), NLB (e.g. analyses IVa, Va, Vc), DKB (e.g. analyses IIa, IIb), XCB (e.g. analyses Ic, Ib), BPL (e.g. analyses Ia, Vb); and
- on **factor 3**: NLB (e.g. analyses IVb, IVc).

In summary, the following morphological features (or groups of inter-related variables) have often been selected for the five types of factor analyses:

- **analysis I**: cranial breadth (XCB, ZYB)

and prognathism (BPL, BNL);
- **analysis II**: vault length (GOL), orbital height (OBH); interorbital (DKB) and frontal (STB) breadths;
- **analysis III**: facial protrusion (BPL), malar (XML) and vault (GOL) length; orbital (OBH) and nasal (NLH) height;
- **analysis IV**: height/length of vault (BBH, GOL) and nose width (NLB); and
- **analysis V**: orbital (OBH) and nasal (NLH) height; nose width (NLB) and facial protrusion (BPL).

These results are again in agreement with previous statistical tests. They support the studies of Howells, (1989) and Froment (1992a, 1992b), who often obtained the same variables for expressing geographical differences on a worldwide scale.

4.1.1.3.1. Sex differences

According to the independent sample t-tests (Table 15, p. 84), sex differences performed on regression factor scores are always significant at the highest levels of probability on both factors.

Scatterplots for factors 1 and 2 of the five analyses (Ia-Va) are drawn in order to show group clustering by sex evaluations for each analysis (Figure 26, p. 81). Ellipses of confidence (with a probability of 0.9) are drawn, in order to measure the overlapping variability (in terms of distance or standard deviations) between the two groups (Defrise-Gussenhoven, 1955). Marked sex differences especially in relation to vault and face size, are observed along factor 1 of two analyses (IIIa, IVa), as it is suggested by the highest t-values (>15).

4.1.1.3.2. Geographical differences

Geographical differences were tested by performing one-way analyses of variance on the regression factor scores obtained from the five types of analyses (I-V). The results showed that, they are always significant at the highest level of probability (Table 16, p. 85).

A slight decrease in the F-values is often noticed when sexes are considered separately. This suggests that sexual dimorphism might slightly increase geographical differences. However, post hoc multiple comparisons tests showed that, the localization of the differences are similar when either pooling or separating sexes (like for previous univariate analyses).

Scatterplots for factors 1 and 2 (or 3) of the five types of analyses were drawn in order to show group clustering by region for both males (analyses Ib-Vb; Figure 27, p. 82) and females (analyses Ic-Vc; Figure 28, p. 83). Therefore, the morphological features are observed more clearly, and they are summarised in Table 17 (p. 86). Three main comments can be made.

Firstly, for both sexes, sub-Saharan African populations are often morphologically very close to the Australians and Melanesians. They both share features such as a prognathic face and a wide nasal aperture. They are often in the highest values and the American Indians in the lowest ones, due to differences in both shape and overall size. Although these facts were suggested previously (Howells, 1989; Froment, 1992a, 1992b), it was observed here that the sub-Saharan African population, which is morphologically close to Australian-Melanesians, corresponded mainly to West Africans.

Secondly, when pooling or analysing sexes separately, both the level of significance and the localization of the geographical differences are very similar, although some slight changes are observed in the F-values. Therefore, as it was suggested previously by Howells (1989) and Lahr (1996), multivariate analyses seem to vary minimally by the bimodality of the variables due to sexual dimorphism.

Thirdly, according to higher F-values, geographical differences are slightly more marked when only female populations are considered.

These results are supported by previous independent sample t-tests that showed marked sex differences for all analyses. For females (Figure 28, p. 83), group clustering by region, is also more evident in the scatterplots. Although this slightly more marked geographical differentiation within female populations may be due to smaller sample size, previous multiple discriminant analyses already suggested a similar observation.

4.1.2. Within sub-Saharan Africa

Geographical differences are explored here within sub-Saharan Africa, with the same subdivisions as those used previously. However, 34 additional variables were analysed in relation not only to the cranium (24 variables) but also to the mandible (10 variables) (Appendix X, p. 205).

Variable entered	Factor 1 eigen-value	Factor 1 (%)	Factor 2 eigen-value	Factor 2 (%)	Cumulative % of variance	N° of factors extracted		Factor loadings after varimax rotation factor 1	Factor loadings after varimax rotation factor 2	% of variance[1] factor 1	% of variance[1] factor 2
GOL, XCB, ZYB, BBH, BPL, NPH, BNL, NLB							GOL	0.495	0.577		
							XCB	**0.896**	-0.119	**80.28**	
							ZYB	**0.765**	0.371	**58.52**	
Analysis Ia (pooled sexes) (N=1497)	4.05	(50.56)	1.06	(13.22)	**63.78**	2	BBH	0.599	0.437		
							BPL	0.202	**0.826**		**68.22**
							NPH	0.594	0.387		
							BNL	0.486	**0.737**		**54.32**
							NLB	0.018	0.655		
Analysis Ib (only males) (N=903)	3.26	(40.78)	1.21	(15.12)	**55.91**	2	GOL	0.610	0.315		
							XCB	-0.194	**0.870**		**75.69**
							ZYB	0.292	**0.741**		**54.91**
							BBH	0.355	0.560		
							BPL	**0.826**	0.127	**68.23**	
							NPH	0.331	0.521		
							BNL	**0.761**	0.413	**57.91**	
							NLB	0.551	-0.011		
Analysis Ic (only females) (N=509)	3.35	(41.93)	1.22	(15.20)	**57.13**	2	GOL	0.545	0.363		
							XCB	-0.147	**0.903**		**81.54**
							ZYB	0.318	**0.775**		**60.06**
							BBH	0.513	0.443		
							BPL	**0.828**	0.059	**68.56**	
							NPH	0.442	0.470		
							BNL	**0.803**	0.340	**64.48**	
							NLB	0.523	-0.035		
NOL, FMB, STB, IML, NLB, DKB, OBH							NOL	0.670	0.289		
							FMB	0.634	0.591		
							STB	0.095	0.650		
Analysis IIa (pooled sexes) (N=671)	2.75	(39.30)	1.07	(15.30)	**54.60**	2	**IML**	**0.769**	0.024	**59.14**	
							NLB	0.391	0.512		
							DKB	0.149	**0.801**		**64.16**
							OBH	0.659	-0.028		
Analysis IIb (only males) (N=401)	2.46	(35.16)	1.20	(16.50)	**51.64**	2	NOL	0.620	0.232		
							FMB	0.662	0.524		
							STB	-0.217	0.698		
							IML	**0.755**	-0.096	**57.00**	
							NLB	0.385	0.425		
							DKB	0.235	**0.766**		**58.68**
							OBH	0.617	0.010		

[1] Percentage of variance accounted on each factor obtained by squaring the correlation coefficient (or factor loading) for variables which have the highest factor loadings (>0.7).

Table 14. Summary of five types of factor analyses (I-V) performed on both African and non-African samples.

Table 14 continued

Variables entered	Factor 1 eigen-value	Factor 1 (%)	Factor 2 eigen-value	Factor 2 (%)	Factor 3 eigen-value	Factor 3 (%)	Cumulative % of variance	N° of factors extracted		Factor loadings after varimax rotation factor 1	factor 2	factor 3	% of variance[1] factor 1	factor 2	factor 3
(NOL, FMB, STB, IML, NLB, DKB, OBH) **Analysis IIc** (only females) (N=250)	2.44	(34.97)	1.18	(16.84)	1.04	(14.79)	**66.60**	3	NOL	0.191	0.696	-0.014			
									FMB	**0.763**	0.407	0.000	**58.22**		
									STB	0.222	0.310	**0.773**			**59.75**
									IML	0.360	0.395	-0.645			
									NLB	**0.704**	0.047	-0.106	**49.56**		
									DKB	**0.865**	-0.050	0.138	**74.82**		
									OBH	-0.054	**0.817**	0.086		**66.75**	
GOL, BPL, ZMB, XML, NLH, OBH **Analysis IIIa** (pooled sexes) (N=867)	2.97	(49.45)	1.04	(17.26)	-	-	**66.71**	2	GOL	0.741	0.260	-			
									BPL	**0.832**	0.034	-	**69.22**		
									ZMB	0.565	0.447	-			
									XML	**0.765**	0.150	-	**58.52**		
									NLH	0.417	**0.754**	-		**54.85**	
									OBH	0.007	**0.907**	-		**82.26**	
Analysis IIIb (only males) (N=498)	2.57	(42.77)	1.17	(19.42)	-	-	**62.19**	2	GOL	0.691	0.229	-			
									BPL	**0.844**	0.048	-	**71.23**		
									ZMB	0.385	0.552	-			
									XML	**0.717**	0.149	-	**51.41**		
									NLH	0.199	**0.849**	-		**72.08**	
									OBH	0.015	**0.858**	-		**73.62**	
Analysis IIIc (only females) (N=342)	2.51	(41.83)	1.06	(17.63)	-	-	**59.46**	2	GOL	0.667	0.261	-			
									BPL	**0.813**	-0.010	-	**66.10**		
									ZMB	0.528	0.330	-			
									XML	0.694	0.113	-			
									NLH	0.357	**0.752**	-		**56.55**	
									OBH	0.008	**0.905**	-		**81.90**	
FRC, PAC, OCC, BBH, NPH, XFB, NLB, OBB **Analysis IVa** (pooled sexes) (N=1050)	3.47	(43.31)	1.00	(12.54)	-	-	**55.85**	2	**FRC**	**0.754**	0.323	-	**56.85**		
									PAC	0.329	0.592	-			
									OCC	**0.748**	-0.171	-	**55.06**		
									BBH	**0.742**	0.351	-	**55.06**		
									NPH	0.621	0.216	-			
									XFB	0.666	0.147	-			
									NLB	0.008	**0.852**	-		**72.59**	
									OBB	0.555	0.376	-			

Table 14 continued

Variables entered	Factor 1 eigen-value	Factor 1 (%)	Factor 2 eigen-value	Factor 2 (%)	Factor 3 eigen-value	Factor 3 (%)	Cumulative % of variance	N° of factors extracted		Factor loadings after varimax rotation factor 1	factor 2	factor 3	% of variance[1] factor 1	factor 2	factor 3
Analysis IVb (only males) (N=624)	2.71	(33.92)	1.16	(14.49)	1.06	(13.21)	**61.62**	3	**FRC**	**0.702**	0.329	0.200	**49.28**		
									PAC	0.424	-0.129	0.660			
									OCC	**0.708**	0.177	-0.234	**50.13**		
									BBH	**0.835**	0.076	0.256	**69.72**		
									NPH	0.174	**0.739**	-0.008		**54.61**	
									XFB	0.402	0.492	-0.058			
									NLB	-0.110	0.195	**0.787**			**61.94**
									OBB	0.058	**0.787**	0.160		**61.94**	
Analysis IVc (only females) (N=389)	2.79	(34.86)	1.19	(14.83)	1.06	(13.18)	**62.88**	3	FRC	0.650	0.430	-0.105			
									PAC	**0.812**	-0.137	0.203	**65.93**		
									OCC	0.420	0.251	-0.601			
									BBH	**0.804**	0.256	-0.071	**64.64**		
									NPH	0.007	**0.787**	-0.001		**61.94**	
									XFB	0.335	0.487	-0.172			
									NLB	0.237	0.256	**0.779**			**60.68**
									OBB	0.139	**0.765**	0.209		**58.52**	
WCB, BPL, JUB, NLH, NLB, OBH **Analysis Va** (pooled sexes) (N=879)	2.93	(48.81)	1.04	(17.34)	-	-	**66.15**	2	**WCB**	**0.736**	0.235	-	**54.17**		
									BPL	0.189	**0.778**	-		**60.53**	
									JUB	**0.705**	0.524	-	**49.70**		
									NLH	**0.838**	0.206	-	**70.22**		
									NLB	0.063	**0.803**	-		**64.48**	
									OBH	**0.750**	0.054	-	**56.25**		
Analysis Vb (only males) (N=501)	2.61	(43.45)	1.17	(19.45)	-	-	**62.90**	2	**WCB**	**0.720**	0.081	-	**51.84**		
									BPL	0.150	**0.797**	-		**63.52**	
									JUB	**0.738**	0.421	-	**54.46**		
									NLH	**0.838**	0.056	-	**70.22**		
									NLB	0.023	**0.787**	-		**61.94**	
									OBH	**0.736**	0.016	-	**54.17**		
Analysis Vc (only females) (N=350)	2.61	(43.52)	1.09	(18.18)	-	-	**61.70**	2	**WCB**	**0.758**	0.122	-	**57.45**		
									BPL	0.123	**0.744**	-		**55.35**	
									JUB	0.633	0.554	-			
									NLH	**0.804**	0.190	-	**64.64**		
									NLB	0.035	**0.794**	-		**63.04**	
									OBH	**0.721**	-0.030	-	**51.98**		

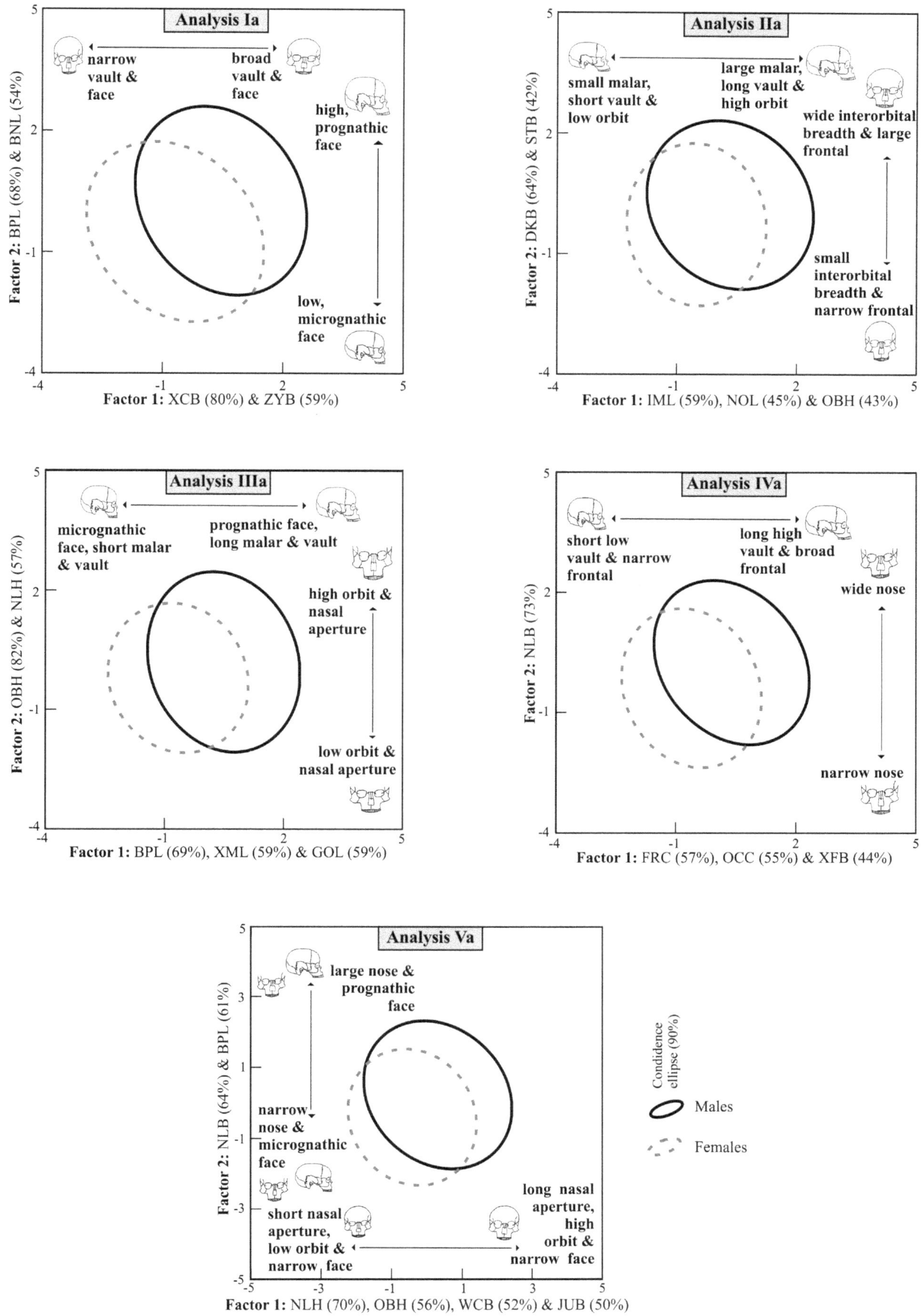

Figure 26. Scatterplots of first and second factors for five factor analyses (Ia-Va) with pooled sexes showing grouping by sex.

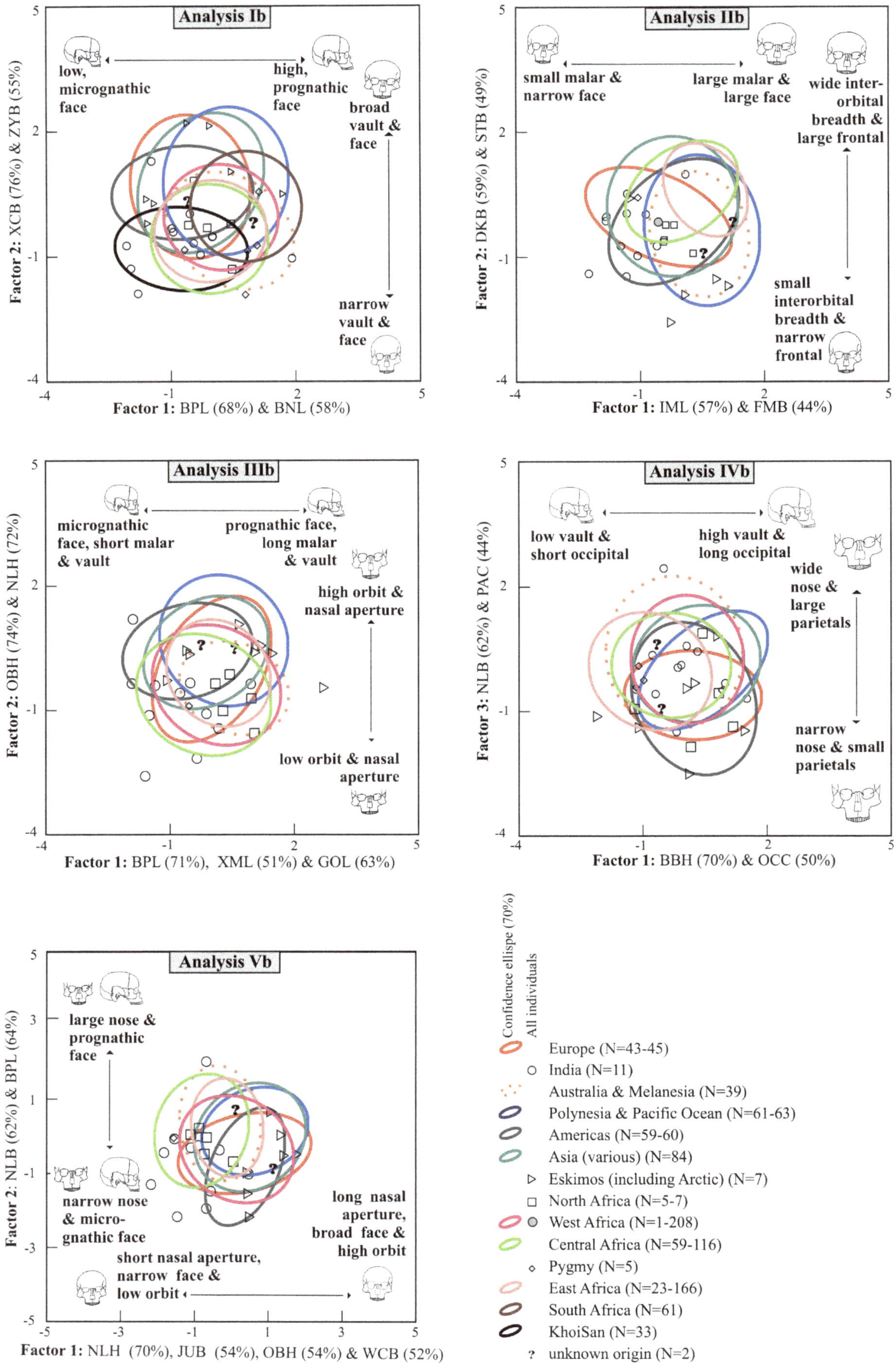

Figure 27. Scatterplots of first and second factors for five factor analyses (Ib-Vb) with only males showing grouping by region.

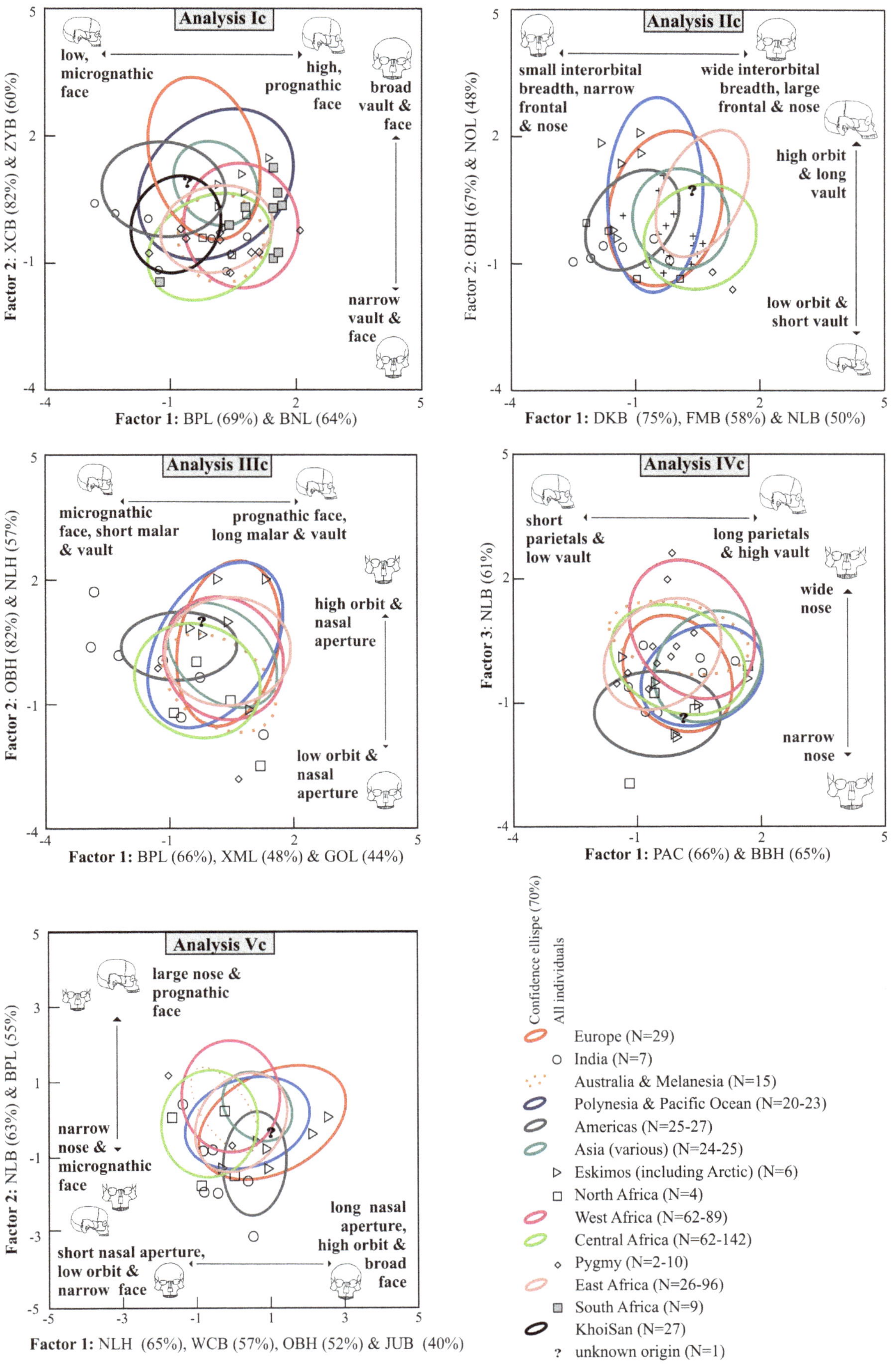

Figure 28. Scatterplots of first and second factors for five factor analyses (Ic-Vc) with only females showing grouping by region.

4.1.2.1. One-way analyses of variance

The 23 variables already used on a worldwide scale were re-analysed for the levels of between to within variance among Africans (with pooled sexes).

This procedure allows us to compare the variation of the latter when adding and/or excluding non-African populations.

The variables are statistically significant at the highest level of probability (P<0.001), except for STB (Appendix XI, p. 206).

Most of the F-values tend to increase except for 9 variables (e.g. XCB, WCB, ZYB, JUB, STB, FMB, ZMB, IML and DKB). The localization of the regional differences (according to post hoc tests) are similar to those previously observed, but more numerous and widespread especially between West Africa and all other regions (e.g. OCC, BBH, XFB). Additional differences also appear within Africans for four variables (e.g. OCC, STB, IML and OBH).

When both sexes are pooled, most of the 34 additional metric traits (except for WBAB) are significantly different within Africa at various levels of probability (Appendix XII, p. 207). According to the highest F-values (>30), the most marked regional differences are observed for the following variables: BIMAS.BR, PMH, PABR, EAM/EAM and ASB.

When analysing only crania with matching mandibles, the morphological differences tend to decrease in comparison to previous results (Appendix XIII, p. 208). This fact suggests that, sample size variability might affect the results.

When both sexes are considered separately (Appendix XII, p. 207), a higher number of variables in relation to various parts of the skull (11 for males; 14 for females) appear not to be significant. According to the F-values, regional differences tend to decrease for most of the variables, except for 7 variables (Q, Q1, PABR, bw, har, mwar and ma) that showed an increase.

The geographical localization of morphological

		Factor	N	Mean	S.D.	Mean diff.	t	Sig.[1]
Analysis Ia	males	1	903	0.391	0.871	0.228	12.76	***
	females	1	509	-0.619	0.901			
	males	2	903	0.262	0.952	0.242	14.76	***
	females	2	509	-0.504	0.906			
Analysis IIa	males	1	401	0.330	0.979	0.191	11.95	***
	females	1	250	-0.521	0.820			
	males	2	401	0.189	0.987	0.118	6.34	***
	females	2	250	-0.307	0.948			
Analysis IIIa	males	1	498	0.439	0.891	0.181	17.49	***
	females	1	342	-0.620	0.818			
	males	2	498	0.181	1.056	0.055	6.24	***
	females	2	342	-0.236	0.869			
Analysis IVa	males	1	624	0.420	0.898	0.211	19.22	***
	females	1	389	-0.631	0.813			
	males	2	624	0.260	0.955	0.149	10.94	***
	females	2	389	-0.409	0.931			
Analysis Va	males	1	501	0.335	0.984	0.101	12.19	***
	females	1	350	-0.436	0.851			
	males	2	501	0.301	0.968	0.120	11.03	***
	females	2	350	-0.421	0.897			

[1] Level of significance: P<0.001***.

Table 15. Independent sample t-tests on the regression factor scores of five analyses (Ia-Va) for sex differences.

differences (according to post hoc tests) is similar for both pooled and separated sexes. They are localized as follows:

- for 4 variables (PAR, TOTARCH, MAL, PABR), between all regions;
- for 7 variables (FAR, PAR, WFB, MAB, MAL, PAL, PABR), between South Africa (Bantu-speaking and/or KhoiSan) and most of the remaining regions; and
- for 6 variables (CC, ASB, PAR, EKB, MAB, PABR), between West Africa and most of the remaining regions.

In summary, in comparison to previous studies (Howells, 1989; Froment, 1992a, 1992b), a higher number of inter-regional differences are found within Africa, especially between West Africa and other regions. This has also been currently observed when using mandibular variables.

Analysis n°	Type	Factor n°	Compared groups[1]	F	Sig.[2]	L.[3]
Ia	pooled	1	**1, 2, 3, 4, 6, 7, 8, 9, 10**	55.82	***	***
	sexes	2		40.44	***	*
Ib	only	1	1, 2, 3, 4, 6, 7, 8, 9, 10	20.09	***	-
	males	2		27.09	***	***
Ic	only	1	1, 2, 3, 4, 6, 7, 8, 10	15.32	***	-
	females	2		31.49	***	***
IIa	pooled	1	**1, 2, 3, 4, 6, 7, 8**	16.73	***	***
	sexes	2		17.88	***	-
IIb	only	1	1, 2, 3, 4, 6, 7, 8	12.30	***	**
	males	2		10.05	***	*
IIc	only	1	1, 2, 3, 4, 6, 7, 8	26.80	***	-
	females	2		7.74	***	***
IIIa	pooled	1	**1, 2, 3, 4, 6, 7, 8**	13.54	***	-
	sexes	2		17.23	***	***
IIIb	only	1	1, 2, 3, 4, 6, 7, 8	6.23	***	-
	males	2		10.86	***	*
IIIc	only	1	1, 2, 3, 4, 6, 7, 8	7.13	***	-
	females	2		6.45	***	*
IVa	pooled	1	**1, 2, 3, 4, 6, 7, 8**	25.94	***	*
	sexes	2		27.32	***	*
IVb	only	1	1, 2, 3, 4, 6, 7, 8	10.25	***	-
	males	3		13.10	***	**
IVc	only	1	1, 2, 3, 4, 6, 7, 8	6.11	***	-
	females	3		19.01	***	-
Va	pooled	1	**1, 2, 3, 4, 6, 7, 8**	52.14	***	***
	sexes	2		24.70	***	-
Vb	only	1	1, 2, 3, 4, 6, 7, 8	21.83	***	***
	males	2		16.85	***	-
Vc	only	1	1, 2, 3, 4, 6, 7, 8	21.62	***	**
	females	2		15.31	***	-

1 **1**=Europe, **2**=Asia (various, India), **3**=Australia (Melanesia, Polynesia, Pacific Ocean), **4**=Americas (including Eskimos), **5**=North Africa, **6**=West Africa, **7**=Central Africa (including Pygmies), **8**=East Africa, **9**=Bantu-speaking South Africa, **10**=KhoiSan;

2 Levels of significance: P<0.05*, P<0.01**, P<0.001***;

3 Significance of Levene's test of homogeneity of variance.

Table 16. One-way analyses of variance on the regression factor scores obtained from eight types of analyses (I-V).

Regions	Vault	Face	Nose	Orbit
1. Europe	broad, variable length	broad, low, micrognathic	narrow	variable, small interorbital breadth
2. Asia (various)	broad	high, broad, prognathic	wide	large interorbital breadth
(2.) India	narrow, low, short	narrow, micrognathic, small malars	low nose	low orbits, small interorbital breadth
3. Australia & Melanesia	narrow	narrow, prognathic, high, large malars	wide, low	low orbits, large interorbital breadth
(3.) Polynesia & Pacific Ocean	broad, long, high	broad	wide, high	high orbits, interorbital breadth variable
4. Americas	narrow	high, narrow, micrognathic	narrow, high	high orbits, low interorbital breadth
(4.) Eskimos	broad, long	broad, high, large malars	variable width, high	high orbits, small interorbital breadth
5. North Africa	medium-size, high	variable prognathism, variable facial breadth	variable width, low	low orbits
6. West Africa	narrow	narrow, prognathic, high, large malars	wide, low	low orbits, large interorbital breadth
7. Central Africa Bantu-speaking & Pygmy	short, narrow	narrow, prognathic	wide	low orbits, large interorbital breadth
8. East Africa	narrow, long, low	narrow, prognathic	wide, high	less low than previous ones
9. South Africa Bantu-speaking	narrow	narrow, prognathic	-	-
10. KhoiSan	narrow	narrow, micrognathic	-	-

Table 17. Summary of the morphological trends observed from factor analyses on the crania of worldwide origin.

4.1.2.2. Multiple discriminant analyses

In order to establish to what extent the metric traits are good predictors of regional differences within Africa, several multiple discriminant analyses were performed on both cranial and mandibular variables separately (Appendix XIV, p. 209).

Eleven analyses using different groups of variables are presented here, as they showed both the highest and lowest rates of classification. They were performed systematically with pooled sexes, as well as with both male and female samples separately. Three types (analyses 2, 4 and 5) had been already performed on a worldwide scale. However, as the different analyses involved a variable number of groups, they are not always comparable with each other.

According to the highest standardized canonical discriminant function coefficients, the variables which showed most frequently regional differences within Africa are:

- for the vault: BBH, XCB, GOL, Q1 and ASB;
- for the face: ZYB, NLB, EKB, OBB and PABR; and
- for the mandible: sh.

Although these results are also based on additional variables, they agree with previous results on a worldwide scale. However, the metric traits appear to be even more variable in predicting geographical differences within Africa. This fact is suggested by the following inter-regional total correct classification rates:

- for pooled sexes: from 30% up to 71.8%;
- for males: from 33% up to 79%; and
- for females: from 42.6% up to 70.6%.

The low rates result from similar mis-classifications when both pooling and separating

Variables entered	Groups compared	Total rates of correct classification[1] Initial sample	First random set: ± 50%	Second random set: ± 50%	Third random set: ± 75%
VAULT					
Analysis 1: CC, ASB, Q1, FAR, PAR	II, IV, VI	**69.0%** (336)	**65.5%** (165)	-	**54.5%** (112)
Analysis 2: GOL, XCB, PAC, OCC, BBH, WCB, BNL	I, II, IV, VI	**42.6%** (1389)	**41.2%** (738)	**43.7%** (701)	**41.8%** (316)
BASE					
Analysis 3: BIMAS.BR, EAM/EAM, FOL, FOB	I, II, IV	**64.0%** (347)	**50.0%** (290)	**46.4%** (276)	**40.6%** (128)
FACE					
Analysis 4: BPL, ZYB, JUB, STB, XFB, FMB, ZMB	I, II, IV	**44.5%** (596)	**51.8%** (274)	**44.1%** (345)	-
Analysis 5: IML, XML, NLH, NLB, DKB, OBB, OBH	I, II, IV	**39.3%** (909)	**42.6%** (575)	**40.6%** (571)	**52.7%** (404)
MANDIBLE					
Analysis 6: war, bw, sh, ma, bb, har	I, II, IV	**60.9%** (294)	**58.5%** (135)	**54.9%** (153)	-
Analysis 7: war, bw, sh, ma, bb, har	I, II	**71.8%** (245)	**68.8%** (112)	**68.3%** (123)	-
VARIOUS					
Analysis 8: BBH, GOL, BNL, ZYB, XCB, BPL, NPH	I, II, IV-VI	**54.7%** (1086)	**54.2%** (537)	**54.6%** (564)	**57.9%** (235)
Analysis 9: BBH, GOL, ZYB, XCB, BPL, NPH, BNL, WFB	I, II, IV-VI	**56.7%** (934)	**57.8%** (464)	**55.0%** (564)	**48.6%** (288)
Analysis 10: FMB, OBH, Q1, DKB, NLB	I, II, IV	**60.3%** (453)	**61.5%** (239)	**56.2%** (226)	**61.2%** (98)
Analysis 11: ASB, Q1, TOTARCH, FOL, FOB, EKB, WFB, PABR	I, II, IV-VI	**54.6%** (668)	**59.0%** (278)	**55.3%** (340)	**56.2%** (146)

[1] Sample size is put into brackets.

Table 18. Summary of multiple discriminant analyses performed with various random samples originating from sub-Saharan Africa (pooled sexes).

sexes (Appendix XV, p. 210). The most frequently misclassified groups are:

- KhoiSan with Central Africans (Bantu-speaking), West Africans and East Africans;
- East Africans with Central Africans (Bantu-speaking) and West Africans; and Central Africans (Bantu-speaking) with West Africans.

In fact, individuals originating from small samples (e.g. KhoiSan) tend to be misclassified into larger samples (e.g. West Africans, Central Africans). Nevertheless, for all samples, the mandible seems to express better the geographical differences within Africa: it provided the highest range of correct classification (54.1%-79%), in contrast to the face (range: 33%-44.5%). In order to test if these results are affected by random factors and/or sample size, a randomization procedure is achieved for eleven analyses with pooled sexes (Table 18, p. 87). Two random samples (around 50% smaller than the initial one) were generated by the approximate procedure of SPSS 9.0 (1999a, 1999b). A further random sample (one around 75% smaller) was generated when there were at least 20 individuals.

For the correct geographical classification, only slight changes (either a decrease or an increase in accuracy) were observed when using different parts of the skull (cranial base: 40.6%-64%; face: 39.3%-52.7%). However, the mandibular variables still appear to be the best predictors of geographical differences within Africa (54.9%-68.8%).

4.1.2.3. Factor analyses

Different series of metric traits are used here to

Variable entered	Factor 1 eigen-value	Factor 1 (%)	Factor 2 eigen-value	Factor 2 (%)	Cumulative % of variance	N° of factors extracted		Factor loadings after varimax rotation factor 1	Factor loadings after varimax rotation factor 2	% of variance[1] factor 1	% of variance[1] factor 2
GOL, XCB, ZYB, BBH, BPL, NPH, BNL, NLB **Analysis VIb** (only males) (N=611)	3.24	(40.44)	1.07	(13.40)	**53.85**	2	GOL	0.352	0.563		
							XCB	-0.260	**0.860**		**73.96**
							ZYB	0.438	0.581		
							BBH	0.338	0.527		
							BPL	**0.762**	0.251	**58.06**	
							NPH	0.417	0.454		
							BNL	**0.735**	0.428	**54.02**	
							NLB	0.625	-0.034		
Analysis VIc (only females) (N=405)	3.28	(41.03)	1.16	(14.47)	**55.50**	2	GOL	0.287	0.680		
							XCB	-0.191	**0.861**		**74.13**
							ZYB	0.515	0.538		
							BBH	0.374	0.600		
							BPL	**0.805**	0.118	**64.80**	
							NPH	0.576	0.320		
							BNL	0.671	0.407		
							NLB	0.589	-0.044		
GOL, BPL, ZMB, XML, NLH, OBH **Analysis VIIa** (pooled sexes) (N=441)	2.74	(45.73)	1.14	(19.07)	**64.80**	2	**GOL**	**0.740**	0.250	**54.76**	
							BPL	**0.811**	0.004	**65.77**	
							ZMB	0.585	0.375		
							XML	**0.768**	0.044	**58.98**	
							NLH	0.435	**0.717**		**51.41**
							OBH	-0.058	**0.915**		**83.72**
Analysis VIIb (only males) (N=185)	2.30	(38.39)	1.36	(22.59)	**60.98**	2	GOL	0.614	0.236		
							BPL	**0.822**	-0.028	**67.57**	
							ZMB	0.512	0.473		
							XML	**0.758**	-0.037	**57.46**	
							NLH	0.252	**0.815**		**66.42**
							OBH	-0.133	**0.862**		**74.30**
Analysis VIIc (only females) (N=229)	2.51	(41.84)	1.05	(17.41)	**59.26**	2	**GOL**	**0.711**	0.275	**50.55**	
							BPL	**0.780**	-0.002	**60.84**	
							ZMB	0.590	0.251		
							XML	0.636	0.108		
							NLH	0.372	**0.753**		**56.70**
							OBH	0.019	**0.912**		**83.17**

[1] Percentage of variance accounted on each factor obtained by squaring the correlation coefficient (or factor loading) for variables which have the highest factor loadings (>0.7).

Table 19. Summary of eight types of factor analyses (VI-XIII) performed on African samples.

Table 19 continued

Variables entered	Factor 1 eigen-value	Factor 1 (%)	Factor 2 eigen-value	Factor 2 (%)	Factor 3 eigen-value	Factor 3 (%)	Cumulative % of variance	N° of factors extracted		Factor loadings after varimax rotation factor 1	Factor loadings after varimax rotation factor 2	Factor loadings after varimax rotation factor 3	% of variance[1] factor 1	% of variance[1] factor 2	% of variance[1] factor 3
FRC, PAC, OCC, BBH, NPH, XFB, NLB, OBB **Analysis VIIIb** (only males) (N=321)	2.53	(31.64)	1.14	(14.26)	1.01	(12.62)	**58.52**	3	FRC	0.657	0.302	0.268			
									PAC	0.303	-0.102	**0.745**			**55.50**
									OCC	**0.753**	0.078	-0.183	**56.70**		
									BBH	**0.799**	0.060	0.352	**63.84**		
									NPH	0.065	0.627	0.258			
									XFB	0.387	0.620	-0.138			
									NLB	-0.085	0.242	0.618			
									OBB	0.050	**0.767**	0.038		**58.83**	
Analysis VIIIc (only females) (N=281)	2.79	(34.87)	1.19	(14.86)	-	-	**49.73**	2	**FRC**	**0.748**	0.246	-	**55.95**		
									PAC	0.614	0.104	-			
									OCC	0.690	-0.129	-			
									BBH	**0.803**	0.192	-	**64.48**		
									NPH	0.135	0.686	-		**47.06**	
									XFB	0.513	0.203	-			
									NLB	-0.013	0.674	-		**45.43**	
									OBB	0.260	0.692	-		**47.89**	
WCB, BPL, JUB, NLH, NLB, OBH **Analysis IXa** (pooled sexes) (N=432)	2.87	(47.90)	1.01	(16.77)	-	-	**64.67**	2	WCB	0.646	0.357	-			
									BPL	**0.789**	-0.102	-	**62.25**		
									JUB	**0.786**	0.313	-	**61.78**		
									NLH	0.475	0.678	-			
									NLB	0.654	0.192	-			
									OBH	0.015	**0.915**	-		**83.72**	
Analysis IXb (only males) (N=184)	2.51	(41.83)	1.15	(19.18)	-	-	**61.01**	2	WCB	0.505	0.481	-			
									BPL	**0.838**	-0.148	-	**70.22**		
									JUB	**0.759**	0.278	-	**57.61**		
									NLH	0.260	**0.801**	-		**64.16**	
									NLB	0.481	0.370	-			
									OBH	-0.043	**0.848**	-		**71.91**	

Table 19 continued

Variables entered	Factor 1 eigen-value	Factor 1 (%)	Factor 2 eigen-value	Factor 2 (%)	Factor 3 eigen-value	Factor 3 (%)	Cumulative % of variance	N° of factors extracted		Factor loadings after varimax rotation factor 1	factor 2	factor 3	% of variance[1] factor 1	factor 2	factor 3
bb, bw, sh, har, war, ma															
Analysis Xa (pooled sexes) (N=300)	2.35	(39.20)	1.24	(20.58)	-	-	**59.78**	2	bb	0.649	0.403	-			
									bw	**0.817**	-0.102	-	**66.75**		
									sh	**0.709**	0.313	-	**50.27**		
									har	0.420	0.501	-			
									war	0.250	**0.720**	-		**51.84**	
									ma	0.165	**-0.888**	-		**78.85**	
Analysis Xb (only males) (N=158)	1.80	(30.04)	1.31	(21.89)	-	-	**51.93**	2	bb	0.500	0.567	-			
									bw	0.127	**0.793**	-		**62.88**	
									sh	-0.106	0.661	-			
									har	0.424	0.092	-			
									war	**0.703**	0.133	-	**49.42**		
									ma	**-0.827**	0.260	-	**68.39**		
Analysis Xc (only females) (N=113)	2.05	(34.16)	1.34	(22.38)	-	-	**56.53**	2	bb	0.610	0.339	-			
									bw	**0.811**	-0.022	-	**65.77**		
									sh	0.633	-0.105	-			
									har	0.544	0.492	-			
									war	0.150	**0.706**	-		**49.84**	
									ma	0.179	**-0.862**	-		**74.30**	
NPH, WFB, NLB, bb, sh															
Analysis XIb (only males) (N=157)	2.01	(40.23)	1.05	(20.97)	-	-	**61.20**	2	**NPH**	0.171	**0.838**	-		**70.22**	
									WFB	**0.792**	0.039	-	**62.72**		
									NLB	0.672	0.087	-			
									bb	0.665	0.252	-			
									sh	0.094	**0.851**	-		**72.42**	
Analysis XIc (only females) (N=157)	1.93	(38.66)	1.12	(22.40)	-	-	**61.06**	2	**NPH**	**0.864**	0.029		**74.65**		
									WFB	-0.047	**0.843**			**71.06**	
									NLB	0.523	0.384				
									bb	0.200	0.697				
									sh	**0.802**	0.051		**64.32**		

Table 19 continued

Variables entered	Factor 1 eigen-value	Factor 1 (%)	Factor 2 eigen-value	Factor 2 (%)	Factor 3 eigen-value	Factor 3 (%)	Cumulative % of variance	N° of factors extracted		Factor loadings after varimax rotation factor 1	factor 2	factor 3	% of variance[1] factor 1	factor 2	factor 3
ASB, Q1, EKB, WFB, PABR, TOTARCH **Analysis XIIa** (pooled sexes) (N=541)	3.06	(50.99)	1.05	(17.53)	-	-	**68.52**	2	**ASB**	**0.721**	0.086	-	**51.98**		
									Q1	**0.803**	0.118	-	**64.48**		
									EKB	0.513	0.697	-			
									WFB	**0.735**	0.295	-	**54.02**		
									PABR	-0.006	**0.914**	-		**83.54**	
									TOT-ARCH	**0.841**	0.076	-	**70.73**		
Analysis XIIb (only males) (N=275)	2.47	(41.16)	1.14	(19.00)	1.05	(17.41)	**77.57**	3	ASB	0.575	0.339	-0.316			
									Q1	0.143	**0.872**	0.136		**76.04**	
									EKB	**0.837**	0.024	0.398	**70.06**		
									WFB	**0.831**	0.189	-0.010	**69.06**		
									PABR	0.066	0.128	**0.931**			**86.68**
									TOT-ARCH	0.173	**0.897**	0.014		**80.46**	
Analysis XIIc (only females) (N=236)	2.58	(43.03)	1.25	(20.80)	-	-	**63.83**	2	ASB	0.648	0.125				
									Q1	**0.791**	0.023		**62.57**		
									EKB	0.319	**0.839**			**70.39**	
									WFB	0.652	0.424				
									PABR	-0.081	**0.832**			**69.22**	
									TOT-ARCH	**0.811**	0.010		**65.77**		
GOL, XCB, PAC, OCC, BBH, WCB, BNL **Analysis XIIIb** (only males) (N=197)	2.72	(38.84)	1.15	(16.37)	-	-	**55.21**	2	GOL	0.593	0.520	-			
									XCB	0.035	**0.747**	-		**55.80**	
									PAC	**0.723**	0.048	-	**52.27**		
									OCC	0.606	0.108	-			
									BBH	**0.819**	-0.001	-	**67.08**		
									WCB	0.109	**0.807**	-		**65.12**	
									BNL	0.552	0.372	-			
Analysis XIIIc (only females) (N=239)	3.03	(43.27)	1.04	(14.89)	-	-	**58.16**	2	GOL	0.684	0.437	-			
									XCB	0.073	**0.834**	-		**69.56**	
									PAC	**0.790**	0.046	-	**62.41**		
									OCC	0.275	0.582	-			
									BBH	0.646	0.420	-			
									WCB	0.132	**0.724**	-		**52.42**	
									BNL	**0.726**	0.091	-	**52.71**		

		Factor	N	Mean	S.D.	Mean diff.	t	Sig.[1]
Analysis VIIa	males	1	185	0.608	0.698	1.08	12.76	***
	females	1	229	-0.476	0.828			
	males	2	185	0.140	1.08	0.25	2.48	*
	females	2	229	-0.106	0.934			
Analysis IXa	males	1	184	0.581	0.901	1.04	11.97	***
	females	1	220	-0.461	0.846			
	males	2	184	0.130	1.08	0.22	2.20	*
	females	2	220	0.093	0.929			
Analysis Xa	males	1	275	0.427	0.865	0.94	8.76	*
	females	1	236	-0.510	0.870			
	males	2	275	0.258	0.988	0.22	2.20	*
	females	2	236	-0.284	0.896			
Analysis XIIa	males	1	275	0.529	0.850	1.11	14.90	***
	females	1	236	-0.587	0.838			
	males	2	275	0.222	0.978	0.46	5.35	***
	females	2	236	-0.242	0.981			

1 Level of significance (P<0.05*; P<0.01**; P<0.001***).

Table 20. Independent sample t-tests on the regression factor scores with analyses for sex differences.

perform eight types of factor analyses (VI-XIII). They were also analysed systematically in three steps:

- analysis a: pooled sexes;
- analysis b: only males; and
- analysis c: only females.

However, in order to be able to compare a minimum amount of variance in a two-dimensional space, only results that provided a model with at least two factors, are presented here (Table 19, p. 88-91).

For the first four types of analyses (VI-IX), the groups of variables entered are similar to those used previously on a worldwide scale:

- for **analysis VI** (similar to I): GOL, XCB, ZYB, BBH, BPL, NPH, BNL and NLB;
- for **analysis VII** (similar to III): FRC, PAC, OCC, BBH, NPH, XFB, NLB and OBB;
- for **analysis VIII** (similar to IV): GOL, BPL, ZML, XML, NLH and OBH; and
- for **analysis IX** (similar to V): WCB, BPL, JUB, NLH, NLB and OBH.

For the four other types of factor analyses (X-XIII), new combinations of variables using not only the cranium but also the mandible were explored:

- for **analysis X:** bb, bw, sh, har, war and ma;
- for **analysis XI**: NPH, WFB, NLB, bb and sh;
- for **analysis XII**: ASB, Q1, EKB, WFB, PABR and TOTARCH; and
- for **analysis XIII**: GOL, XCB, PAC, OCC, BBH, WCB and BNL.

For all analyses with pooled and/or separated sexes, the variables that are present most frequently on the factors with the highest loading (≥0.7) are:

- for factor 1: BPL (e.g. analyses VIa VIb, VIIa, VIIb, VIIc, IXa, IXb), BBH (e.g. analyses VIIIa, VIIIb, XIIIa) and WFB (e.g. analyses XIa, XIIa, XIIb); and
- for factor 2: OBH (e.g. analyses VIIa, VIIb, VIIc, IXa, IXb), NLH (e.g. analyses VIIa, VIIb, VIIc, IXb) and XCB (e.g. analyses VIa, VIb, XIIIa, XIIIb).

For the first four types of factor analyses (VI-IX), the selected morphological features (or groups of inter-related variables) are similar to previous analyses obtained on a worldwide scale (Table 14, p. 78-80). For the other four types of factor analyses (X-XIII), they are the following:

- for **analysis X**: mandible breadth (bb, bw), chin height (sh); mandibular angle size (ma) and ascending ramus width (war);

- for **analysis XI**: breadth of frontal (WFB), nose (NLB) and mandible (bb); height of chin (sh) and face (NPH);
- for **analysis XII**: length/height of vault (TOTARCH, Q1) and frontal breadth (WFB); breadth of palate (PABR) and face (EKB); and
- for **analysis XIII**: length/height of vault (BBH, PAC) and breadth of vault (WCB, XCB).

Geographical differences are expressed here through morphological trends (e.g. X, XI) that are different from previous analyses (see Table 14, p. 78-80).

4.1.2.3.1. Sex differences

According to independent sample t-tests (Table 20, p. 92), sex differences are significant at various levels of probability. They are more marked along factor 1, which is related mainly to the size of the skull (especially vault: analyses VIIa, XIIa).

Scatterplots for factors 1 and 2 (Figure 29, p. 94) show the overlapping variation of the regression factor scores for each sex. They underline the fact that, the mandible (analysis Xa) is slightly less sexually dimorphic than other parts of the skull, as it is also suggested by the t-values and their low level of significance.

4.1.2.3.2. Geographical differences

In order to explore in more detail the geographical differences within Africa, two types of tests were performed on the regression factor scores (Table 21, p. 95):

- firstly, one-way analyses of variance with three categorical variables such as broad regions, latitude areas and longitude areas; and
- secondly, non-parametric correlation (Spearman Rank) with two continuous variables such as latitude and longitude coordinates.

4.1.2.3.2.1. One-way analyses of variance

Scatterplots for factors 1 and 2 (or 3) obtained from the various previous analyses (see Table 19, p. 88-91) were drawn, in order to show within Africa the group clustering by regions for both male and female populations (Figures 30-33, p. 96-99).

In order to test the geographical differences obtained from previous regression factor scores (see Table 19, p. 88-91), one-way analyses of variance were performed again. They showed that, the regional differences within Africa are significant at various levels of probability, except for factor 2 of four analyses (e.g. VIIb, IXb, Xb and Xc) (Appendix XVI, p. 211).

In general, differences between regions are found to be statistically significant more often for factor 1 (16/19 tests) than factor 2 (12/19 tests). Therefore, they are mainly related to size and shape of vault, as the latter is expressed through factor 1 in five types of analyses (e.g. VI, VII, VIII, XII and XIII).

For the first four types of factor analyses (VI-IX) (Figures 30-31, p. 96, 97), the morphological differences observed within Africa are similar to previous ones (Table 17, p. 86). However, they appear to be slightly different, as groups were not compared on a worldwide scale (Table 22, p. 100). For example, African crania appear to have wider vaults, when compared with each other. Furthermore, with the four other types of factor analyses (X-XIII) (Figures 32-33, p. 98, 99), additional morphological differences were observed, especially in relation to the mandible.

Post hoc multiple comparisons tests allowed us to localize differences more precisely. Although broad differences between regions such as West Africa, Central Africa, East Africa and/or South Africa can vary slightly, they tend to be the same when analysing both pooled and separated sexes.

In general, differences were always present between two extremes: the West Africans (broad vault, prognathic face, high nose and orbit, high chin) and South African KhoiSan (narrow vault, micrognathic face, low nose and orbit, low chin).

Due to small sample size (N≤10), the Pygmies are difficult to localize within African variation. However, they often appear to be well within the Central African group. Sometimes they were morphologically close to both East Africans (analysis VIIIc; Figure 31, p. 97) and West Africans (analysis Xc; Figure 33, p. 99).

Although regional differences are also similar for both sexes, they seemed to be slightly more marked in females, especially for analysis VIIc (factor 2). This fact was already suggested by independent sample t-tests. The first two factors of analysis VIIc that appeared significantly different on a regional level for females (Table 21, p. 95). These differences reflect morphological variation in relation to prognathism, vault size, orbital and nose height.

The broad regions used here do not always express marked geographical differences. It seems that the use of latitude and/or longitude areas can often provide additional information on this respect. For example, regression factor scores that appear not significant between regions were instead significant between either latitude and/or longitude areas for certain analyses.

Post hoc multiple comparisons for both latitude and longitude areas reflect similar differences as the previous ones on an inter-regional level in this section. However, they are expressed in a more

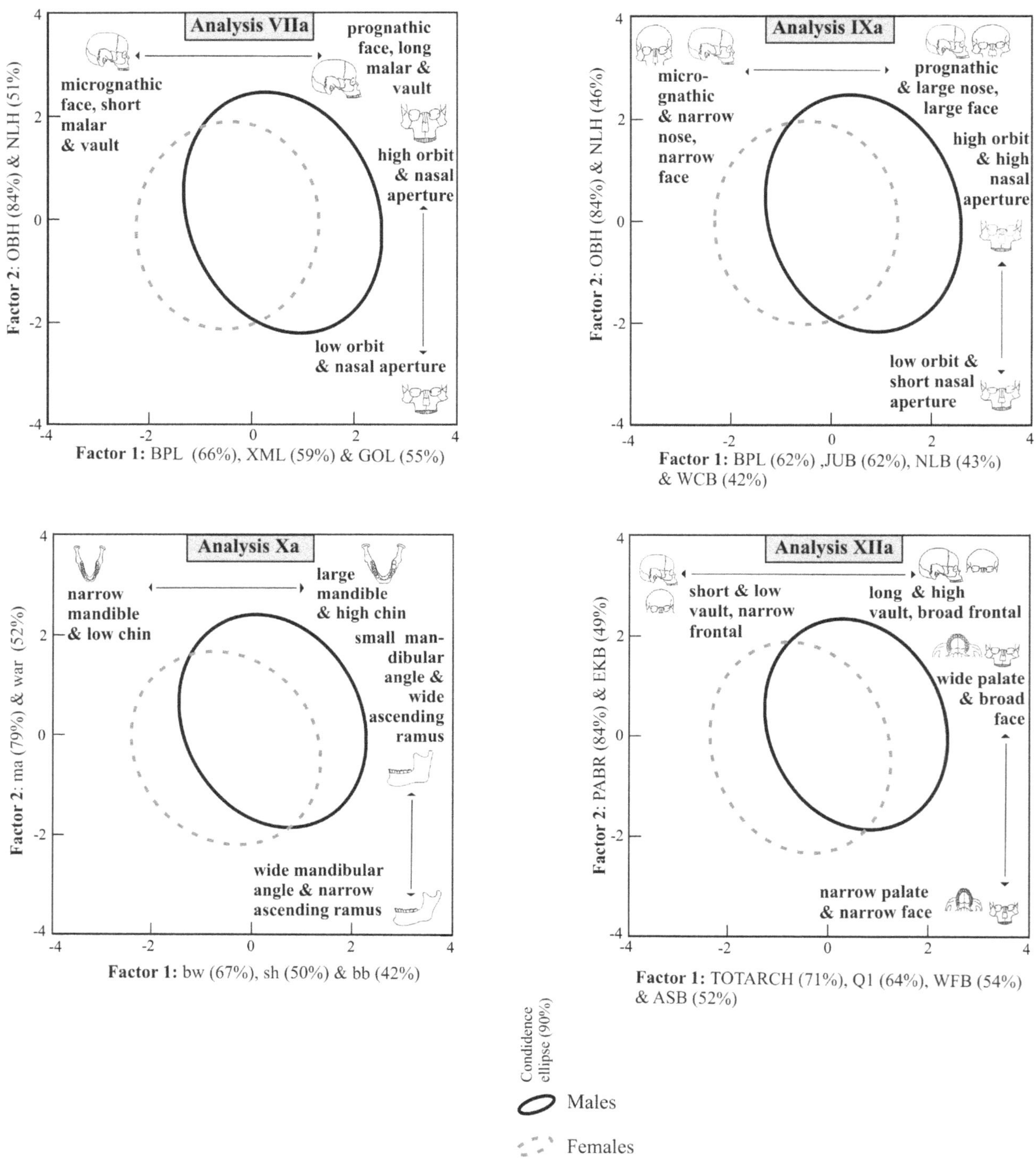

Figure 29. Scatterplots of first and second factors for four factor analyses (VIIa-XIIa) with pooled sexes showing grouping by sex.

			Levels of significance of one-way analyses of variance[1]			Spearman Rank correlation coefficients[1]	
Analysis n°	Type	Factor n°	Broad regions	Latitude areas	Longitude areas	Latitude[2]	Longitude[3]
VIb	only	1	***	***	***	0.009	**-0.178****
	males	2	***	***	***	0.026	-0.031
VIc	only	1	***	***	***	**0.377****	**-0.264****
	females	2	***	***	***	0.017	0.050
VIIa	pooled	1	***	***	***	**0.218****	**-0.081**
	sexes	2	***	**	***	**0.147****	-0.075
VIIb	only	1	-	-	-	0.103	-0.017
	males	2	-	-	-	0.120	-0.070
VIIc	only	1	***	***	***	**0.273****	**-0.089**
	females	2	***	**	***	**0.165***	-0.045
VIIIb	only	1	***	***	***	**0.259****	**-0.316****
	males	3	-	*	***	**0.129***	**-0.070****
VIIIc	only	1	-	*	***	0.029	**-0.136***
	females	2	***	***	***	**0.388****	**-0.217****
IXa	pooled	1	***	**	***	**0.315****	**-0.194****
	sexes	2	***	**	***	**0.147****	-0.030
IXb	only	1	**	***	*	**0.176***	-0.095
	males	2	-	-	-	0.145	-0.083
Xa	pooled	1	-	-	*	0.014	0.004
	sexes	2	-	*	-	0.032	0.018
Xb	only	1	*	-	-	0.087	-0.084
	males	2	-	-	-	0.059	-0.057
Xc	only	1	-	*	***	0.180	-0.142
	females	2	-	-	-	0.127	-0.014
XIb	only	1	***	-	*	-0.031	-0.150
	males	2	-	-	*	0.066	0.044
XIc	only	1	***	***	***	**0.338****	**-0.269****
	females	2	-	-	*	0.013	-0.038
XIIa	pooled	1	***	***	***	**-0.243****	**0.173****
	sexes	2	***	***	***	**0.382****	**-0.207****
XIIb	only	1	***	***	***	**-0.241****	**0.254****
	males	3	***	***	***	0.040	**-0.209****
XIIc	only	1	**	***	**	-0.126	**0.227****
	females	2	***	***	***	**0.402****	**-0.146****
XIIIb	only	1	***	***	***	**0.295****	**-0.348****
	males	2	***	***	***	**0.161***	0.124
XIIIc	only	1	***	***	***	**0.332****	**-0.251****
	females	2	**	***	***	**0.148***	0.146

1 Levels of significance: P<0.05*, P<0.01**, P<0.001***;
2 Latitude expresses here the distance from the equator (a south-north or north-south trend starting from 0°);
3 Longitude expresses here the distance from 0° which reflects mainly a west-east trend.

Table 21. Relationship between various geographical variables and the regression factor scores obtained from eight types of analyses (VI-XIII).

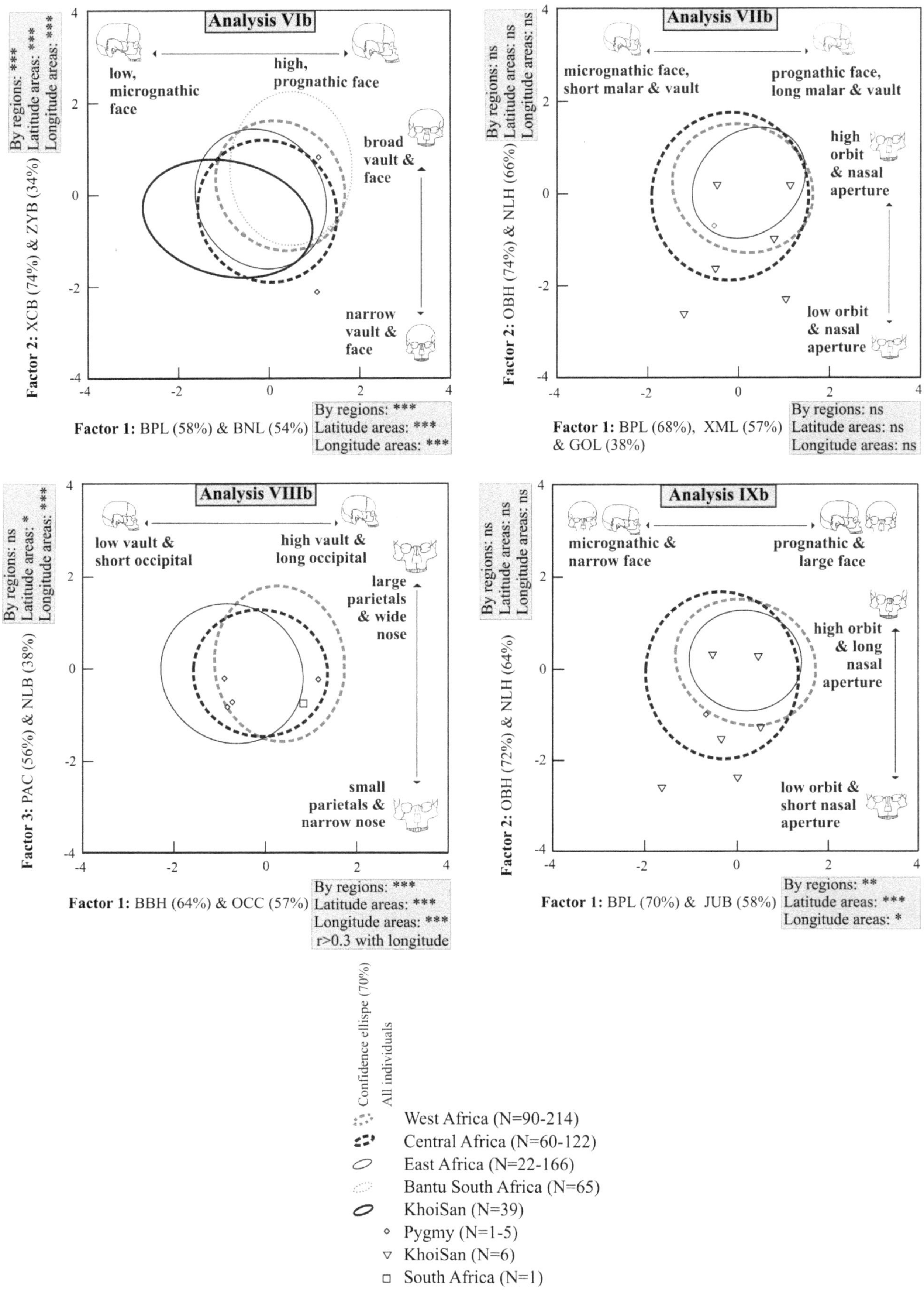

Figure 30. Scatterplots of first and second factors for four factor analyses (VIb-IXb) with only males showing grouping by region. The results of various tests (analysis of variance and Spearman Rank correlation) are mentioned for each factor. Levels of significance: ns=not significant, P<0.05*, P<0.01**, P<0.001***.

detailed manner, as the areas are subdivided along a grid having either a north-south direction (latitude areas) or an east-west direction (longitude areas).

4.1.2.3.2.2. Spearman Rank correlations

In order to test more precisely the correlation between regression factor scores and geographical coordinates, non-parametric tests (Spearman Rank correlation) were performed. They were used here, because geographical coordinates do not present a normal distribution even after logarithmic transformation.

According to the highest values (r≥0.3), correlation coefficients for latitude are often relatively higher than those for longitude (Table 21, p. 95). Therefore, on a geographical level, morphological differences are more marked when looking in a

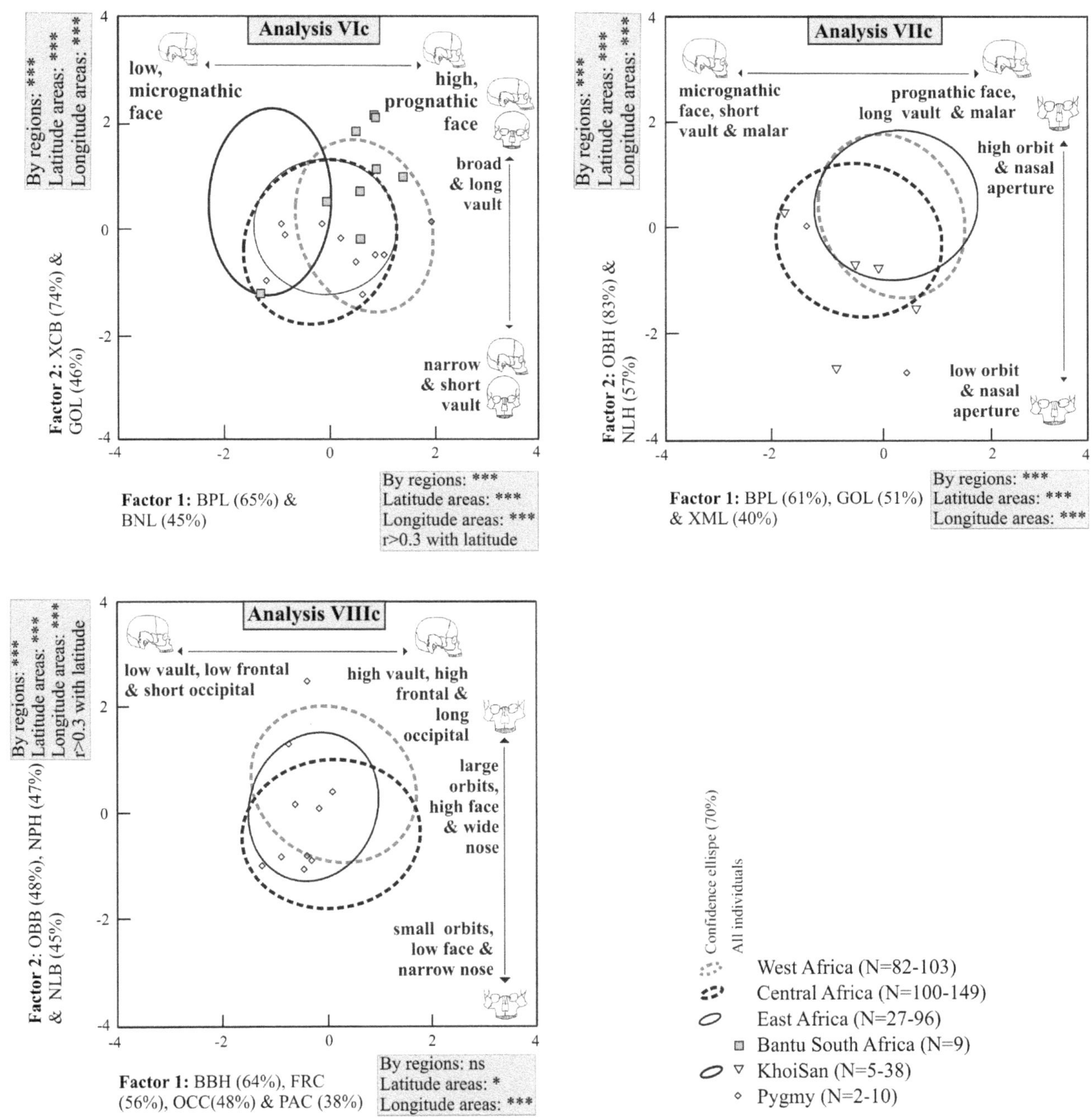

Figure 31. Scatterplots of first and second factors for three factor analyses (VIc-IXc) with only females showing grouping by region. The results of various tests (analysis of variance and Spearman Rank correlation) are mentioned for each factor. Levels of significance: ns=not significant, P<0.05*, P<0.01**, P<0.001***.

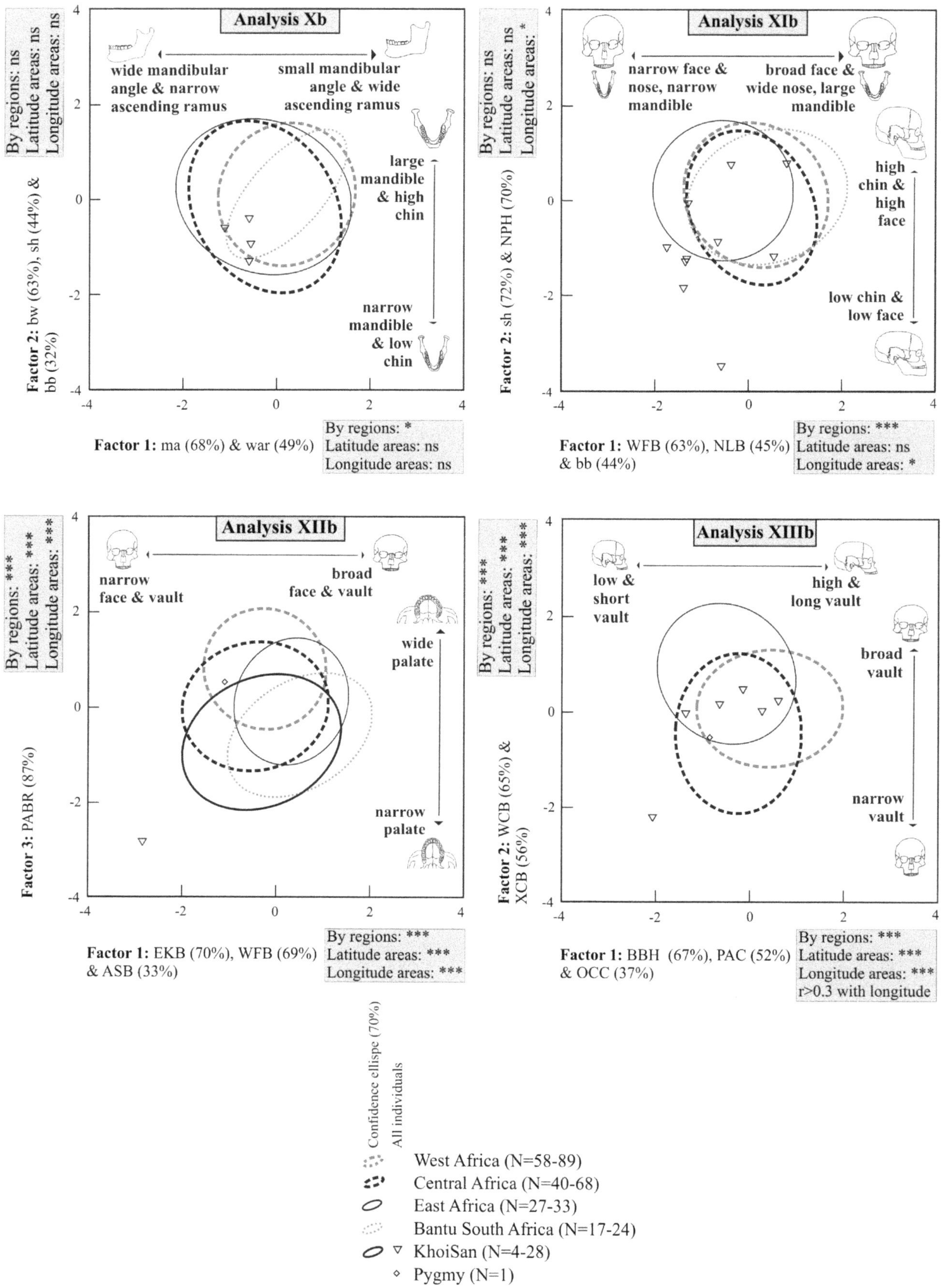

Figure 32. Scatterplots of first and second factors for four factor analyses (Xb-XIIIb) with only males showing grouping by region. The results of various tests (analysis of variance and Spearman Rank correlation) are mentioned for each factor. Levels of significance: ns=not significant, P<0.05∗, P<0.01∗∗, P<0.001∗∗∗.

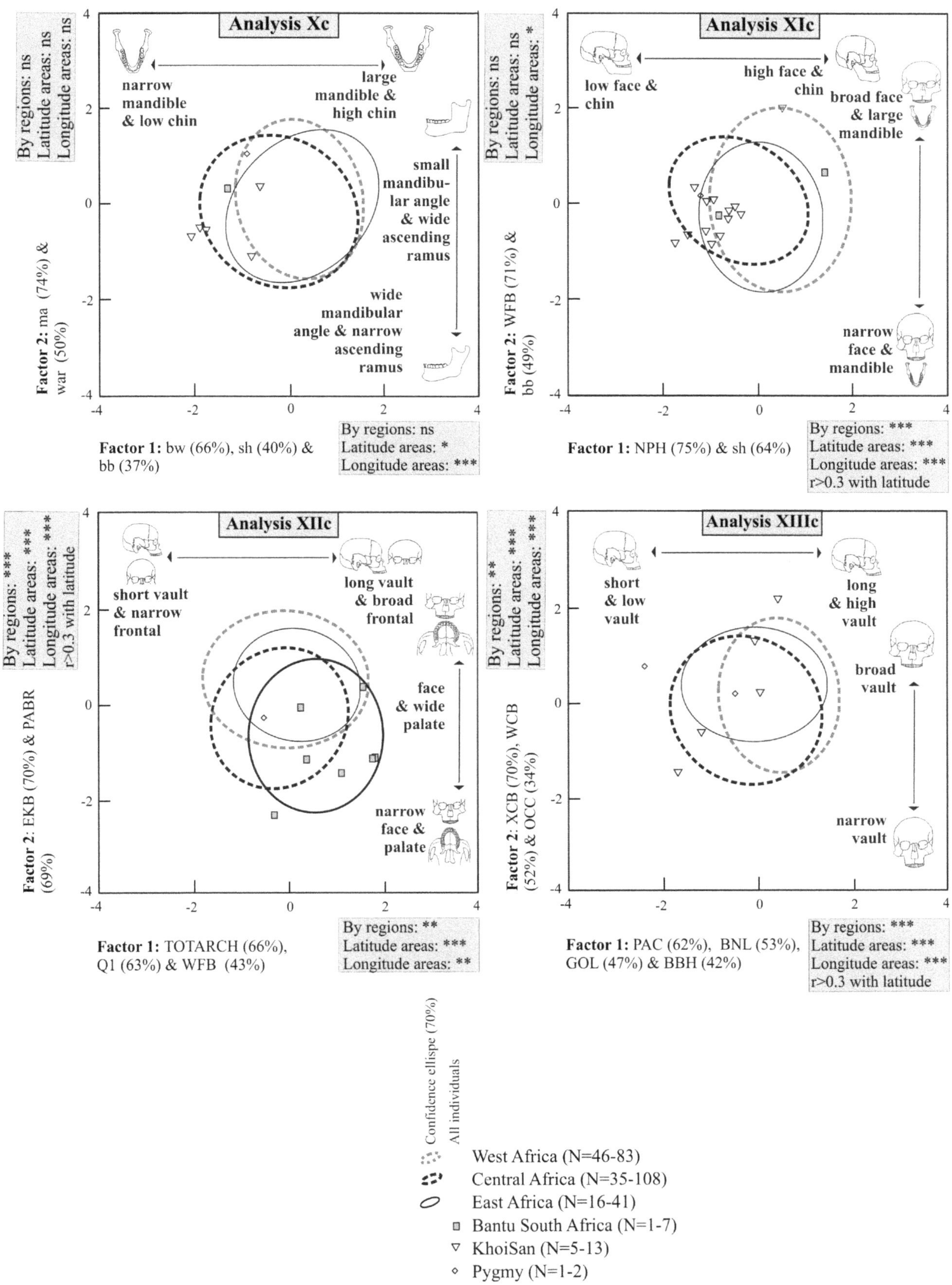

Figure 33. Scatterplots of first and second factors for four factor analyses (Xc-XIIIc) with only males showing grouping by region. The results of various tests (analysis of variance and Spearman Rank correlation) are mentioned for each factor. Levels of significance: ns=not significant, P<0.05∗, P<0.01∗∗, P<0.001∗∗∗.

Regions	Vault	Face	Nose	Orbit	Mandible
I. West Africa	broad, long, high	broad, prognathic, high, large malars, wide palate	high, wide	high, wide	large mandible, high chin
II. Central Africa (Bantu-speaking)	medium-size, moderately high	moderately broad, high & prognathic, moderately narrow palate	moderately wide	moderately high	low chin
III. Pygmy (Central Africa)	medium-size, moderately high	narrow, prognathic	variable width	variable size	
IV. East Africa	medium-size, long, moderately low	moderately broad, high, prognathic, large malars & wide palate	wide, high	variable height	moderately high chin
V. South Africa (Bantu-speaking)	broad	high, broad, prognathic, narrow palate		wide	large mandible, high chin
VI. KhoiSan (South Africa)	narrow	narrow, low, micrognathic, narrow palate	low	low	low chin

Table 22. Summary of the morphological trends observed from factor analyses on the skulls originating only from sub-Saharan Africa.

latitudinal way especially along the equator (north-south or south-north direction) than in a longitudinal way (or mainly west-east direction). This fact was observed especially in female populations. In contrast, geographical differentiation of males follows slightly more often a longitudinal direction.

4.1.3. Conclusions

In answering our two initial questions about geography (and sex) as a source of diversity, the following comments can be made.

1. In comparison to previous studies (Howells, 1989; Froment, 1992a, 1992b), both univariate and multivariate tests showed similar morphological trends amongst modern populations on a worldwide scale. Geographical affinities such as between Australian-Melanesians and Africans that had already been observed have been further specified here. The latter correspond mainly to West Africans, with a prognathic face and a wide nose. As it has been previously observed on both living (Hiernaux and Froment (1976) and skeletal (Howells, 1989) populations, these features are often encountered in Australian-Melanesians and Asians.

   As new combinations of variables were tested here on a large sub-Saharan African sample, additional information is provided especially for inter-regional differences within Africa. For example, both mandibular (e.g. sh) and facial traits (e.g. NPH, PABR and EKB) reflect differences in size and shape between two extremes: West Africans (largest skull, prognathic face, high orbit, high chin) and KhoiSan (smaller skull, micrognathic, low orbit, low chin). Between these two extreme groups, Central Africans and East Africans often vary a lot and overlap each other.

2. On a worldwide scale, geographical differences are similar for both sexes, as was already suggested by Howells (1989) and Lahr (1996). Although the variation due to sexual dimorphism is often significant, the localization of the geographical differences does not change. However, within Africa, slightly more marked geographical differentiation of female populations is observed when using both the vault and the face. They are expressed especially in a latitudinal way (north-south or south-north direction along the equator), in contrast to males, whose regional differences are slightly more visible in a longitudinal (or mainly west-

east) direction. These different geographical trends for males and females are difficult to interpret. The variability of social systems among the populations represented in our sample precludes possible associations between specific patterns of male and female spatial mobility and the morphological ones observed.

In conclusion, geography is the key-element in shaping the morphological differences between African (and non-African) populations. However, although sex differences are much smaller than the geographical ones, a separate analysis for both males and females was worthwhile to complete. It provided information on the degrees of geographical differentiation rather than on the changes of basic morphological differences.

4.2. Climate and skull morphology: the re-evaluation of a possible relationship

In addition to geography, population differentiation can be also affected by additional factors such as both selective pressures (and history), as it has been previously suggested (Hiernaux and Froment, 1976; Hiernaux, 1974, 1976a; Rightmire, 1972). However, the effects of these factors are difficult to identify and to distinguish between each other, especially when they follow the same geographical trend. Nevertheless, having approached sexual dimorphism and geography, it is also worthwhile here to attempt to understand skull morphology within sub-Saharan Africans, in relation to the ecological background.

Concerning the influence of ecology on head shape, previous literature showed that several features especially facial traits might be under the control of strong selective pressures. For example, on a worldwide scale, Hiernaux and Froment (1976) identified a relatively high correlation between nasal breadth and average rainfall (r=0.4). However, Howells (1989) suggested that, a broader range of features are correlated to climate, such as not only the face (e.g. prognathism, facial height) but also the vault (e.g. breadth). Within sub-Saharan Africa, Hiernaux (1976a), Vecchi *et al.* (1987), Roseman and Weaver (2004) and Hájek *et al.* (2008) also determined a west-east gradient of morphologies (especially in relation to nasal breadth) that could reflect climatic adaptations.

Although a causal relationship between ecology and skull morphology is difficult to demonstrate, all these results suggest a geographical distribution or differentiation. However, in comparison to the studies cited in the paragraph above, a wider range of ecological variables (mainly related to climate) and much larger sample of sub-Saharan African populations will be utilized here. Therefore, it will allow us to confirm and/or provide additional information on previous hypotheses.

Therefore, in order to explore in more detail sub-Saharan African diversity, one main question is addressed here. Is it possible to determine the influence of selective pressures, by looking at the skull morphology in parallel to various ecological variables? Or, more precisely, are some of the facial features (especially nasal breadth) more affected by climate than other parts of the skull, as it has been already suggested by Hiernaux and Froment (1976)?

As a starting point, the relationship between geography and/or the ecological variables used here is tested, in order to see to what extent the latter are also correlated to latitude and longitude. Thereafter, the relationship between various ecological variables and the metric traits of the skull is explored, in order to attempt to reply to the question above.

4.2.1. Ecology and geography

As all the variables do not follow a normal distribution, only non-parametric tests are used here. The Spearman Rank correlations between the various continuous geographical and ecological variables are tested here (Table 23, p. 102). The results show that most of the correlation coefficients although very variable, are relatively high (up to 0.8), especially between mean annual temperature and longitude. This supports the fact that, marked differences in temperature in sub-Saharan Africa follow mainly a west-east axis even more than a north-south axis.

Relatively high correlations (r=0.3) are also observed between mean annual rainfall (or pluviometry) and mean annual temperature. This also indicates that, climatic variables such as pluviometry reflect the importance, within sub-Saharan Africa, of equatorial areas as characterized by tropical climate.

Therefore, both pluviometry (reflecting ambient humidity) and especially temperature are correlated to geographical variables, although each one in a different manner.

	GEOGRAPHY[1]		ECOLOGY[1]	
	Latitude[2]	Longitude[3]	Mean annual temperature	Total annual mean rainfall
Mean annual temperature	0.497**	**-0.777****		
Total annual mean rainfall	-0.171**	-0.245**	0.307**	

[1] Level of significance: P<0.05*, P<0.01** and P<0.001***;
[2] Latitude expresses here the distance from the equator (a south-north or north-south trend starting from 0°);
[3] Longitude expresses here the distance from 0° which reflects mainly a west-east trend.

Table 23. Spearman Rank correlation coefficients between various geographical and ecological variables.

	GEOGRAPHY[1]		ECOLOGY[1]	
	Latitude[2]	Longitude[3]	Mean annual temperature	Total annual mean rainfall
Vegetation (1-6)	516.44***	**964.42*****	**1379.54*****	**1206.04*****
Humidity in January (3-8)	215.66***	444.49***	809.40***	848.35***
Humidity in July (3-9)	**887.28*****	716.24***	992.389***	456.62***

[1] Level of significance: P<0.05*, P<0.01** and P<0.001***;
[2] Latitude expresses here the distance from the equator (a south-north or north-south trend starting from 0°);
[3] Longitude expresses here the distance from 0° which reflects mainly a west-east trend.

Table 24. Kruskal-Wallis one-way analysis of variance between geographical and ecological variables.

The Kruskal-Wallis one-way analysis of variance will be performed to test if five (geographical and/or ecological) variables differ according to some ecological parameters. They will allow us to see whether, for example, latitude or temperature variability is strongly dependent on ecological zones such as vegetation and mean annual extremes of humidity.

Table 23 shows that all the results are significant at the highest level of probability, and according to the highest F-values, the following comments can be made:

- temperature, pluviometry and longitude are strongly dependent on vegetation areas; this fact suggests that the variable vegetation itself reflects in a synthetic manner various climatic parameters;
- latitude is strongly dependent on humidity in July; this also suggests that, humidity at its highest peak reflects marked geographical differences along the equator and in a north-south direction.

Therefore, both vegetation and humidity variables appear to be the result of various ecological parameters that in turn are also closely related to geography.

4.2.2. Ecology and skull morphology

In order to explore whether the metrical variables of the skull are either dependent or independent of ecological parameters, two kinds of different tests were done:

- the Spearman Rank correlation (using 5 continuous variables, 2 ecological and 2 geographical); and
- the one-way analyses of variance (using 4 categorical ecological variables).

However, as significant correlations do not always indicate a causal relationship, the interpretations of the result will be done with caution.

4.2.2.1. Spearman Rank correlations

Non-parametric correlations (Spearman Rank) are used, as both geographical and ecological variables do not present normal distribution even after attempting logarithmic transformation. They are tested on both individual metric traits of the skull and regression factor scores obtained from the previous factor analyses completed in section 4.1.2.3. (p. 87).

4.2.2.1.1. Individual metric traits

The correlation coefficients between the ecological (and geographical) variables and the individual metric variables of the skull vary between 0.003 and 0.424 (Appendix XVII, p. 212).

According to relatively high correlation coefficients (r≥0.3), four morphological trends are evident:

1. concerning the nose shape, with increasing distance from the equator, a nasal aperture lengthening (NLH); and, in a longitudinal (or west-east) direction, a nasal widening (NLB);
2. along a longitudinal gradient (mainly west-east), a lengthening of the vault (NOL);
3. with rising mean annual temperature, a shortening and broadening of the vault (NOL, Q); and
4. with increasing pluviometry and decreasing distance from the equator, a narrowing of the cranial base (EAM/EAM).

For the first morphological trend about nose shape, the lengthening of the latter agrees with Hiernaux and Froment (1976). In addition, these authors observed the same trend when both humidity decreases and distance from the equator increases. As we saw previously, humidity is in fact highly correlated to distance from equator (Table 24, p. 102).

However, the increase of nasal breadth in a west-east direction does not entirely support the high correlation of variable NLB with humidity (Hiernaux and Froment, 1976). In fact, it should appear in an opposite (east-west) direction along the equator, in order to support the hypothesis of the nose shape working as a humidifier.

The second morphological trend could agree in part with the results of Hiernaux (1974). The latter observed an elongation of the body morphology from west-central wet environments (e.g. forest) to eastern drier areas (e.g. savanna). However, in his study, the skull tended to elongate more in height than in length.

The third morphological trend is completely in contradiction with the hypothesis of Beals (1974), who predicts an increasing brachycephalism with decreasing temperature. As it was already suggested by Pales (1953) and Chabeuf (1959), this phenomenon of brachycephalism is present in several areas of Africa.

The last morphological trend related to the base of the cranium could be correlated with a phenomenon of general size reduction resulting from an adaptation to forest. However, according to Hiernaux (1974), skull morphology is not affected by this size reduction in contrast to the rest of the body.

In addition, as the NOL variable presents the highest number of high values (r≥0.3), it seems therefore that length of the skull is strongly dependent on both ecological and geographical parameters. However, another measurement such as GOL which is similar to NOL provided much lower correlation coefficients. Therefore, it is questioned whether the nature (and the size) of the sample can affect the results and interpretations cited above.

4.2.2.1.2. Regression factor scores

The correlation coefficients between the ecological variables and the regression factor scores of the various factor analyses vary between 0.004 and 0.488 (Table 25, p. 104).

According to relatively high correlation coefficients (r≥0.3), several morphological trends are clear, for example:

1. with rising mean annual temperature, increasing prognathism (BPL, BNL) (analysis VIc, factor 1), increasing total facial height (NPH, sh) (analysis XIc, factor 1), narrowing of vault and face (EKB, WFB, ASB) (analysis XIIb, factor 1), broadening of palate (PABR) (analysis XIIb, factor 2), and increasing length and height of vault (BBH, PAC, OCC) (analysis XIIIb, factor 1); and
2. with increasing pluviometry, increasing facial and mandibular breadth (WFB, bb) (analysis XIc, factor 2).

As is summarised in Table 26 (p. 105), both prognathism (analysis VIc, factor 1) and total facial height (analysis XIc, factor 1), increase according to increasing distance from equator. Increasing length and height of vault (analysis XIIIb, factor 1) also follows a longitudinal direction (mainly west-east one).

Analysis n°	Type	Factor n°	Geography[1] Latitude[2]	Longitude[3]	Climate[1] Annual mean temperature	Total mean annual rainfall
VIb	only	1	0.009	**-0.178****	**0.091***	-0.019
	males	2	0.026	-0.031	-0.067	**0.102***
VIc	only	1	**0.377****	**-0.264****	**0.313****	0.048
	females	2	0.017	0.050	-0.042	**0.101***
VIIa	pooled	1	**0.218****	**-0.081**	**0.134****	0.016
	sexes	2	**0.147****	-0.075	0.075	0.063
VIIb	only	1	0.103	-0.017	0.105	0.066
	males	2	0.120	-0.070	0.069	0.058
VIIc	only	1	**0.273****	**-0.089**	**0.165***	-0.023
	females	2	**0.165***	-0.045	0.062	0.074
VIIIb	only	1	**0.259****	**-0.316****	**0.264****	**0.219****
	males	3	**0.129***	**-0.070****	**0.200****	**-0.125***
VIIIc	only	1	0.029	**-0.136***	**0.199****	**0.198****
	females	2	**0.388****	**-0.217****	**0.212****	**-0.147***
IXa	pooled	1	**0.315****	**-0.194****	**0.281****	0.034
	sexes	2	**0.147****	-0.030	0.067	0.073
IXb	only	1	**0.176***	-0.095	**0.195****	0.104
	males	2	0.145	-0.083	0.072	0.068
Xa	pooled	1	0.014	0.004	0.030	0.004
	sexes	2	0.032	0.018	-0.092	0.096
Xb	only	1	0.087	-0.084	-0.021	0.060
	males	2	0.059	-0.057	0.050	-0.044
Xc	only	1	0.180	-0.142	0.190*	0.042
	females	2	0.127	-0.014	-0.014	0.138
XIb	only	1	-0.031	-0.150	-0.018	0.134
	males	2	0.066	0.044	-0.008	-0.090
XIc	only	1	**0.338****	**-0.269****	**0.320****	0.039
	females	2	0.013	-0.038	0.017	**0.340****
XIIa	pooled	1	**-0.243****	**0.173****	**-0.284****	**-0.160****
	sexes	2	**0.382****	**-0.207****	**0.269****	0.021
XIIb	only	1	**-0.241****	**0.254****	**-0.376****	-0.137
	males	3	0.040	**-0.209****	**0.488****	**0.258****
XIIc	only	1	-0.126	**0.227****	**-0.188****	0.027
	females	2	**0.402****	**-0.146****	**0.263****	-0.051
XIIIb	only	1	**0.295****	**-0.348****	**0.386****	0.048
	males	2	**0.161***	0.124	-0.074	0.059
XIIIc	only	1	**0.332****	**-0.251****	**0.270****	-0.102
	females	2	**0.148***	0.146	0.029	**0.148***

[1] Level of significance: P<0.05*, P<0.01**, P<0.001***;
[2] Latitude expresses here the distance from the equator (a south-north or north-south trend starting from 0°);
[3] Longitude expresses here the distance from 0° which reflects mainly a west-east trend.

Table 25. Spearman Rank correlation coefficients between (geographical and) ecological variables and regression factor scores of eight types of factor analyses (VI-XIII).

			Geography[1]			Ecology[1]			
Analysis n°	**Type**	**Factor n°**	**Broad regions**	**Latitude areas**	**Longitude areas**	**Areas of distance from equator**	**Vegetation areas**	**Humidity areas in January**	**Humidity areas in July**
VIb	only	1	***	***	***	***	***	-	**
	males	2	***	***	***	***	***	***	***
VIc	only	1	***	***	***	***	***	**	***
	females	2	***	***	***	***	***	***	***
VIIa	pooled	1	***	***	***	**	***	***	***
	sexes	2	***	**	***	**	-	-	-
VIIb	only	1	-	-	-	-	*	*	-
	males	2	-	-	-	-	-	-	-
VIIc	only	1	***	***	***	-	***	***	**
	females	2	***	**	***	**	*	*	*
VIIIb	only	1	***	***	***	-	***	***	***
	males	3	-	*	***	**	***	-	*
VIIIc	only	1	-	*	***	-	***	**	***
	females	2	***	***	***	**	**	***	***
IXa	pooled	1	***	**	***	**	***	***	***
	sexes	2	***	**	***	***	*	-	-
IXb	only	1	**	***	*	-	**	**	**
	males	2	-	-	-	-	-	-	-
Xa	pooled	1	-	-	*	-	-	**	-
	sexes	2	-	*	-	-	**	-	*
Xb	only	1	*	-	-	*	*	-	-
	males	2	-	-	-	-	-	-	-
Xc	only	1	-	*	***	-	-	-	-
	females	2	-	-	-	-	-	-	-
XIb	only	1	***	-	*	-	***	*	-
	males	2	-	-	*	-	-	-	-
XIc	only	1	***	***	***	**	-	-	-
	females	2	-	-	*	-	-	-	-
XIIa	pooled	1	***	***	***	***	***	***	***
	sexes	2	***	***	***	***	***	***	***
XIIb	only	1	***	***	***	***	***	***	***
	males	3	***	***	***	***	***	***	***
XIIc	only	1	**	***	**	***	*	**	*
	females	2	***	***	***	***	***	***	**
XIIIb	only	1	***	***	***	*	***	***	***
	males	2	***	***	***	***	***	***	-
XIIIc	only	1	***	***	***	-	*	*	***
	females	2	**	***	***	***	*	***	*

[1] Level of significance of F-values: P<0.05*; P<0.01**; P<0.001***.

Table 26. Summary of one-way analyses of variance performed on the regression factor scores obtained from eight types of analyses (VI-XIII), and various ecological (and geographical) grouping variables.

These results support the observations of Hiernaux (1974) in relation to increasing prognathism and temperature. The latter considered facial protrusion as a result of an adaptation to hot conditions. Increasing palate breadth might also be related to the same phenomenon, as it showed a similar trend to prognathism.

Broadening of the vault and mandible increases with pluviometry (or ambient humidity). This supports the hypothesis of Hiernaux (1974) that a narrower vault is favoured in drier, hot conditions.

However, the narrowing of the vault (e.g. WFB, ASB) according to increasing temperature seems to be in contradiction with those results obtained on other individual variables (e.g. Q) (see section 4.2.2.1.1., p. 103). It suggests that, vault breadths taken on different areas of the skull could reflect different trends.

4.2.2.2. One-way analyses of variance

A series of one-way analyses of variance were performed on the regression factor scores of various analyses, using various ecological grouping variables.

The F-values were significant at various levels of probability, except always for factor 2 of four analyses (VIIb, IXb, Xb and Xc), as it was previously observed for the geographical parameters (regions, longitude and latitude) (Table 26, p. 105). This again supports the fact that, the nature of ecological variables is closely related to geography.

For five analyses (factor 1: VIIc, VIIIa, VIIIb, IXb, XIIIb), no differences were observed on the regression factor scores for the variable named distance from the equator. However differences were significant when using vegetation and humidity. One analysis (factor 1: XIb) that appeared significant for the regression scores using distance areas from equator, did not provide significant results when using other ecological groupings.

However, no particular differences were noted in the morphological trends. Concerning the regression factor scores of various analyses having high loadings for NLB and NLH (see Table 19, p. 88-91), the levels of significance of the F-tests is also very variable.

For the factor analyses involving nasal breadth (factor 2: VIIIb; factor 1: XIb), the regression factor scores appear to be more dependent of longitude (expressing mainly a west-east trend) and vegetation than the other ones. As it was already suggested by previous non-parametric analyses of variance (see Table 24, p. 102), these results indicate again two facts:

- variables such as longitude and vegetation are highly correlated; and
- nasal widening is dependent of longitude, as it follows a west-east trend.

For the factor analyses involving nasal height (factor 2: VIIb, VIIc and IXb), the regression factor scores are either not significant or related to various geographical and ecological variables. However, no morphological trend is observed or confirmed from previous tests.

According to post hoc multiple comparison tests when using the ecological groupings, both localization and distribution of differences between various categories are very variable. The differences can appear very spread among most of the categories as well as localized between extreme categories (e.g. VIb, XIIb). However, although same peaks for the F-values were observed for all grouping variables, it is not possible to compare all results. This is mainly due to the variable nature of the sample and number of categories under study.

4.2.3. Conclusions

The possible influence of various selective pressures has been explored here. The results show that, it is difficult to identify a clear causal relationship, as it is not possible to distinguish the effects of each factor separately. Although selection is closely related to geography, various comments can be made about the question initially addressed (see section 4.2, p. 101).

It was shown here that most of the ecological variables are interrelated as well as to geography. Although it was not possible to determine directly the influence of selective pressures on skull morphology, various morphological trends in relation to climatic variables have been observed. Most of them agree with previous work (Hiernaux, 1974; Hiernaux and Froment, 1976; Howells, 1989; Roseman and Weaver, 2004; Smith, 2009). For example:

- with increasing distance from the equator (mainly related to decreasing humidity),

the nose tends to lengthen (Appendix XVII, p. 212);
- with increasing temperature, prognathism tends to increase (e.g. BNL, BPL);
- with increasing distance from the Equator (VIc, Figure 31, p. 97), the vault tends to narrow (e.g. WFB, ASB) mainly in a south-north direction (XIIb, Figure 32, p. 98), and the vault tends to shorten (NOL) mainly in an east-west direction; and
- with increasing pluviometry, skull broadening (e.g. WFB, bb) towards the equator (XIc, Figure 33, p. 99).

Therefore, the facial features do not appear more "affected" by climate than other parts of the skull, in contrast to what it has been already suggested by Hiernaux and Froment (1976).

Furthermore, as nasal breadth tends to increase in a west-east direction, it does not support entirely previous observations in relation to average rainfall (Hiernaux and Froment, 1976).

As vault breadth appears to increase with increasing temperature, this fact also contradicts the hypothesis of Beals (1974), which supports the opposite trend.

In conclusion, geography appears to be the key element in shaping the differences between populations, although the ecological background of the latter could have an impact on the variation. Geography also reflects indirectly history as gene flow is mainly directed by geographical proximity. For example, West Africans are morphologically closer to Central than East Africans. However it is necessary to discuss historical aspects in more detail to better understand this diversity.

CHAPTER 5

EXPLORING HISTORY THROUGH MODERN AND PAST DIVERSITY

Historical aspects such as large and/or small scale population movements are explored below through modern and past diversity. Firstly, the question of the possible effects of the Bantu-speakers expansion on skull morphology is approached in detail, using a large modern sample. Secondly, geography and history are explored more generally, as key factors of morphological evolution in various West and Central African archaeological samples, mostly dated to the Holocene period.

5.1. Possible effects of the Bantu-speakers expansion on skull morphology

This section explores the question of the effects on skull morphology of large-scale historical events within sub-Saharan Africa such as the Bantu-speakers expansion. The variation of sub-Saharan Africans (of known ethnic origin) is analysed in order to identify processes of both homogenization and diversification that could be related to such migratory events.

The Bantu-speakers expansion, that started at least 3,000 years ago, is at the centre of the debate. According to Hiernaux (1976a), its biological effects are visible through a general process of morphological homogenization. This fact was also observed by Froment (1992a, 1992b) who analysed a wider range of craniometric features amongst both Bantu-speaking populations and other linguistic groups.

Furthermore, through the morphological affinities of populations of known ethno-historical background, Rightmire (1976a) could identify the advance of the Bantu-speakers expansion in both East Africa and Southern Africa. However, morphological diversity between the two main streams of the Bantu-speakers dispersal (e.g. West-Central and Eastern) has not yet been explored in relation to the ethno-linguistic background.

Using our corpus of data based on three main ethnic groups (e.g. Bahutu, Basuku, Ashanti), it will allow us to re-evaluate the differences and/or similarities within the same linguistic family (Niger-Khordofanian), between and/or within various sub-groups (e.g. Bantu, Kwa).

Two main questions are addressed here.

1. Does skull morphology support processes of homogenization resulting from the Bantu-speakers expansion, as it has been already suggested by Hiernaux (1976a) and Froment (1992a, 1992b, 1998)?
2. Despite this apparent homogenization, are differences within Bantu-speaking populations sufficient enough to be able to support a two streams phenomenon through west-central and eastern regions?

5.1.1. Variation within only Bantu-speaking populations: analyses of variance

Using only Bantu-speaking populations, one-way analyses of variance have been performed on 35 variables previously analysed (see section 4.1.2., p. 77) (Appendix X, p. 205). Regional differences are tested again with five groups (West Africa, non-Pygmy Central Africa, Pygmy Central Africa, East Africa and South Africa).

The results show that the metric traits are significantly different within each broad region (Appendix XVIII, p. 213). However, the levels of significance (when pooled sexes) is slightly lower than previous analyses that also included non-Bantu-speaking populations. For example, two additional variables (PALBA, mpl) are not significant, and two others (NASGNA, bw) have a lower level of significance. For the vault, the F-values are increasing systematically, but for other parts of the skull, they are either increasing and/or decreasing.

Although sample size variation might be responsible for these facts, it seems that there is slightly less variation within only the Bantu-speaking populations. Intense gene flow within the latter could also be responsible for this slight homogenization process. However, this question remains one hypothesis and can be contradicted by the fact that, for the vault, regional differences increase slightly.

5.1.2. Variation of Bantu-speaking populations in comparison to other ethno-linguistic groups

In order to test morphological differences and/or

similarities of Bantu-speaking populations in comparison to various other linguistic groups, the variation is approached here on two different levels of analysis:

- firstly, using broad (ethno-) linguistic classifications; and
- secondly, focusing only on a few groups of well-known ethno-linguistic background.

5.1.2.1. Testing different ethno-linguistic classifications: analyses of variance

One-way analyses of variance were performed on the regression factor scores previously analysed (Table 14, p. 78) (Appendix XIX, p. 214). For this purpose, two cultural grouping variables already presented previously (Table 3, p. 52) are used: they are based on either only linguistic data (Greenberg, 1964) or both linguistic and ethnographic data (Murdock, 1959).

Four analyses (the same as previously when testing both geographical and ecological variables) (see section 4.2.1, p. 101) are not significant. They are: VIIb (factors 1 & 2), IXb (factor 2) and Xc (factor 2). This fact supports again a relationship not only between geographical and ecological variables, but also with cultural variables (Appendix XIX, p. 214).

However, in contrast to previous results, there are a few changes: analysis VIIIb (factor 3: PAC, NLB) is significant as well as analysis XIc (factor 2: WFB, bb) for both cultural grouping variables.

Post hoc multiple comparisons tests were also performed in order to localize the morphological differences. For linguistic differences, the latter are mainly localized between the following groups: West Atlantic group (1) (with widest nose and narrower skull) and two Bantu-speaking ones (West & Central: 3; Eastern: 4) (with narrower nose and broader skull). For ethno-linguistic differences, the latter are mainly localized between the following groups: Yam belt region (11) (with widest nose) and two Bantu-speaking ones (North West: 2; Cameroon Highlands: 1) (with narrower nose).

Most of the (ethno-) linguistic differences and/or similarities reflect those observed previously on a geographical basis (Table 22, p. 100). For example, the KhoiSan group (with micrognathism, small nose and short vault) appears to be often in the lowest values of the sub-Saharan African variation again. West Africans corresponding to the West Atlantic group are also more prognathic than the other Bantu-speaking groups (e.g. analysis VI, factor 1). Furthermore, Eastern Bantu-speakers present wider palate than both Southern Bantu-speakers and KhoiSan (analysis XII, factor 3).

However, differences within the Bantu-speaking populations are better seen with the cultural grouping variables, especially the Murdock's ethno-linguistic units. In fact, the latter provide the Bantu-speaking group with a higher number of clear subdivisions (1 to 9) than the geographical ones (I partially, II, III, IV and V).

For example, Central Bantu-speakers (group 6) appear to be often in the lowest values of the variation, in comparison to other Bantu-speaking groups (Eastern and Southern: 8 & 9) and non-Bantu-speaking ones: they have smaller and narrower vault (analysis VI, XII, XIII, factor 1 &/or 2), less prognathic face, lower orbits and nose (analyses VII & IX, factors 1 & 2), narrower nose (analysis VIII, factor 2) and smaller palate (analysis XII, factor 2).

5.1.2.2. Analysing variation within three known ethnic groups

Both univariate and multivariate analyses were performed here in order to analyse the variation within three known ethnic groups (Bahutu, Basuku, Ashanti), especially in relation to Bantu-speaking populations (Bahutu, Basuku) (Appendix XX, p. 215).

5.1.2.2.1. Analyses of variance

When focusing on differences between only the three ethnic groups (with pooled sexes), the F-tests for 16 measurements also provide similar results to most previous analyses (see section 4.1.2.1, p. 84) (Table 27, p. 110). The levels of significance are however more variable than when comparing the broad regional groupings, and an additional variable such as OCC is not significant.

The post hoc multiple comparison tests showed where the differences are localized:

- for 8 variables (e.g. GOL, WCB, BNL, BPL, ZYB, JUB, FMB, ZMB) between the Basuku and the other 2 groups;
- for 2 variables (e.g. XCB, XFB) between the Bahutu and the other 2 groups;
- for 2 variables (e.g. NLB, BBH) between all groups;
- for one variable (e.g. PAC) between

	F	Sig.[1]	L.[2]
GOL	16.87	***	-
XCB	23.69	***	-
PAC	4.65	*	-
OCC	2.22	-	-
BBH	22.35	***	-
WCB	9.04	***	-
BNL	12.72	***	-
BPL	7.72	***	-
ZYB	22.71	***	-
JUB	21.40	***	-
STB	0.88	-	-
XFB	6.04	**	-
FMB	13.14	***	***
ZMB	5.86	**	-
NLB	11.55	***	-
OBH	9.26	***	-

[1] Level of significance (P<0.05*; P<0.01**; P<0.001***);
[2] Levene's test of homogeneity of variances.

Table 27. One-way analyses of variance for comparing 16 cranial metric traits between three main ethnic groups (Bahutu, Basuku and Ashanti).

Ashanti and Basuku; and
- for another variable (e.g. OBH) between Basuku and Bahutu.

Therefore, most of these cranial metric traits appear to be markedly different between a large Bantu-speaking group, the Basuku, and two other groups, both Bantu-speaking (Bahutu) and non-Bantu-speaking (Ashanti) ones.

5.1.2.2.2. Multiple discriminant analyses

Multiple discriminant analyses are performed in order to see whether ethno-linguistic affiliation is a good predictor for cranial morphology.

Five groups of variables were entered, such as (Table 28, p. 111):
- for the vault: GOL, XCB, PAC, OCC, BBH, WCB and BNL;
- for the cranial base: BIMAS.BR, EAM/EAM, FOL and FOB;
- for the face: BPL, ZYB, JUB, STB, XFB, FMB and ZMB;
- for both the vault and face: BBH, OBB, GOL, DKB, EKB, ZYB and FMB; and
- for both the vault and face: FMB, OBH, Q1, DKB and NLB.

The total correct rates of classification are higher for the vault (69.9%-71.1%) than for the face (61.1%-64.2%) and the base (68.5%). When entering both vault and face variables together, the range of total correct rates of classification tend to decrease (65.2%-66.2%). However, these results show that, inter-ethno-linguistic differences are in general slightly higher than inter-regional ones (Table 18, p. 87).

According to highest standardized canonical coefficients (≥0.7), the variables which are most frequently effective in "classifying" groups are:
- for the vault: BBH, BNL and GOL;
- for the cranial base: BIMAS.BR and EAM/EAM;
- for the face: ZYB and BPL; and
- for both vault and face: BBH, NLB and DKB.

For the Bahutu, 40% to 72% of the cases are correctly classified, all the remaining ones being misclassified into another Bantu-speaking group (Basuku).

For the Basuku, 80% to 93% of the cases are correctly classified, the remaining ones being misclassified mainly into another Bantu-speaking

Variables entered, total sample size & groups analysed	N	Classification results: Bahutu	Basuku	Ashanti	Total correct rate	Standardized canonical discriminant function coefficients: Selected variables	function 1	function 2
VAULT: GOL, XCB, PAC, OCC, BBH, WCB, BNL								
						GOL	0.492	-0.105
Bahutu (Rwanda)	50	**36**	14	0		XCB	0.635	0.089
Basuku (D.R.C.)	120	6	**111**	3	**77.1%**	**BBH**	**-1.119**	0.317
Ashanti (Ghana)	31	1	22	**8**		**BNL**	0.313	**0.818**
BASE: BIMAS.BR, EAM/EAM, FOL, FOB								
Bahutu (Rwanda)	49	**27**	22	0				
Basuku (D.R.C.)	120	12	**108**	0	**68.5%**	**BIMAS.BR**	**0.641**	-1.212
Ashanti (Ghana)	28	5	23	**0**		**EAM/EAM**	0.445	**1.297**
FACE: BPL, ZYB, JUB, STB, XFB, FMB, ZMB								
Bahutu (Rwanda)	50	**20**	30	0				
Basuku (D.R.C.)	120	11	**109**	0	**64.2%**	**ZYB**	**1.000**	-
Ashanti (Ghana)	31	6	25	**0**				
VAULT & FACE: BBH, OBB, GOL, DKB, EKB, ZYB, FMB								
Bahutu (Rwanda)	50	**32**	18	-				
Basuku (D.R.C.)	120	8	**112**	-	**65.2%**	**BBH**	**1.000**	-
VAULT & FACE: FMB, OBH, Q1, DKB, NLB								
						FMB	0.741	0.598
Bahutu (Rwanda)	50	**30**	20	0		OBH	0.239	0.380
Basuku (D.R.C.)	120	19	**96**	5	**66.2%**	Q1	-0.482	0.387
Ashanti (Ghana)	31	1	23	**7**		**DKB**	0.271	**-0.623**
						NLB	**-0.766**	0.202

Table 28. Summary for five multiple discriminant analyses on the cranium: predicted group membership for ethnic affiliation.

group (Bahutu) (5%-16%) and only a few into a non-Bantu-speaking one (Ashanti) (0%-4%).

For the Ashanti, a non-Bantu-speaking group, only 0% to 26% of the cases are correctly classified, all the remaining ones being misclassified into the two Bantu-speaking groups, mainly the Basuku (71%-87%), and a few into the Bahutu (4%-13%).

Therefore, according to these results, Bantu-speaking groups appear to be more similar to each other than with non-Bantu-speaking ones. This fact also supports homogeneity within the Bantu-speakers. Nevertheless, they are slightly different especially for the vault, with the following trends:

- the Bahutu have relatively long, broad and low vault; and
- the Basuku have relatively narrow, short and high vault.

However, the variation of the West Atlantic-speaking group (Ashanti) is closer to the Bantu-speaking Central rather than the Eastern African group. Therefore, it supports again that, geographical proximity is a key-element in morphological affinities, although variable sample size might have slightly affected all these observations.

5.1.2.2.3. Factor analyses

Three types of factor analyses (pooled sexes) are performed here. The series of variables used (that are similar to previous factor analyses) (Table 19, p. 88-91), are:

- for **analysis XIV** (similar to I & VI, but

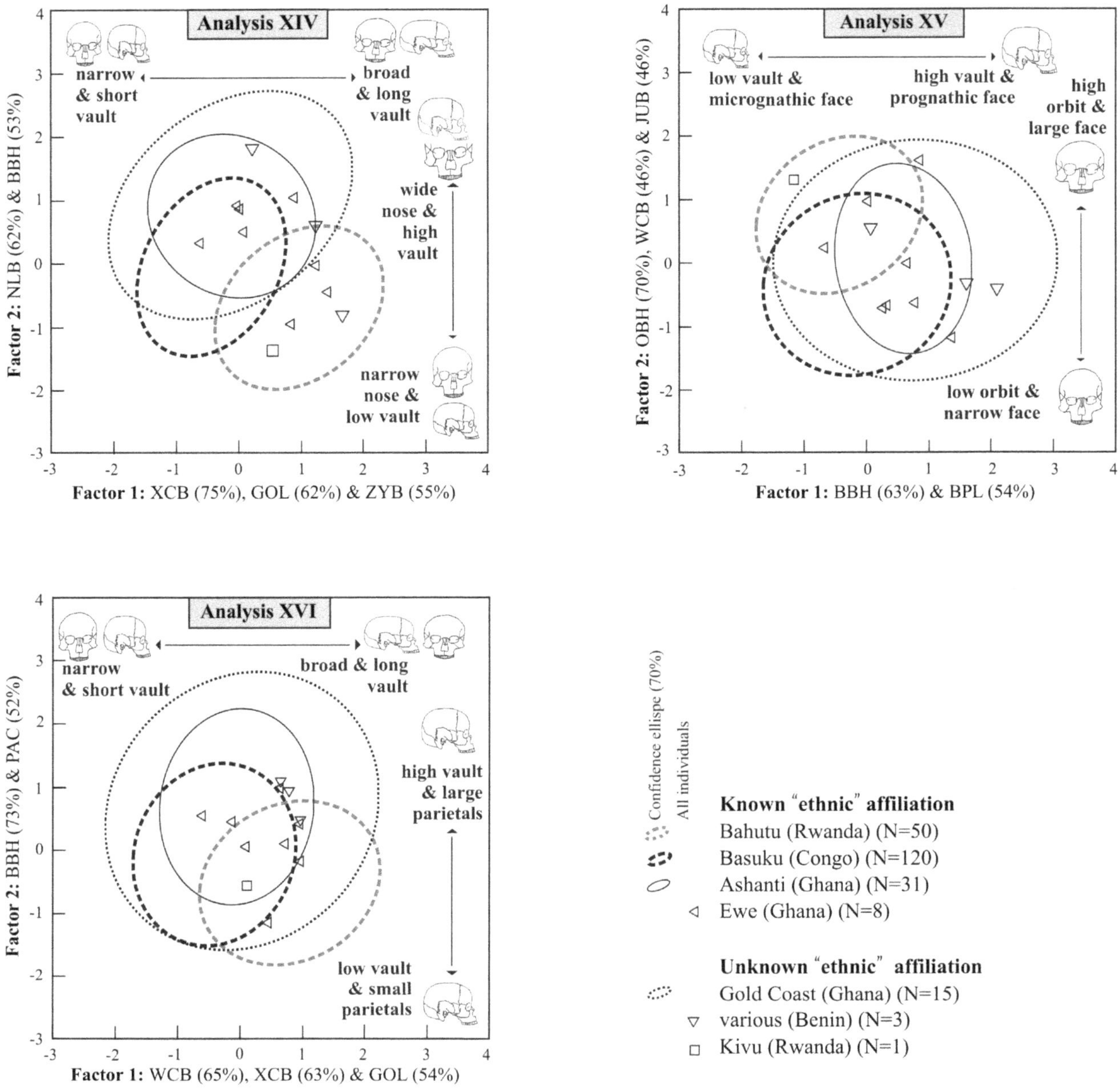

Figure 34. Scatterplots of first and second factors for three analyses (XIV-XVI) with pooled sexes, showing grouping by ethnic affiliation.

without NPH): GOL, XCB, ZYB, BBH, BPL, BNL and NLB;
- for **analysis XV** (similar to V & IX, but BBH used instead of NLH): WCB, BPL, JUB, BBH, NLB and OBH; and
- for **analysis XVI** (similar to XIII): GOL, XCB, PAC, OCC, BBH, WCB and BNL.

These three analyses that summarised different combinations of variables were selected, as the varimax rotation phase could also be achieved (with two factors having eigenvalues greater than one).

In addition to the three previous ethnic groups (N=228), a few individuals of both known (Ewe, Ghana; N=8) and unknown (Ghana, Benin, Rwanda; N=19) ethnic affiliation were included in the analysis.

In comparison to previous analyses (Table 19, p. 88-91), the highest factor loadings (>0.7) are observed here often for the same variables (Table 29, p. 113). A few additional variables presented highest loading such as:
- for **analysis XIV**: ZYB (factor 1) and NLB (factor 2);
- for **analysis XV**: NLB (factor 2); and

Variables entered	Factor 1 eigen-value	Factor 1 (%)	Factor 2 eigen-value	Factor 2 (%)	Cumulative % of variance	N° of factors extracted	Factor loadings after varimax rotation	factor 1	factor 2	% of variance[1] factor 1	% of variance[1] factor 2
VAULT & FACE											
GOL,XCB,							**GOL**	**0.786**	0.360	**61.78**	
ZYB,BBH,							**XCB**	**0.866**	-0.081	**75.00**	
BPL, BNL,	3.71	(53.03)	1.05	(15.02)	**68.05**	2	**ZYB**	**0.742**	0.384	**55.06**	
NLB							**BBH**	0.203	**0.731**		**53.44**
Analysis							BPL	0.520	0.583		
XIV							BNL	0.549	0.673		
(N=228)							**NLB**	0.004	**0.788**		**62.09**
BASE & FACE											
WCB, BPL,							WCB	0.439	0.679		
JUB, BBH,							**BPL**	**0.733**	0.214	**53.73**	
NLB, OBH	2.70	(44.92)	1.13	(18.76)	**63.68**	2	JUB	0.516	0.675		
Analysis							NLH	0.671	0.226		
XV							**NLB**	-0.146	**0.838**		**70.22**
(N=228)							**OBH**	**0.794**	-0.089	**63.04**	
VAULT & BASE											
GOL, XCB,							**GOL**	**0.736**	0.502	**54.17**	
PAC, OCC,							**XCB**	**0.793**	0.039	**62.88**	
BBH, WCB,							**PAC**	0.243	**0.718**		**51.55**
BNL	3.36	(48.01)	1.10	(15.68)	**63.69**	2	OCC	0.033	0.648		
Analysis							**BBH**	0.174	**0.855**		**73.10**
XVI							**WCB**	**0.808**	0.103	**65.29**	
(N=228)							BNL	0.596	0.508		

[1] Percentage of variance accounted on each factor obtained by squaring the correlation coefficient (or factor loading) for variables which have the highest factor loadings (>0.7).

Table 29. Summary of three types of factor analyses (XIV-XVI) performed on modern African populations using only main ethnic groups.

- for **analysis XVI**: GOL (factor 1).

Scatterplots for factor 1 and 2 are drawn in order to show group clustering by ethnic origin (Figure 34, p. 112). In order to explore in more detail, whether the morphological differences are significant between the main ethnic groups, one-way analyses of variance were performed on the regression factor scores. The results were all significant at the highest level of probability (Table 30, p. 114). Post hoc multiple comparison tests also provided information that confirmed the localization of the main morphological differences.

For example, in analysis XIV, differences were observed along factor 1 (53% of variance) that is related to breadth and length of vault, especially between Bahutu (higher values) and Basuku (lower values). More marked differences especially between all three main ethnic groups (Ashanti being in highest values) are also observed along factor 2 (15% of variance). The latter is however related to facial features such as nasal breadth and vault height. For both factors, the group of unknown ethnic affiliation originating from Ghana (Gold Coast) is highly variable, but it clearly overlaps the variation of the Ashanti.

In analysis XV, differences are especially present between the Ashanti (higher values) and two other groups (Bahutu and Basuku being in lowest values) along factor 1 (45% of the variance). The latter is related to vault height and degree of facial prognathism. Differences are however distributed in a different manner along factor 2 (19% of variance) that is related to orbital height and facial

		Factor	N	Mean	S.D.	F	Sig.[1]	L.[2]
Analysis XIV	Bahutu	1	48	0.815	0.918	35.031	***	-
	Basuku	1	120	-0.398	0.838			
	Ashanti	1	31	-0.038	0.773			
	Bahutu	2	48	-0.512	0.815	16.708	***	-
	Basuku	2	120	-0.062	0.920			
	Ashanti	2	31	0.690	0.963			
Analysis XV	Bahutu	1	48	0.996	0.851	55.861	***	-
	Basuku	1	120	-0.436	0.761			
	Ashanti	1	31	-0.111	0.833			
	Bahutu	2	48	-0.700	0.810	26.783	***	-
	Basuku	2	120	-0.059	0.896			
	Ashanti	2	31	0.749	0.802			
Analysis XVI	Bahutu	1	48	-0.425	0.839	12.722	***	-
	Basuku	1	120	-0.141	0.960			
	Ashanti	1	31	0.597	0.682			
	Bahutu	2	48	0.749	0.778	26.643	***	-
	Basuku	2	120	-0.351	0.912			
	Ashanti	2	31	0.058	0.928			

1 Level of significance: P<0.05*; P<0.01**; P<0.001***;
2 Levene's test of homogeneity of variances.

Table 30. One-way analyses of variance on the regression factor scores for ethnic group differences.

breadth: the Bahutu having the highest values, are very different from the two other groups (Basuku and Ashanti). For both factors, the Ghanaian group is highly variable again, but it overlaps the variation of both Ashanti and Basuku.

In analysis XVI, the differences are very similar to those observed in analysis XIV especially for factor 1. They are marked between all groups along factor 2 which is related to vault height and parietal size. The Ashanti variation falling within the highest range is overlapped again by the highly variable Ghanaian group.

In all analyses, the position of the Ewe is also very variable. However, the individual originating from Kivu is always localized within 70% of Bahutu variation.

In summary, these morphological differences are summarised in Table 31 (p. 115). They confirm the fact that, the two Bantu-speaking ethnic groups (Bahutu, Basuku), although slightly variable, share common features. Marked differences are always observed between the Bahutu and a non-Bantu-speaking group (Ashanti).

5.1.3. Multiple discriminant analyses compared to RMET analyses: correcting for genetic drift?

Statistics of population genetics have been applied here, using the RMET software (e.g. Relethford-Blanguero analysis) of Relethford (1994) available online. They have the advantage of identifying levels of group differentiation through (R-)matrices of distances (D^2) like in multiple discriminant analyses. However, in contrast to the latter, it can also correct for the effect of genetic drift in small populations (Relethford *et al.*, 1997). Therefore it was useful to apply this technique as the sample under study needed to be assessed in terms of both differentiation as well as homogenization. Furthermore, it contained low density populations such as Pygmies and KhoiSan foragers whose extreme variation could affect final results as we saw in Figures 30-33 (p. 96-99).

The modern African sample (pooled sexes) was subdivided in the same four regions and foraging groups as discussed in **Chapter 4** (Table 22, p. 100). It was also grouped into three different groups: all groups, Bantu-speakers and non-Bantu-speakers (with Pygmies). Only 14 variables have

Groups	Vault	Face	Nose	Orbit
Bahutu	broad, long, low	large, relatively less prognathic	relatively narrow	high
Basuku	relatively narrow, short, low & medium size height	relatively narrow & less prognathic	medium size width	low
Ashanti	narrow, high, moderate length	prognathic & medium size breadth	broad	medium size height

Table 31. Summary of the morphological trends observed from factor analyses on the crania of three main ethnic groups.

Types of analysis	12 variables			9 variables		
	N	Group number	%	N	Group number	%
All groups	785	6	54	842	6	53
Bantu-speakers (without Pygmies)	503	4	71	555	4	70
Non-Bantu-speakers (with Pygmies)	282	4	80	287	4	73

Table 32. Multiple discriminant analyses achieved on two series of variables for the modern sub-Saharan African sample: total correct rates (%) of classification for three analyses.

been used here, as they maximised sample sizes especially for the Pygmies. They correspond to the vault (GOL, XCB, and BBH) and the face (BNL, BPL, NPH, WFB, FMB, ZYB, DKB, OBB, OBH, NLB, and NLH). Two analyses have been performed with two different series of variables: 12 variables (all cited above except for FMB and DKB) and 9 variables (face only).

Both multiple discriminant analyses and RMET analyses were performed, providing therefore population differentiation with biological distances (D^2).

The results obtained from the first multivariate analyses are presented in Tables 32 and 33 (p. 115 and 116) with the various rates of classification, followed by the scatterplots of the discriminant scores (first two functions with highest standardized canonical discriminant coefficients) (Figure 35, p. 117).

The total rates of classification (Table 32) appear to be very variable ranging from 41% up to 80%. These percentages tend to decrease when sample size and/or number of variables used increase. Group differentiation also increases when non-Bantu-speakers only are analysed. In general, the groups are better classified into their original category when they are not analysed altogether, and especially when non-Bantu-speakers only are considered (Table 33, p. 116).

Figure 35 (p. 117) supports these results showing increasing population differentiation when Bantu-speakers are not considered. Although this fact might partly be due to sample size, it reflects more homogeneity within the Bantu-speakers (always having West and Central Africa close together) than in the other groups.

The results obtained from the second multivariate analyses are presented in two ways, with or without correction for population size (N_e). The analyses were repeated systematically with N_e=1 (equal population weights for all groups) and

Regions or groups	All groups		Bantu-speakers (without Pygmies)		Non-Bantu-speakers (with Pygmies)	
	12 variables	9 variables	12 variables	9 variables	12 variables	9 variables
West Africa	41	37	6	6	92	91
Central Africa	65	74	86	91	0	0
East Africa	44	11	49	16	-	-
South Africa	63	64	75	71	-	-
Pygmies	24	14	-	-	38	5
KhoiSan	54	42	-	-	82	65

Table 33. Multiple discriminant analyses achieved on two series of variables for the modern sub-Saharan African sample: correct rates (%) of classification for each group.

N_e= 0.2 (unequal population weights for both Pygmies and KhoiSan). These two values were chosen only, as no accurate demographic information was available for the sample of historic foraging groups under study.

An average heritability of 0.55 was used for the craniometric traits as recommended by Relethford (1994). All the results (plots of the two first principal coordinates of distance matrices) obtained from the different population weights are presented for comparison (Figure 36, p. 118).

All RMET analyses show that when unequal population weights are applied to both Pygmies and KhoiSan, the distance between them is slightly lower. This suggests that genetic drift has a slight impact on the degree of population differentiation, but it does not change the nature of the differences between groups themselves.

In fact, these multivariate analyses agree with previous multiple discriminant analyses, on the four following points:

i) the extremes in the variation are often between the KhoiSan and the other Bantu-speaking groups;
ii) West Africa and Central Africa are always very close to each other;
iii) the Pygmies are close to the Bantu-speakers and they are therefore difficult to distinguish from the latter; and
iv) group differentiation is especially high in non-Bantu-speakers; therefore, it is suggested that there is more homogeneity within the Bantu-speakers.

5.1.4. Conclusions

Various comments can be made about the two questions initially addressed (section 5.1.2., p. 108):

1. These univariate and multivariate analyses so far suggest that there is relatively more morphological homogeneity within Bantu-speakers in comparison to other African linguistic groups. Is it the result of a founder effect phenomenon; or is it influenced by the nature and size of the sample and/or the data, and other historical events (e.g. latest population admixture)?

   As was suggested previously (e.g. Hiernaux, 1976a; Froment, 1992a; Ramachandran *et al.*, 2005; Berniell-Lee *et al.*, 2009; Tishkoff *et al.*, 2009; Bryc *et al.*, 2010), processes of homogenization that could result from the Bantu-speakers expansion, might have affected cranial morphology. For example, geographical differences related to the latter appear to be slightly lower when considering only Bantu-speaking populations.

2. Despite this apparent homogenization, slight differences within Bantu-speaking populations are present such as between Central Bantu (e.g. Basuku) and Eastern Bantu (e.g. Bahutu) speakers. They can be as marked as those observed with some other linguistic groups (e.g. West Atlantic-speakers like the Ashanti). However, both univariate and multivariate analyses showed that, differences of Bantu-speaking groups are the most marked with non-Bantu-speakers from both South Africa (KhoiSan) and West Africa (West Atlantic).

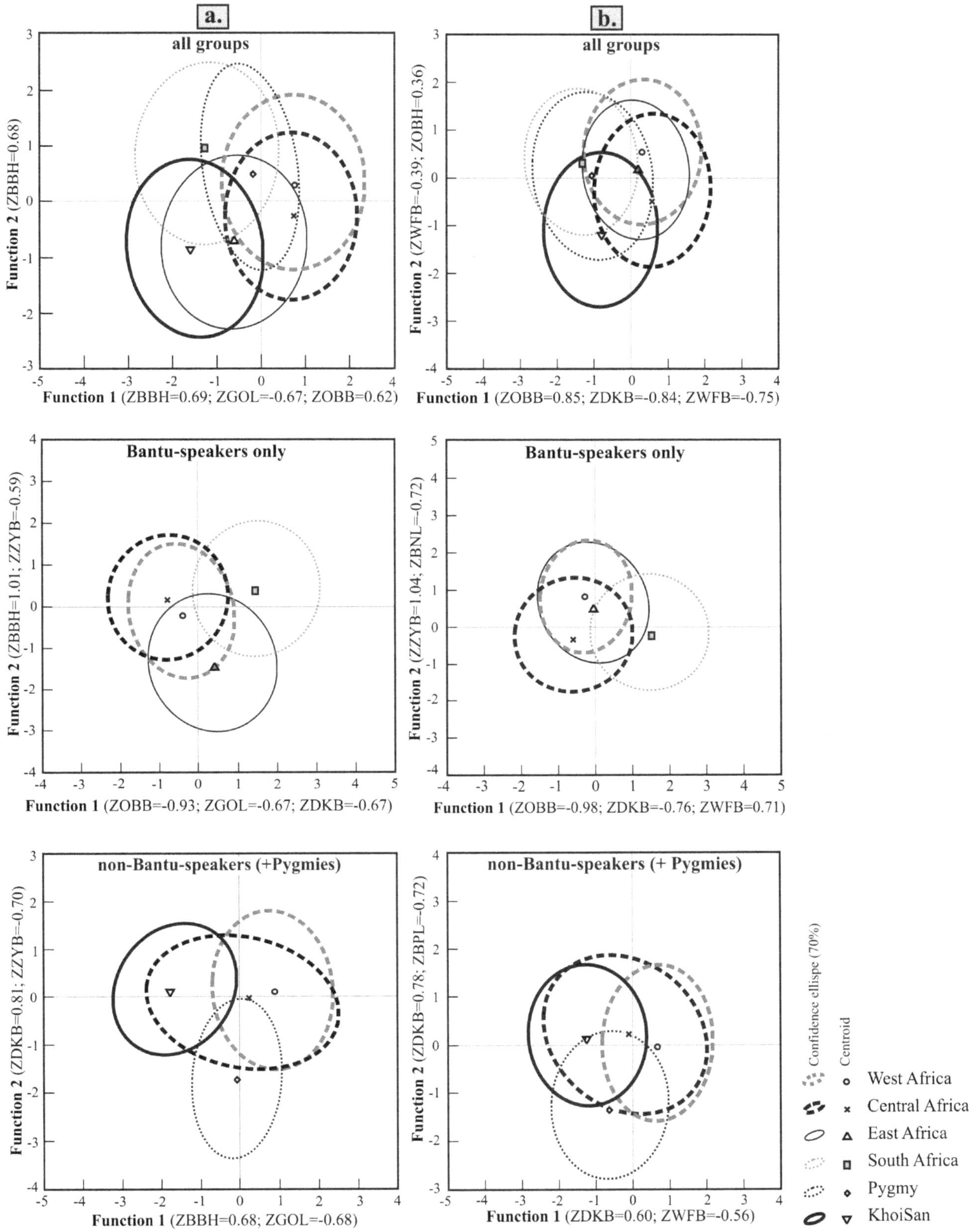

Figure 35. First two functions of discriminant scores for modern sub-Saharan Africans using two series of variables. **a.** 12 variables analysis: all groups (74.5%: variance for first two functions), Bantu-speakers (93%) and non-Bantu-speakers with Pygmies (96.5%); **b.** 9 variables analysis: all groups (90%: variance for first two functions), Bantu-speakers (94.8%) and non-Bantu-speakers with Pygmies (98.7%).

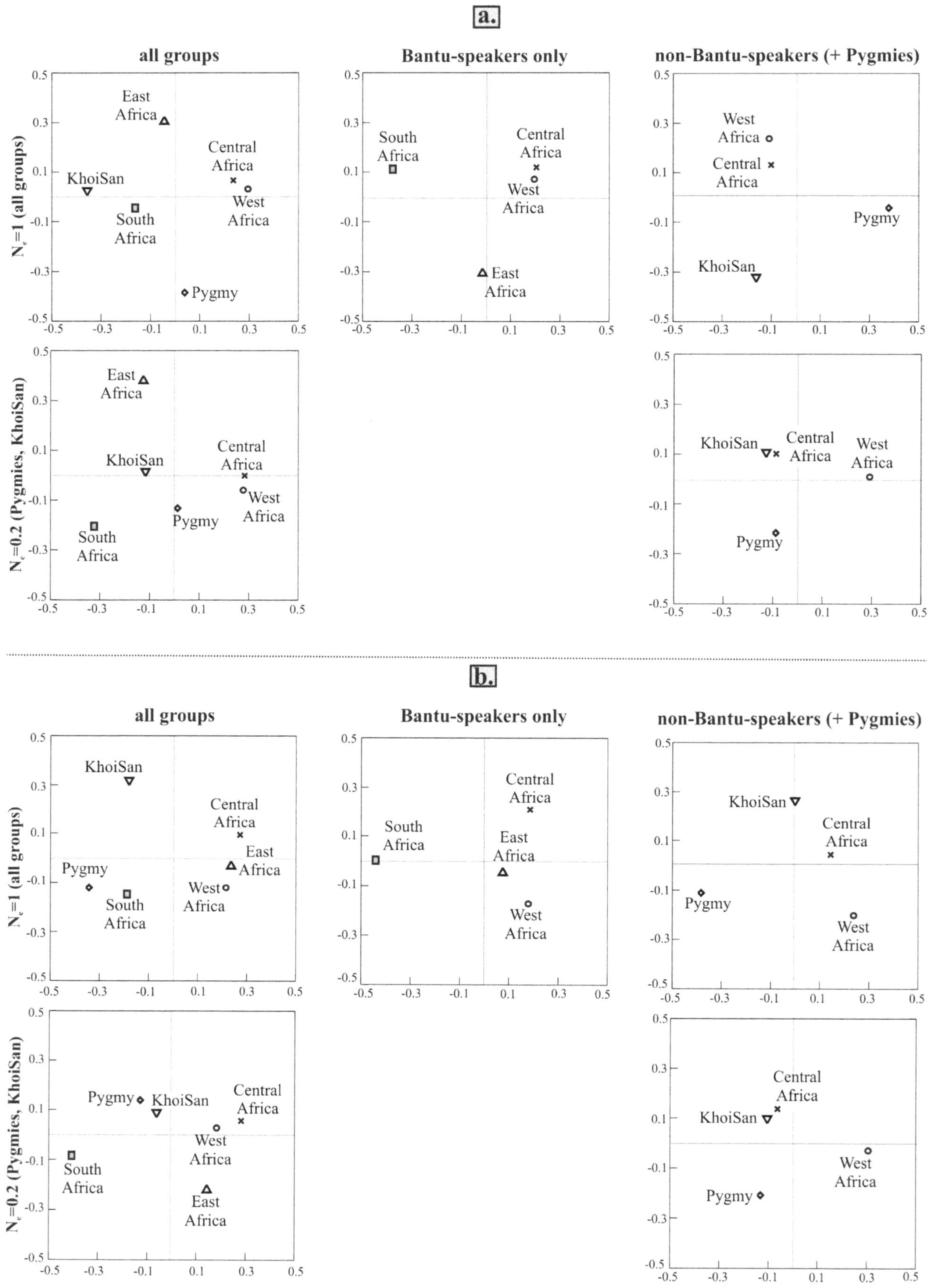

Figure 36. First two principal coordinates (obtained from R-matrices of Relethford-Blanguero analysis) for modern sub-Saharan Africans using two series of variables. **a.** 12 variables analysis with N_e=1 for all groups (up) and N_e=0.2 for both Pygmies and KhoiSan (below); **b.** 9 variables analysis with N_e=1 for all groups (up) and N_e=0.2 for both Pygmies and KhoiSan (below).

This morphological diversification of sub-Saharan Africans is also supported by recent three-dimensional morphometric approach (Franklin *et al.*, 2010). It appears in this study to be higher on an inter-ethnic level than inter-regional one (**Chapter 4**). It could also support indirectly a two streams (at least) phenomena in relation to the Bantu-speakers expansion. However, this hypothesis could also reflect the variable levels of population admixture that are present in each region.

Population admixture with non-Bantu-speakers could have been responsible for this inter-regional diversification. To some extent, the Basuku probably admixed with Pygmy, and the Bahutu with the Tutsi. But it is unknown whether this phenomenon is sufficient enough for producing diversification.

In conclusion, morphological trends that could be directly related to the Bantu-speakers expansion are difficult to identify. Nevertheless, geography appears to be the key element in shaping the differences between populations, although the ecological and cultural backgrounds of the latter also probably have an impact on variation. For example, gene flow is mainly directed by geographical proximity, as West Africans (Ashanti) are morphologically closer to Central Africans (Basuku) than Eastern ones (Bahutu).

On morphology, it is very difficult to separate the effects, of the historical aspects from the geographical and ecological ones (e.g. various selective pressures). Furthermore fluctuating demographic parameters (e.g. initial population size) and natural geographical barriers (as we saw in **Chapter 4**) also could have had an impact on population differentiation. However the impact of genetic drift appears to be minimal as it was probably not the main causal factor of diversity.

5.2. Microevolution through the Holocene period: inter- and intra-regional patterns in West and Central Africa

After having approached the variation of modern sub-Saharan African populations on different levels (geography, ecology and history), some archaeological samples originating especially from West and Central Africa, were analysed here. They were compared to various modern African groups, in order to see whether similarities and/or differences between populations did evolve through time in relation to geography.

The question of morphological continuity and/or differentiation within the African continent has often been previously addressed for the Holocene period. However, studies focused mainly on one region rather than exploring systematically inter-regional patterns (section 2.3.3.2., p. 34).

For Early and Mid-Holocene populations in West and Central Africa, unique combinations of skull features reflecting robusticity (e.g. high broad vault, large nose, broad mandible) have been seen on various specimens (e.g. Asselar, Hassi-el-Abiod, Kobadi, Iwo Eleru). Some (e.g. Asselar) presented strong affinities with Late Pleistocene North Africans (Dutour, 1992), and other (e.g. Ishango) with modern sub-Saharan Africans (Boaz *et al.*, 1990).

However, these regional differences between various Later Stone Age and/or Iron Age African populations need to be re-evaluated with a large comparative modern sample more representative of the sub-continent. The latter will allow us to explore further in West and Central Africa, the biological input originating from both south and north of the Sahara.

So far, through past skeletal morphology, population movements occurring on a micro-scale have been better discovered than those occurring on a macro-scale (e.g. Bantu-speakers expansion). For example, Huizinga *et al.* (1967) observed the successive arrival of different populations (e.g. Tellem, Dogons) in the Middle Niger Valley (Mali), especially during the Late Holocene. However, the progressive advance of the Bantu-speakers expansion in sub-Saharan Africa is more difficult to identify for the two following reasons: (i) it is a very long term phenomenon involving various populations; and (ii) there are too few skeletal remains especially in West-Central Africa for testing the proto-Bantu-speakers dispersal from the Early Iron Age period onwards.

As a consequence, the present section will also focus on the detection of micro-scale events in a region that has not been explored, e.g. in Senegal. The questions are addressed here in two phases: on a broad level and then on a more detailed one.

1. Do geographical differences and/or similarities (in relation to previous regional subdivisions) change through the Holocene in sub-Saharan Africa for both

Later Stone Age (before 1,000 BC) and Iron Age (after 1,000 BC) periods? And do they agree with previous studies (e.g. Brothwell and Shaw, 1971; Dutour, 1992; Sereno *et al.*, 2008)?
2. Is there continuity and/or differentiation between both various Iron Age and modern West African samples?

In particular, this last question is related to the occurrence of population movements on a micro-scale from northern and southern regions, especially in both Senegal (and Mali).

Therefore, large archaeological and modern groups (N>10) originating from West (and North) Africa are the main study samples (Appendix XXI, p. 216-217), including:

- four Senegalese groups dated to both the Iron Age (Dioronboumak, Faboura) and the 20th century AD (Griots Lebu, Griots Serer);
- three Malian groups dated to the Iron Age (Early Tellem, Late Tellem) and the 20th century AD (Nokara); and
- one modern North African group (corresponding to various countries).

Both factor analyses and multiple discriminant analyses are used, as archaeological sample sizes are very variable when addressing the issue of population differentiation through time.

5.2.1. Factor analyses

Factor analyses are done firstly, as they are more adapted to the analysis of small archaeological samples. They will explore variation through time on both continental and regional scales in relation to the two questions cited above.

5.2.1.1. Variation on a continental scale: broad regional differences through time

A maximum number of 16 different archaeological samples have been used, and most of them (10 samples) correspond to West Africans (section 3.3., p. 56).

Four factor analyses are presented here (Table 34, p. 122). According to previous analyses on modern populations that showed marked regional differences (**Chapters 4 and 5**), three series of variables in relation to the skull have been selected, such as:

- for **analysis XVIIa** (similar to I, VI and XIV): GOL, XCB, ZYB, BBH, BPL, NPH, BNL and NLB;
- for **analysis XVIIIa** (similar to III and VIII): FRC, PAC, OCC, BBH, NPH, XFB, NLB and OBB; and
- for **analysis XIX** (b & c) (similar to X): bb, bw, sh, har, war and ma.

The comparative modern sample is also subdivided into main broad regions and a few (ethnic) groups. Sexes are either pooled (e.g. XVIIa, XVIIIa) or separated for both males (e.g. XIXb) and females (e.g. XIXc) in order to increase the archaeological sample under study.

In comparison to previous analyses (Table 19, p. 88-91), the highest loadings (>0.7) are often observed for the same variables, such as:

- for **analysis XVIIa**: XCB, GOL, BBH, ZYB (factor 1) and NLB, BPL (factor 2);
- for **analysis XVIIIa**: BBH, FRC (factor 1); and NLB, OBB (factor 2); and
- for **analysis XIX** (b & c): ma, war (factor 1) and bw, sh (factor 2) .

Scatterplots for factor 1 and 2 are drawn in order to show group clustering for each analysis (Figures 37-39, p. 123-125).

For analysis XVIIa (Figure 37a & b, p. 123), marked morphological differences between modern populations of various regions are observed along both factor 1 (e.g. breadth, length and height of vault) and factor 2 (e.g. nose width, facial protrusion).

Although Iron Age West Africans (especially originating from Senegal) are very variable, they often fall within the highest values of the modern variation (Figure 37c, p. 123). For example, Dioronboumak and Walalde individuals have a relatively large vault, wide nose and prognathic face like the modern West Africans and Bantu-speaking South Africans. However, a few other Senegalese specimens (e.g. Cap des Biches) are less prognathic despite a large vault, and therefore they are closer to the variation of modern North Africans.

In addition, both Later Stone Age East Africans and Iron Age Central (and North) Africans tend to be localized in the lower values of the modern variation especially along factor 2 (Figure 37d, p. 123). For example, Bromhead individuals who have a relatively narrow nose and micrognathic face, fall quite well within the variation of not only some modern East Africans and North Africans but also Iron Age Sudanese.

For analysis XVIIIa (Figure 38a-c, p. 124), the morphological trends observed within modern and past populations are similar to the previous ones. However, they provide additional information especially along factor 2 that is related not only to nasal width but also orbital breadth. Although Later Stone Age and/or Iron Age West Africans are the most variable, especially in relation to vault height (factor 1), they still tend to have the widest nose and orbits such as the modern West Africans (Figure 38c).

Other archaeological specimens also appear to have morphological affinities with modern populations. For example, a Later Stone Age East African specimen (e.g. Gamble Cave II) is located within the middle of the modern variation, and an Iron Age North African one (e.g. Attiri) within the lowest values.

For analyses XIXb (with only males) and XIXc (with only females), morphological differences are related only to the mandible (Figure 39a & b, p. 125). Slight regional differences between modern populations are observed on both factor 1 (e.g. mandibular angle, ascending ramus width) and factor 2 (e.g. mandibular breadth, chin height). However, when analysing only females, they are more marked along factor 2.

For analysis XIXb (Figure 39a), a male Later Stone Age Central African (e.g. Ishango) is located in the highest values of the modern variation such as the Bantu-speaking South Africans (with largest mandible). However, another male Later Stone Age East African (e.g. Gamble Cave II) and Iron Age Senegalese tend to have more gracile mandible, and therefore they fall within the variation of North Africans (e.g. Attiri).

For analysis XIXc (Figure 39b), a female (Later Stone Age-) Iron Age West African (e.g. Shum Laka) has a very narrow mandible and low chin. It therefore appears to be an outlier along factor 2. All the other archaeological specimens originating from various regions tend to be in the lowest values of modern variation.

In summary, Later Stone Age and/or Iron Age specimens originating from West Africa despite their high variability still tend to overlap the variation of their region of origin. In contrast, both East African and North African skulls appear to be, in general, smaller and more gracile, and therefore closer morphologically to modern North Africans. As a consequence, morphological continuity in craniometric terms is variable through time (around last nine millennia), but more visible on an intra-regional level within West Africa.

5.2.1.2. Variation on a regional scale: focusing on Iron Age West African populations

The relationship of West African Iron Age samples with various modern African populations, are explored in detail. Nine factor analyses using different sets of variables (that were also used previously for only modern samples) are presented below. They provided additional information about affinities of Iron Age West Africans (Table 35, p. 126-127).

The variables entered are related to the whole skull, and they are:

- for **analyses XXa and XXa bis** (similar to III & VIII): GOL, BPL, ZMB, XML, NLB and OBB;
- for **analyses XXIb and XXIc** (similar to XII): ASB, Q1, EKB, WFB, PABR and TOTARCH;
- for **analyses XXIIb and XXIIc** (similar to I & VI): WCB, BPL, JUB, NLB, NLH and OBB; and
- for **analyses XXIIIa, XXIIIb and XXIIIc** (similar to XI, but with ma instead of sh): NPH, WFB, NLB, bb and ma.

Most of the factor analyses (except for XXa & XXa bis) were performed by separating sexes in order to explore in more detail the archaeological sample under study. The Iron Age specimens under study corresponded not only to Senegalese (N=15), but also to individuals originating from various other countries (N=4). The sample of modern Africans that was used for comparison is representative of either the broad main regions (XXa, XXIb, XXIc, XXIIIb & XXIIIc) or a few ethnic groups (XXa bis, XXIIb, XXIIc, XXIIIa). Two samples were composed of pooled ethnic groups because of either lack of information (e.g. Ugandese) or too small sample size (e.g. North West Cameroonians).

In comparison to previous analyses (Table 19, p. 88-91), the highest loadings are here observed often for the same variables (Table 35, p. 126-127):

- for **analyses XX** (a & a bis): BPL, XML and GOL (factor 1); OBH and NLH (factor 2);
- for **analyses XXI** (b & c): WFB (factor 1) and PABR (factor 2).

Variable entered	Factor 1 eigen-value	Factor 1 (%)	Factor 2 eigen-value	Factor 2 (%)	Cumulative % of variance	N° of factors extracted		Factor loadings after varimax rotation factor 1	Factor loadings after varimax rotation factor 2	% of variance[1] factor 1	% of variance[1] factor 2
VAULT & FACE											
GOL, XCB, ZYB, BBH, BPL, NPH, BNL, NLB							**GOL**	**0.752**	0.230	**56.55**	
							XCB	**0.788**	-0.231	**62.09**	
							ZYB	**0.700**	0.395	**49.00**	
Analysis XVIIa (pooled sexes) (N=1336)	3.98	(49.74)	1.03	(12.92)	**62.66**	2	**BBH**	**0.701**	0.244	**49.14**	
							BPL	0.419	**0.716**		**51.27**
							NPH	0.614	0.327		
							BNL	0.687	0.516		
							NLB	0.011	**0.782**		**61.15**
FRC, PAC, OCC, BBH, NPH, XFB, NLB, OBB							**FRC**	**0.781**	0.241	**61.00**	
							PAC	0.559	0.215		
							OCC	0.671	-0.083		
Analysis XVIIIa (pooled sexes) (N=729)	3.33	(41.67)	1.01	(12.61)	**54.28**	2	**BBH**	**0.821**	0.211	**67.40**	
							NPH	0.422	0.508		
							XFB	0.639	0.244		
							NLB	-0.067	**0.824**		**67.90**
							OBB	0.386	0.635		
MANDIBLE											
bb, bw, sh, har, war, ma							bb	0.392	0.672		
							bw	0.113	**0.784**		**61.47**
Analysis XIXb (only males) (N=179)	2.00	(33.38)	1.22	(20.30)	**53.68**	2	sh	-0.147	0.658		
							har	0.554	-0.005		
							war	0.655	0.260		
							ma	**-0.853**	0.037	72.76	
							bb	0.486	0.490		
							bw	0.118	**0.816**		**66.59**
Analysis XIXc (only females) (N=122)	1.99	(33.19)	1.36	(22.74)	**55.93**	2	**sh**	-0.051	**0.713**		**50.84**
							har	0.635	0.307		
							war	0.656	0.085		
							ma	**-0.815**	0.299	**66.42**	

[1] Percentage of variance accounted on each factor obtained by squaring the correlation coefficient (or factor loading) for variables which have the highest factor loadings (>0.7).

Table 34. Summary of three types of factor analyses (XVII-XIX) performed on modern African samples and various Later Stone Age and Iron Age specimens.

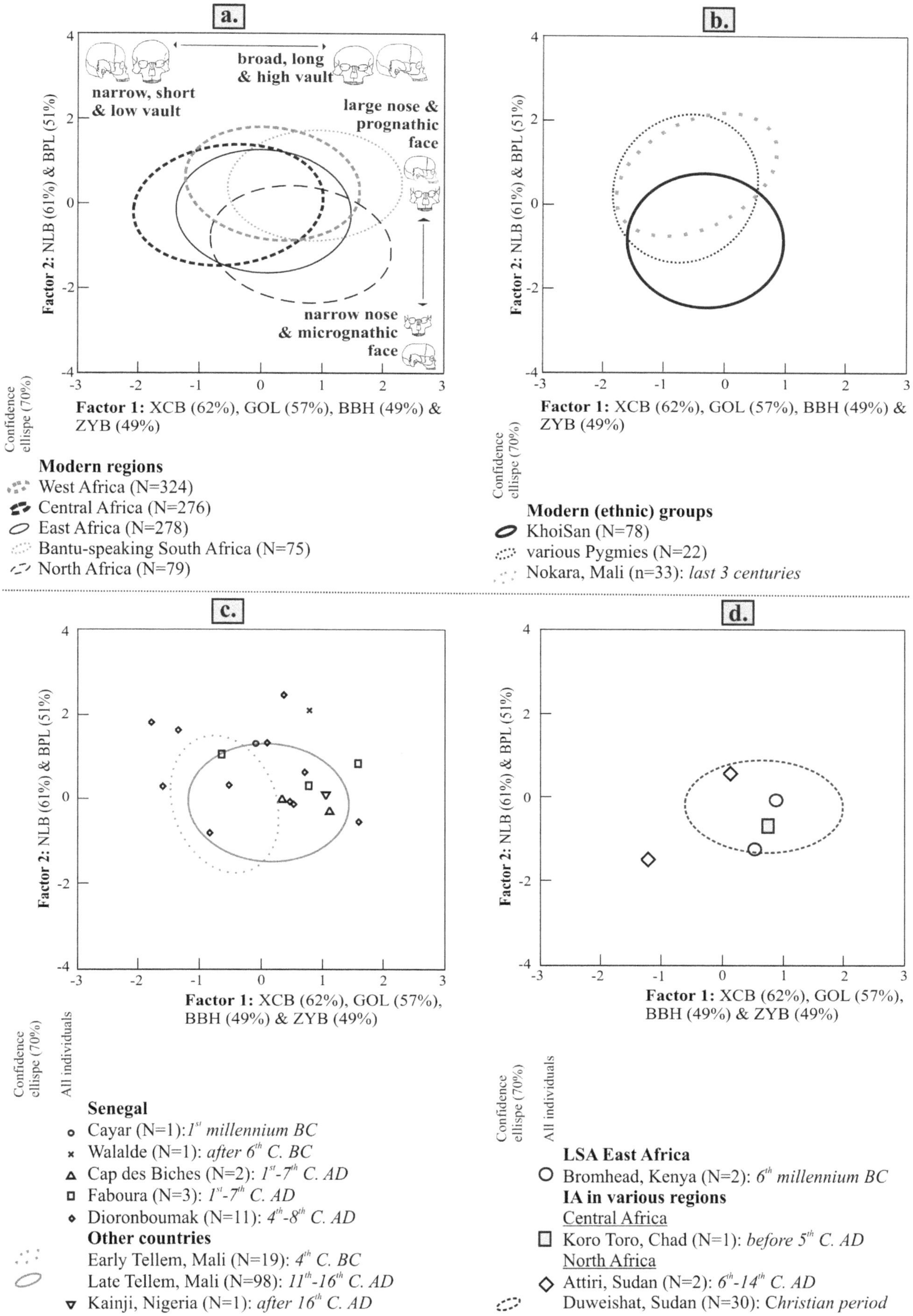

Figure 37. Scatterplots of first and second factors for analysis XVIIa (pooled sexes) expressing 63% of total variance. **a.** Only modern regions; **b.** Only modern (ethnic) groups; **c.** Iron Age (IA) West Africans; **d.** Later Stone Age (LSA) and IA specimens.

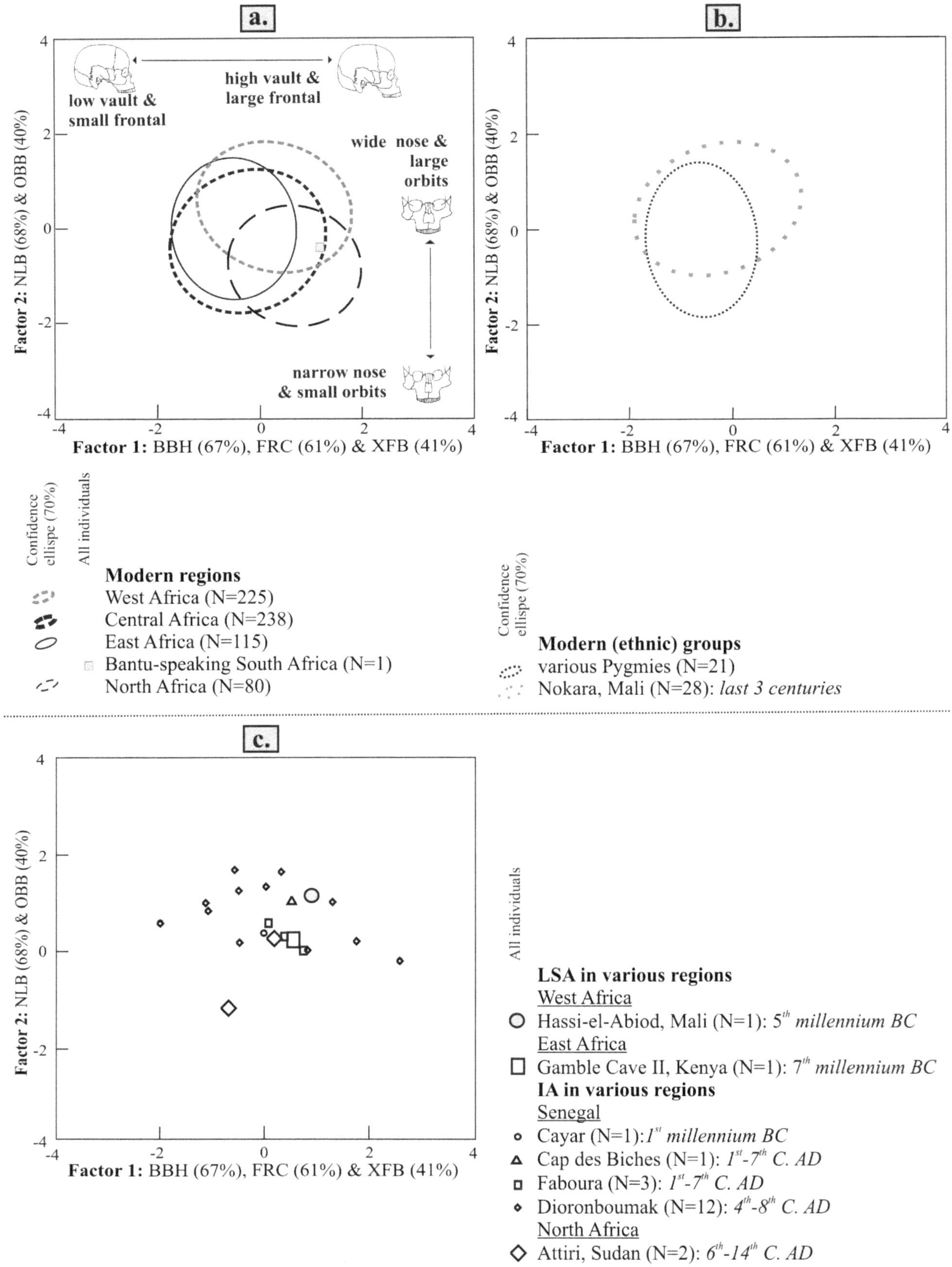

Figure 38. Scatterplots of first and second factors for analysis XVIIIa (pooled sexes) expressing 54% of total variance. **a.** Only modern regions; **b.** Only modern (ethnic) groups; **c.** Later Stone Age (LSA) and Iron Age (IA) specimens.

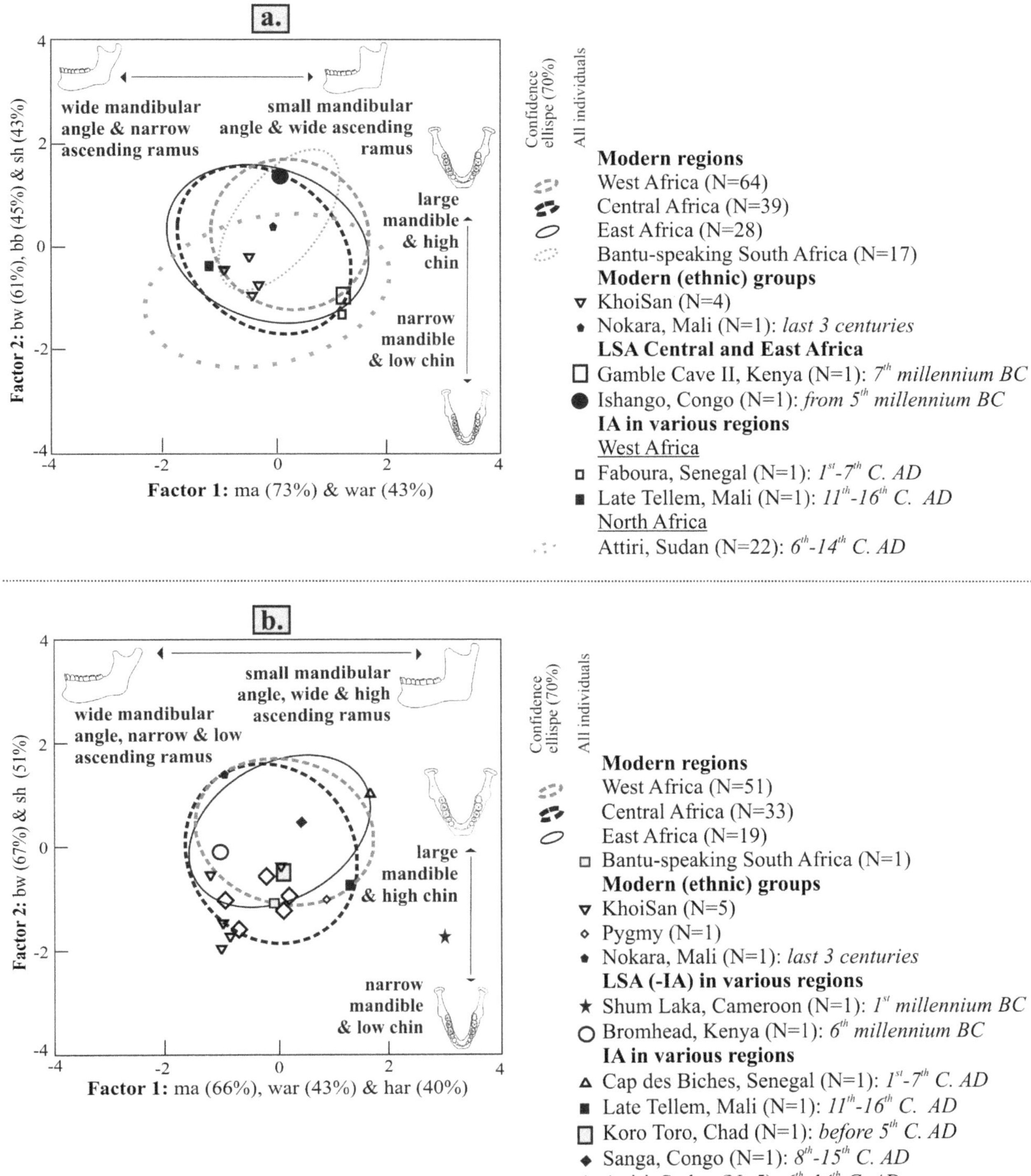

Figure 39. Scatterplots of first and second factors for analyses XIX (b & c) using males and females separately, modern regions, Later Stone Age (LSA) and Iron Age (IA) Africans. **a.** Analysis XIXb (with only males) expressing 54% of total variance; **b.** Analysis XIXc (with only females) expressing 56% of total variance.

Variables entered	Factor 1 eigen-value	Factor 1 (%)	Factor 2 eigen-value	Factor 2 (%)	Factor 3 eigen-value	Factor 3 (%)	Cumulative % of variance	N° of factors extracted		Factor loadings after varimax rotation factor 1	factor 2	factor 3	% of variance[1] factor 1	factor 2	factor 3
VAULT & FACE															
GOL, BPL, ZMB, XML,									**GOL**	**0.720**	0.282	-	**50.40**		
NLH, OBH									**BPL**	**0.803**	0.002	-	**64.48**		
Analysis XXa	2.66	(44.27)	1.14	(18.96)	-	-	**63.23**	2	ZMB	0.604	0.319	-			
(pooled sexes)									**XML**	**0.752**	0.011	-	**56.55**		
main modern regions									**NLH**	0.423	**0.715**	-		**51.12**	
(N=520)									**OBH**	-0.056	**0.908**	-		**82.45**	
									GOL	**0.739**	0.298	-	**54.61**		
Analysis XXa bis									**BPL**	**0.811**	0.016	-	**65.77**		
(pooled sexes)	2.78	(46.33)	1.07	(17.89)	-	-	**64.22**	2	ZMB	0.562	0.386	-			
modern ethnic groups									**XML**	**0.737**	0.068	-	**54.32**		
(N=272)									**NLH**	0.437	**0.719**	-		**51.70**	
									OBH	-0.031	**0.915**	-		**83.72**	
ASB, Q1, EKB, WFB, PABR,									ASB	0.551	0.390	-0.309			
TOTARCH									**Q1**	0.140	**0.864**	0.143		**74.65**	
Analysis XXIb									**EKB**	**0.844**	0.047	0.376	**71.23**		
(only males)	2.50	(41.73)	1.12	(18.73)	1.04	(17.30)	**77.76**	3	**WFB**	**0.841**	0.187	-0.036	**70.73**		
main modern regions									**PABR**	0.070	0.121	**0.936**			**87.61**
(N=275)									**TOT-ARCH**	0.171	**0.895**	0.015		**80.10**	
									ASB	0.647	0.152	-			
Analysis XXIc									**Q1**	**0.792**	0.010	-	**62.73**		
(only females)									**EKB**	0.332	**0.831**	-		**69.06**	
main modern regions	2.58	(42.94)	1.25	(20.80)	-	-	**63.74**	2	WFB	0.661	0.413	-			
(N=233)									**PABR**	-0.088	**0.833**	-		**69.39**	
									TOT-ARCH	**0.804**	-0.010	-	**64.64**		

[1] Percentage of variance accounted on each factor obtained by squaring the correlation coefficient (or factor loading) for variables which have the highest factor loadings (>0.6).

Table 35. Summary of four types of factor analyses (XX-XIII) performed on modern African samples and various Iron Age specimens.

Table 35 continued

Variable entered	Factor 1 eigen-value	Factor 1 (%)	Factor 2 eigen-value	Factor 2 (%)	Cumulative % of variance	N° of factors extracted	Factor loadings after varimax rotation	factor 1	factor 2	% of variance[1] factor 1	% of variance[1] factor 2
FACE											
WCB, BPL, JUB,							WCB	0.619	0.356		
NLH, NLB, OBH							**BPL**	-0.049	**0.878**		**77.09**
Analysis XXIIb							JUB	0.330	0.648		
(only males)	2.37	(39.48)	1.06	(17.63)	**57.11**	2	**NLH**	**0.795**	0.158	**63.20**	
modern ethnic groups							NLB	0.430	0.336		
(N=100)							**OBH**	**0.808**	0.069	**65.29**	
							WCB	0.696	0.094		
Analysis XXIIc							BPL	0.497	0.163		
(only females)	2.53	(42.24)	1.01	(16.82)	**59.06**	2	**JUB**	**0.748**	0.233	**55.95**	
main ethnic groups							**NLH**	**0.799**	0.119	**63.84**	
(N=141)							**NLB**	0.211	**0.886**		**78.50**
							OBH	**0.706**	-0.428	**49.84**	
FACE & MANDIBLE											
NPH, WFB, NLB, bb, ma							NPH	0.656	0.467		
Analysis XXIIIb							WFB	0.674	-0.192		
(only males)	1.77	(35.34)	1.04	(20.80)	**56.14**	2	NLB	0.515	-0.374		
main modern regions							**bb**	**0.735**	-0.079	**54.02**	
(N=155)							**ma**	-0.142	**0.833**		**69.39**
							NPH	-0.036	**0.895**		**80.10**
Analysis XXIIIc							WFB	0.621	0.149		
(only females)	1.62	(32.38)	1.08	(21.67)	**54.05**	2	NLB	0.427	0.486		
main modern regions							bb	0.650	0.217		
(N=115)							**sh**	**-0.728**	0.276	**53.00**	

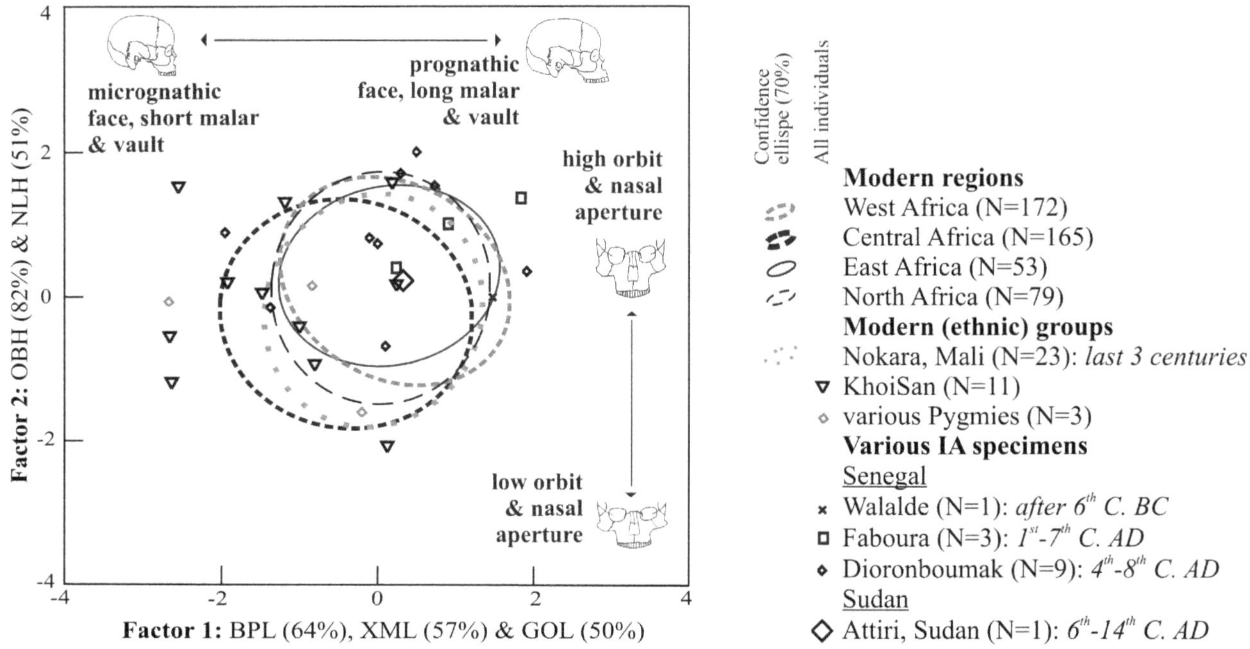

Figure 40. Scatterplot of first and second factors for analysis XXa (pooled sexes, Iron Age (IA) specimens and modern regions) expressing 63% of total variance.

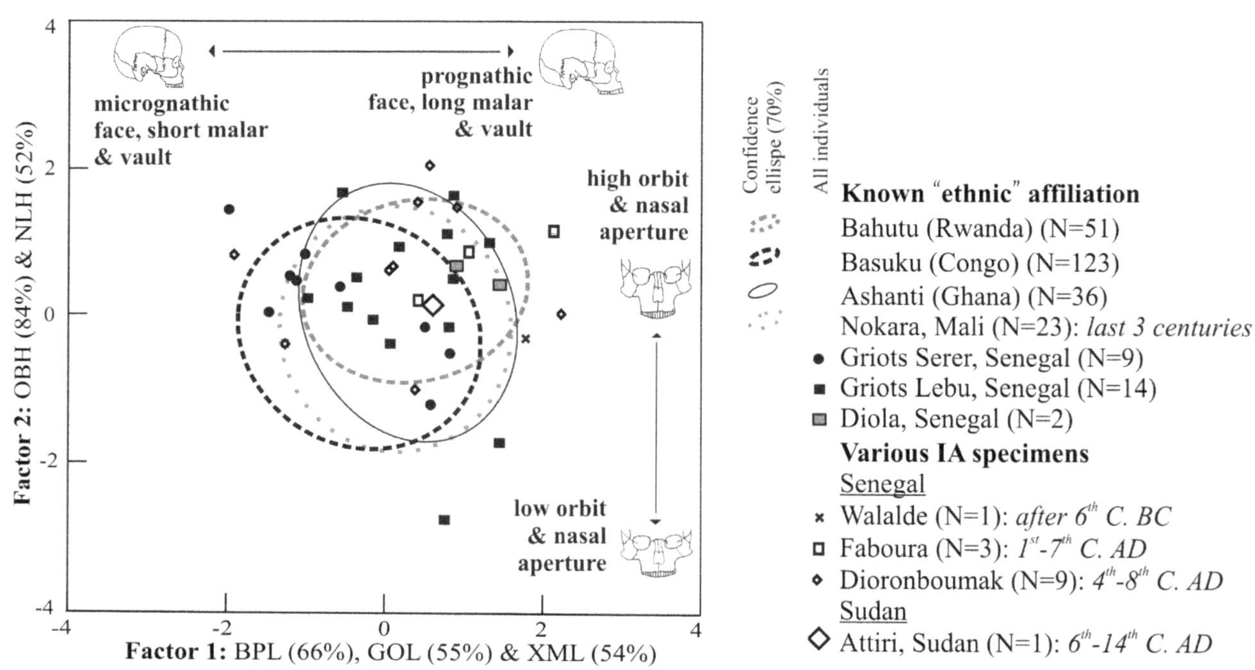

Figure 41. Scatterplot of first and second factors for analysis XXa bis (pooled sexes, Iron Age (IA) specimens and modern ethnic groups) expressing 64% of total variance.

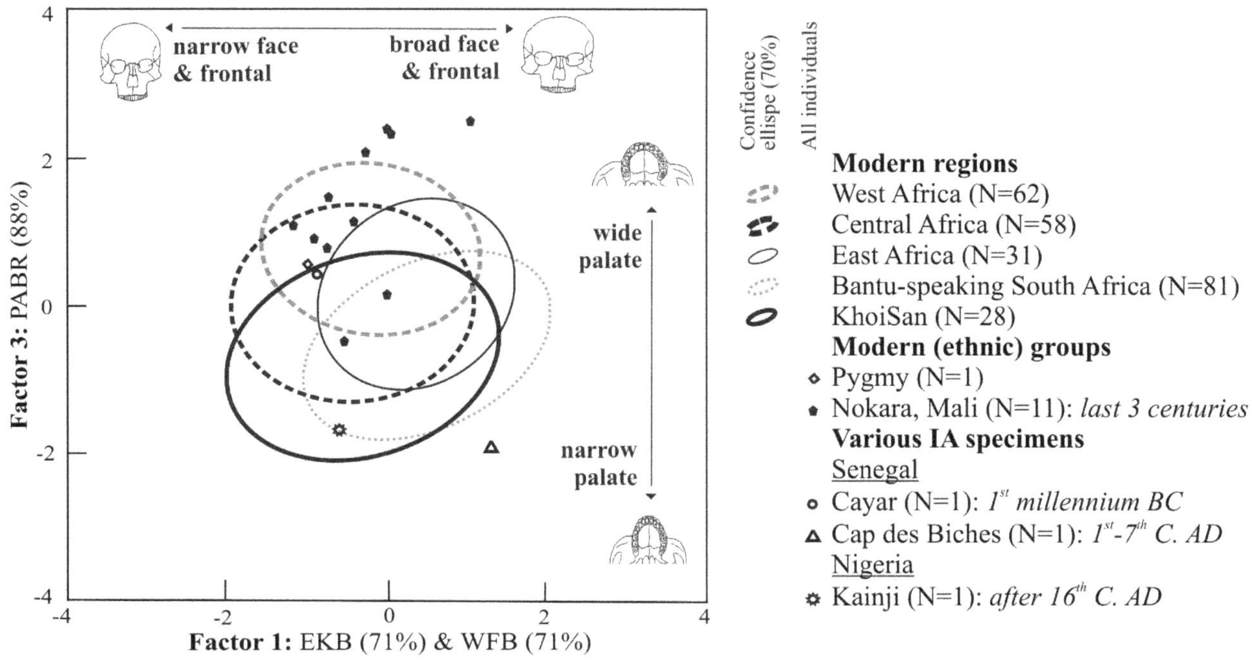

Figure 42. Scatterplot of first and second factors for analysis XXIb (only males, Iron Age (IA) specimens and modern regions) expressing 78% of total variance.

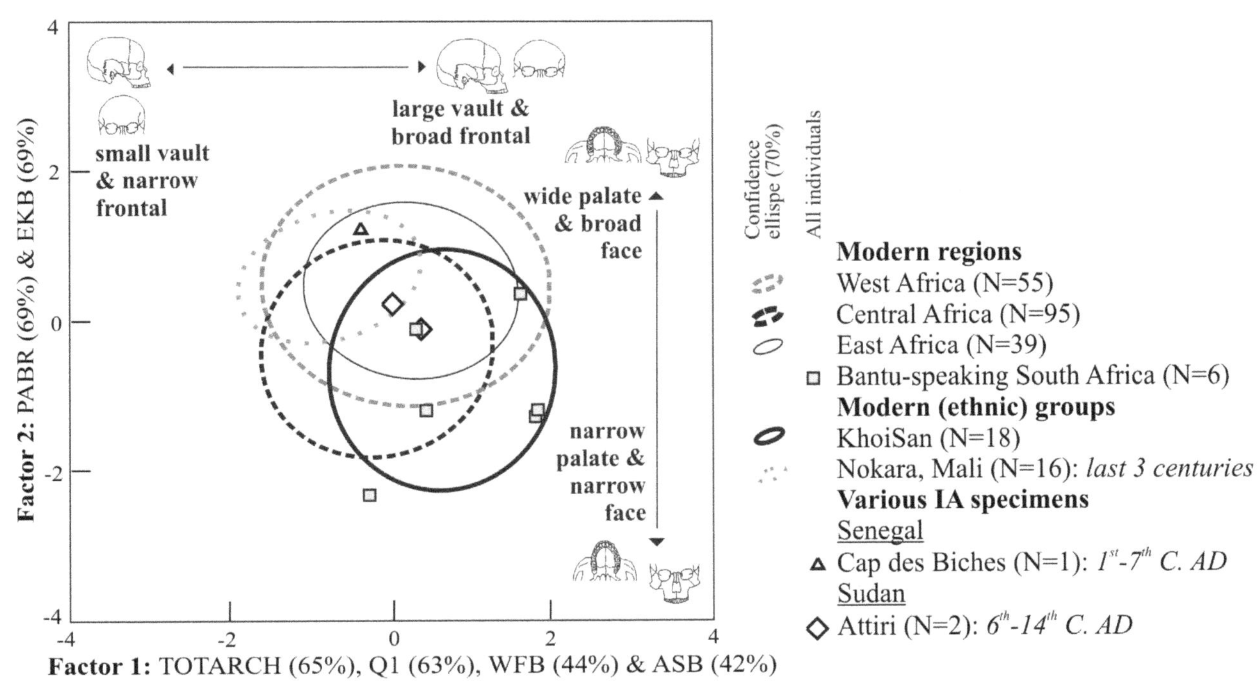

Figure 43. Scatterplot of first and second factors for analysis XXIc (only females, Iron Age (IA) specimens and modern regions) expressing 64% of total variance.

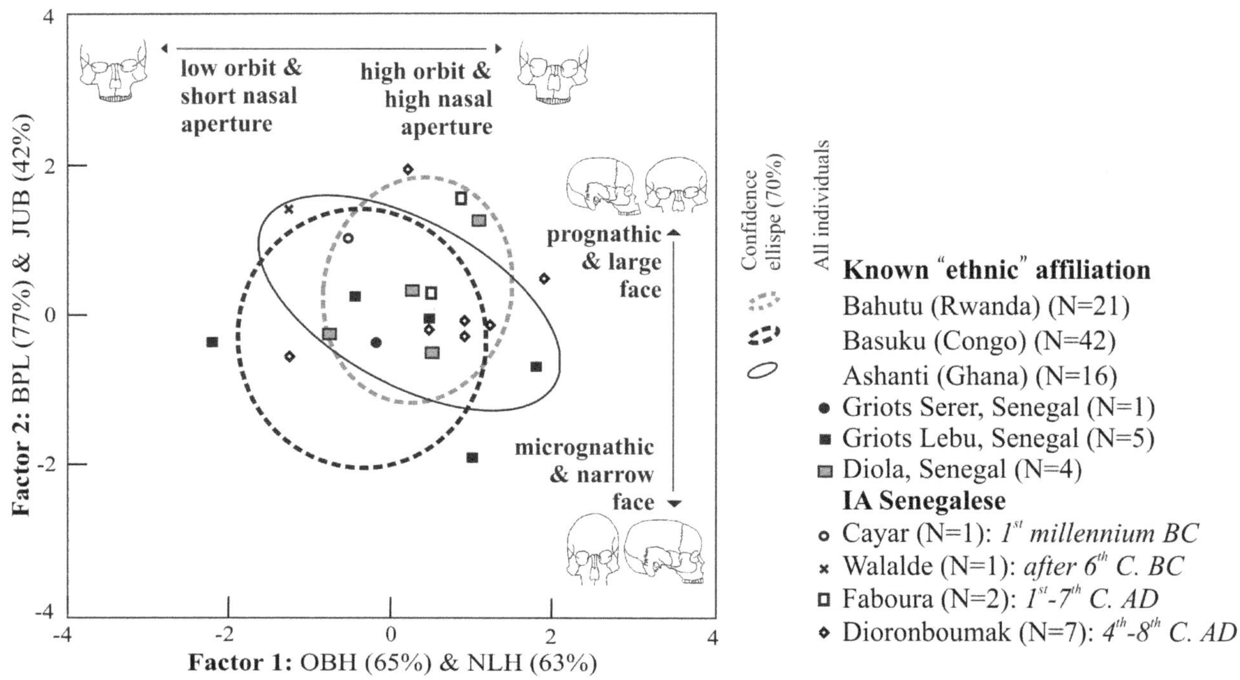

Figure 44. Scatterplot of first and second factors for analysis XXIIb (only males, Iron Age (IA) Senegalese specimens and modern ethnic groups) expressing 57% of total variance.

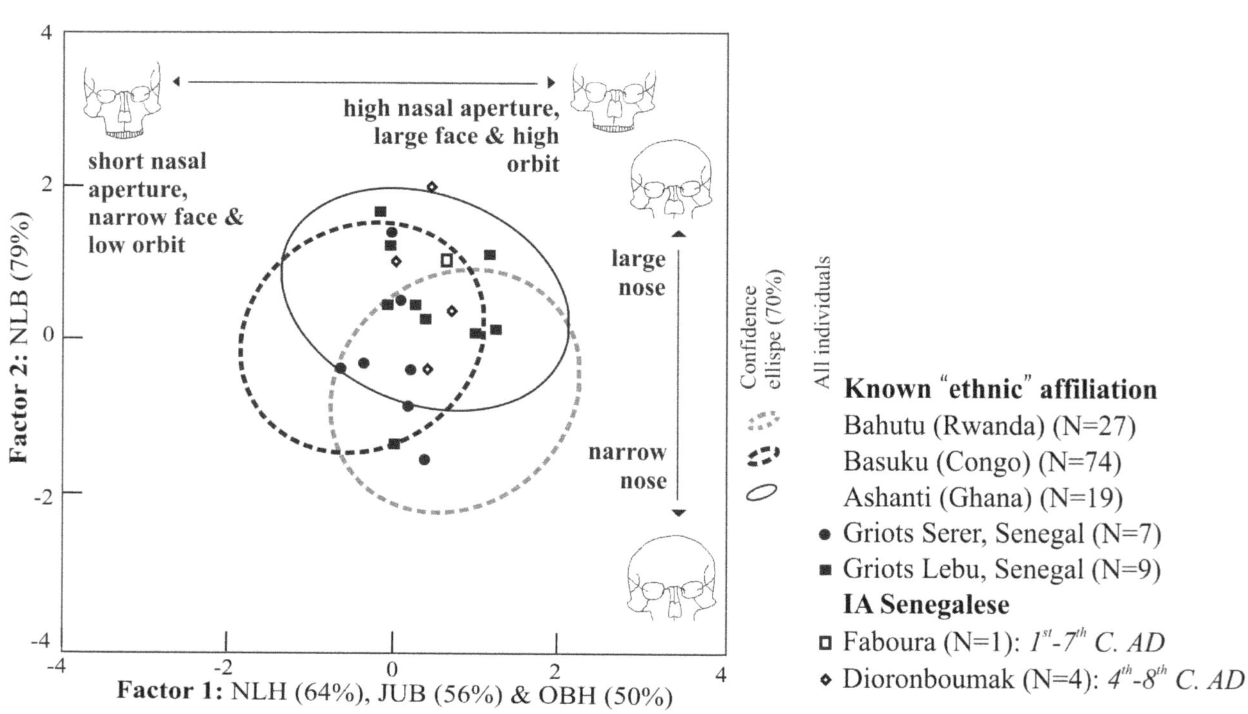

Figure 45. Scatterplot of first and second factors for analysis XXIIc (only females, Iron Age (IA) Senegalese specimens and modern ethnic groups) expressing 59% of total variance.

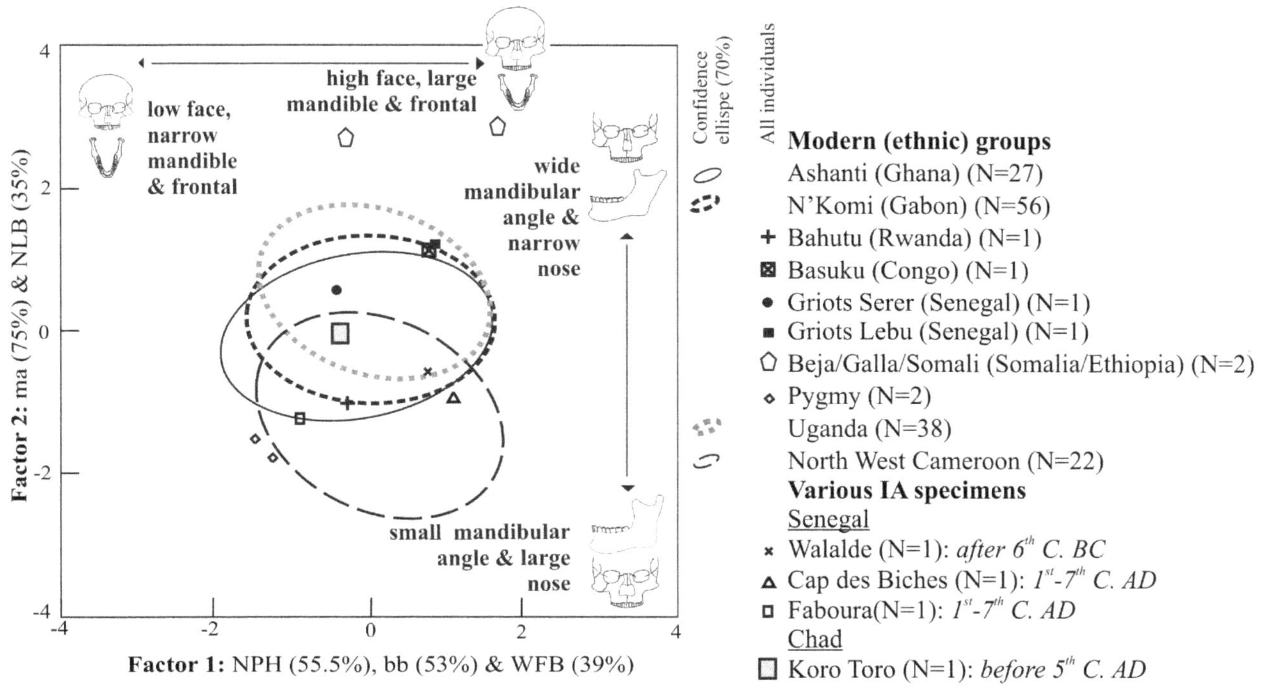

Figure 46. Scatterplot of first and second factors for analysis XXIIIa (pooled sexes, Iron Age (IA) specimens and modern ethnic groups) expressing 58% of total variance.

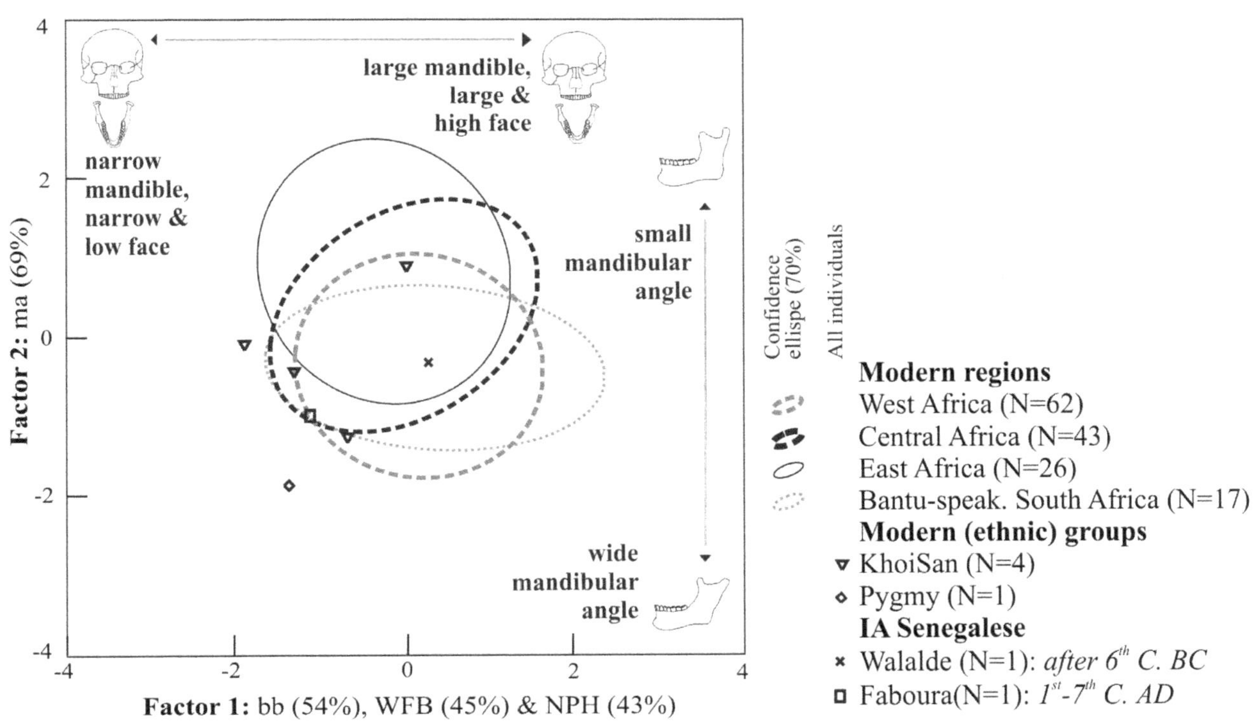

Figure 47. Scatterplot of first and second factors for analysis XXIIIb (only males, Iron Age (IA) specimens and modern regions) expressing 56% of total variance.

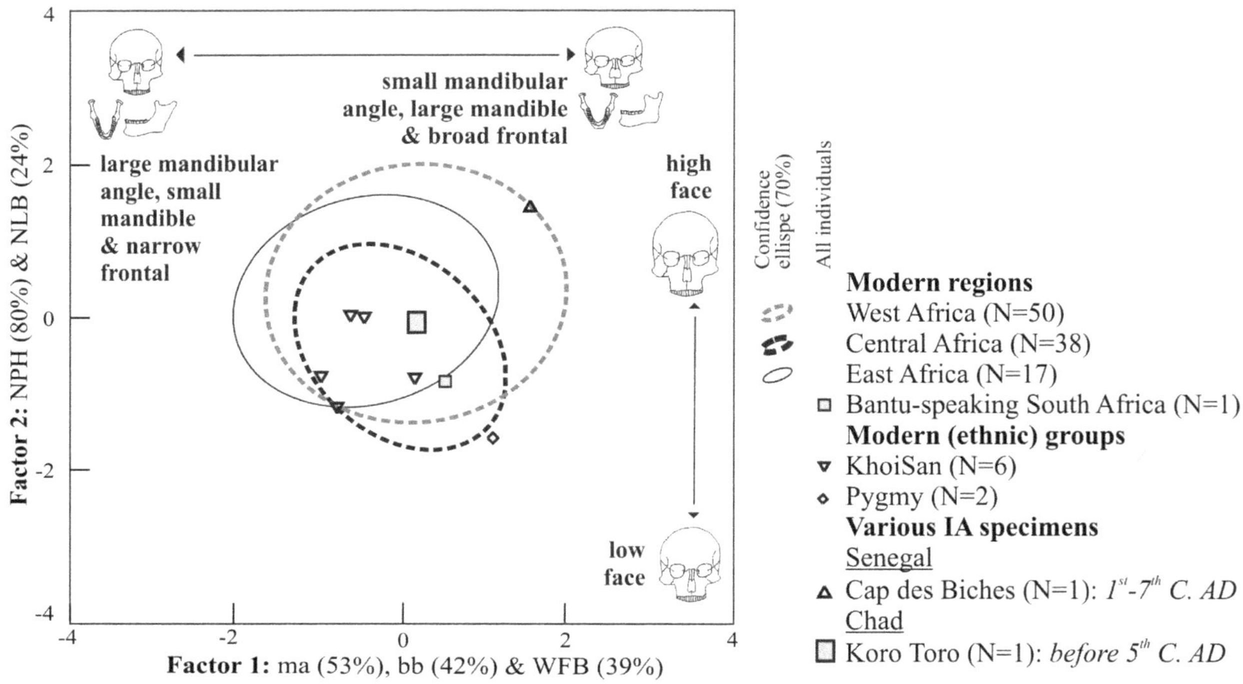

Figure 48. Scatterplot of first and second factors for analysis XXIIIc (only females, Iron Age (IA) specimens and modern regions) expressing 54% of total variance.

- for analyses XXII (b & c): NLH and OBH (factor 1); BPL, JUB and/or NLB (factor 2); and
- for **analyses XXIII** (a, b & c): bb and WFB (factor 1); ma, NPH and/or NLB (factor 2).

Scatterplots for factor 1 and 2 are drawn in order to show group clustering for each analysis (Figures 40-48, p. 128-132). For analyses XX (a & a bis) (Figures 40- 41, p. 128), marked differences are observed between both modern and Iron Age populations. When looking at both broad regional (XXa; Figure 40, p. 128) and ethnic (XXa bis; Figure 41, p. 128) groupings, Dioronboumak and Faboura individuals appear to be very variable, especially for the following morphological trends:

- prognathism degree, length of vault and malar (factor 1); and
- height of orbit and nose (factor 2).

However, Dioronboumak and Walalde specimens (as well as Attiri individual) overlap mainly the variation of West Africans, especially the Ashanti. The Ashanti are characterized by a long vault, a prognathic face, high orbits and a high nasal aperture. The modern Senegalese Griots also appear very variable, but differences are also noted within their group, the Lebu being larger than the Serer. For analyses XXI (b & c) (Figures 42-43, p. 129), marked regional differences are observed, when both males and females are considered separately for morphological trends, such as:

- length and breadth of vault (factor 1); and
- breadth of palate and/or face (factor 2).

When using only male populations (XXIb; Figure 42, p. 129), Cayar individual, who has a moderately broad face and wide palate, falls within the middle of the variation of both West and Central Africans. Cap des Biches and Kainji individuals are closer to the (Bantu-speaking) South African variation reflecting large size for the vault but not for the palate and biorbital breadth.

However, when using only female populations (XXIc; Figure 43, p. 129), regional differences are slightly less marked, especially along factor 1. Another specimen from Cap des Biches appears to be closer to the Nokara group. In fact, Nokara overlaps lower values of the West African variation, tend to have large palate and biorbital breadth and moderately large vault.

For analyses XXII (b & c) (Figures 44-45, p. 130), there is still an overlapping variation between Iron Age Senegalese specimens (especially Dioronboumak) and a large number of ethnic groups (with high orbit and nasal aperture, large and prognathic face, and large nose), e.g. the

Ashanti, the Griots Lebu and the Bahutu. This is observed for both male (XXIIb; Figure 44, p. 130) and female (XXIIc; Figure 45, p. 130) samples. However, when using only females, there are more marked regional differences along factor 1.

In analyses XXIII (a, b & c) (Figures 46-48, p. 131-132), different morphological trends for various archaeological specimens are found on both ethnic and regional levels. When pooling sexes and comparing various groups of both known and unknown ethnic affiliation (analysis XXIIIa; Figure 46, p. 131), the extremes of the variation of the comparative modern sample, are:

- in the lowest values, the Ugandese with various Somalians (e.g. narrow nose, large mandibular angle); and
- in the highest values, the North West Cameroonians (e.g. large nose, small mandibular angle).

Between these last two groups, there is also the variation of both N'Komi and Ashanti groups. Faboura, Cap des Biches and Walalde are located mainly within the North West Cameroonian and Ashanti variation, and a few of the Griots within the Ugandese variation.

The one-way analysis of variance on the regression factor scores shows that, the main modern groups (e.g. N'Komi, Ashanti, Ugandese, and North West Cameroonian) are significantly different at the highest level of probability ($F=24.6$; $P<0.001$). According to the Scheffe's tests, these morphological differences are present especially between the North West Cameroonians and all the remaining groups.

When using only males (XXIIIb; Figure 47, p. 131), results about affinities of Iron Age West Africans are similar to those in Figure 46 (p. 131). Both Walalde and Faboura individuals fall within the variation of the West Africans, as they have a wide mandibular angle (along factor 2).

When using only females (XXIIIc; Figure 48, p. 132), the Cap des Biches specimen is located in the highest values of the West Africans. Although the latter are highly variable, they tend to have a high face and large nose (factor 2). In contrast to the previous ones, the individual from Koro Toro shows an opposite morphological trend, and is therefore closer to Central Africans.

In summary, the Iron Age Senegalese often appear to be close to modern West Africans (especially the Ashanti). Therefore, it seems that during the last three millennia, there is a morphological continuity in relation to geographical differences.

5.2.2. Multiple discriminant analyses

Additional multiple discriminant analyses are performed, to evaluate in more detail the intra-regional patterns within West Africa. Two Iron Age Senegalese samples (Faboura, Dioronboumak) are compared to two series of various regional (and ethnic) groups:

- five modern ethnic groups originating from West Africa (Ashanti, Griots Serer, Griots Lebu), Central Africa (Basuku) and East Africa (Bahutu) (Table 36, p. 134); and
- two modern groups originating from Nokara (Mali) and North Africa, and two other Iron Age Malian groups (Late Tellem, Early Tellem) (Table 37, p. 134).

For this statistical analysis, sexes were pooled, as sample size of both males and females when separated were too small for the comparative modern groups. Different sets of variables in relation to both vault and face are entered for four analyses. According to highest standardized canonical coefficients (≥ 0.7), the variables most frequently effective in "classifying" groups are the followings:

- for the vault: XCB, GOL and BBH; and
- for the face: BPL, BNL and JUB.

The results obtained for the various comparisons (Tables 36-37, p. 134) show that, the total correct rate of classification is quite variable and moderately high (range: 29%-64%). It supports the fact that, the variation of cranial morphologies overlaps considerably between the different groups.

When performing two analyses with the first series of comparative groups (Table 36, p. 134), the total correct rates of classification are higher for the vault (63.6%) than for the face (51.8%). The individuals of Dioronboumak are closer morphologically to Central Africans, as they are often misclassified with the Basuku (25%-67%). In contrast, the individuals of Faboura appear to have more affinity with East Africans, as they are often misclassified with the Bahutu (40%-60%). However, this observation is perhaps biased as sample size is small here ($N=5$). The Griots also appear to have more affinities with Central Africans than Iron Age Senegalese, as they are also very often misclassified with the Basuku (89%).

Variables entered, total sample size & groups analysed		N	Classification results: Bahutu	Basuku	Ashanti	Griots Serer	Griots Lebu	Faboura	Dioron-boumak	Total correct rate	Standardized canonical discriminant function coefficients: Selected variables	function 1	function 2	function 3	function 4
VAULT: GOL, XCB, PAC, OCC, BBH, WCB, BNL															
Rwanda	Bahutu (modern)	50	35	13	0	0	0	1	1	63.6%	GOL	0.488	-0.482	1.160	0.286
D.R.C.	Basuku (modern)	120	5	108	5	0	1	0	1		XCB	0.666	0.315	-0.720	0.540
Ghana	Ashanti (modern)	31	0	24	4	0	1	0	2		BBH	-1.054	0.387	0.242	0.592
Senegal	Griots Serer (modern)	8	0	8	0	0	0	0	0		BNL	0.336	0.864	-0.324	-0.943
	Griots Lebu (modern)	20	3	12	2	0	3	0	0						
	Faboura (IA)	6	2	1	1	0	0	2	0						
	Dioronboumak (IA)	15	2	3	1	0	0	2	7						
FACE: BPL, ZYB, JUB, STB, XFB, FMB, ZMB															
Rwanda	Bahutu (modern)	50	20	30	0	0	0	0	0	51.8%	JUB	1.000	-	-	-
D.R.C.	Basuku (modern)	120	13	107	0	0	0	0	0						
Ghana	Ashanti (modern)	31	6	25	0	0	0	0	0						
Senegal	Griots Serer (modern)	9	1	8	0	0	0	0	0						
	Griots Lebu (modern)	18	2	16	0	0	0	0	0						
	Faboura (IA)	5	3	2	0	0	0	0	0						
	Dioronboumak (IA)	12	4	8	0	0	0	0	0						

Table 36. Summary for two multiple discriminant analyses: predicted group membership for various sub-Saharan African samples dating back to the Iron Age (IA) period.

Variables entered, total sample size & groups analysed		N	Classification results: North Africa	Nokara	Early Tellem	Late Tellem	Faboura	Dioron-boumak	Total correct rate	Standardized canonical discriminant function coefficients: Selected variables	function 1	function 2	function 3
BBH, GOL, BNL, ZYB, XCB, BPL, NPH													
North Africa	various (modern)	70	32	6	1	30	1	0	55.4%	GOL	-0.647	1.122	0.045
Mali	Nokara (modern)	41	4	20	0	17	0	0		XCB	1.080	0.000	0.338
	Tellem (Early IA)	19	4	2	0	13	0	0		BPL	-0.034	-0.754	0.894
	Tellem (Late IA)	98	14	5	0	79	0	0					
Senegal	Faboura (IA)	3	1	1	0	0	1	0					
	Dioronboumak (IA)	11	1	2	0	6	0	2					
BBH, GOL, ZYB, XCB, BPL, NPH, BNL, WFB													
North Africa	various (modern)	70	62	8	-	0	0	0	29.1%	XCB	1.000	-	
Mali	Nokara (modern)	56	35	21	-	0	0	0					
	Tellem (Late IA)	110	104	4	-	0	0	2					
Senegal	Faboura (IA)	10	9	1	-	0	0	0					
	Dioronboumak (IA)	17	14	2	-	0	0	1					

Table 37. Summary for two multiple discriminant analyses: predicted group membership for various North and West African populations dating back to the Iron Age (IA) period.

When performing two other analyses with the second series of comparative groups (Table 37, p. 134), the total correct rates of classification tend to decrease (29.1%-55.4%). For the first analysis, 55% of Dioronboumak individuals, as well as 33% of the Faboura ones, are misclassified with the Tellem group of the late phase. However, for the second analysis (when adding one variable and excluding one group), most of the Senegalese specimens are misclassified with North Africans (Dioronboumak: 82%; Faboura: 90%). Therefore, only very few of them (range for both groups: 0%-33%) are correctly classified into their group of origin.

In summary, Iron Age Senegalese (especially Dioronboumak) are morphologically closer to modern sub-Saharan Africans, especially Congolese (Basuku) as well as to their contemporaneous neighbouring populations (Late Tellem). To some extent, they also share some morphological features with North Africans, but only in relation to vault breadth.

5.2.3. Conclusions

The identification of morphological trends through time, between and within regions, involves hypotheses about the impact of various (macro-scale and micro-scale) phenomena. The latter are mainly related to the geographical differentiation of populations that reflect predominant gene flow within each broad region (**Chapter 4**). Natural geographical barriers as well as historical events such as population movements could be partly responsible for sub-Saharan African diversity.

Although our corpus of data is not fully representative of all modern and past African populations, it is possible to provide the following preliminary comments about the two questions initially addressed in section 5.2. (p. 119):

i) The geographical differences that are present between modern populations on a broad regional level vary through time. Although, these are also observed during the Iron Age period, they are increasingly different during the Later Stone Age. Nevertheless, past West Africans, despite high levels of morphological diversity, appear to be often the closest to modern populations within their region of origin.

   As a series of Later Stone Age and/or Iron Age specimens were compared with the modern African diversity, they showed a gracilization trend of the skull since the mid-Holocene (Dutour, 1992; Henneberg and Steyn 1993; Orban *et al.*, 2001; Sereno *et al.*, 2008). This phenomenon was already observed within Africa by various studies even earlier (Lahr, 1996; Grine *et al.*, 2007; Crevecoeur *et al.*, 2009). However, it did not occur systematically amongst some of the Iron Age West Africans. For example, Cap des Biches and Dioronboumak tend to be in the highest values of the modern variation (e.g. West Africans, Bantu-speaking South Africans) for both vault size and various facial features (prognathism degree, nasal width, orbital breadth). It therefore supports the hypothesis that, the biological input in West Africa originated mainly from sub-Saharan Africa.

   When analysing the mandible morphology, Later Stone Age (and Iron Age) specimens originating from sub-Saharan Africa (e.g. Shum Laka, Ishango, Bromhead) present common features related mainly to robusticity (e.g. very wide ascending ramus, small mandibular angle). For example, although the Shum Laka individual has a very narrow mandible, it also has a very wide ascending ramus. This observation supports previous comments by various authors (Brothwell and Shaw, 1971; Pinhasi and Semal, 2000; Stringer *et al.*, 2010) about the presence of unique mandibular features amongst Later Stone Age West-Central Africans (e.g. Iwo Eleru) (section 2.3.3.2.2., p. 36).

ii) As in (i) above, despite a high variability, there is a morphological continuity between both past and modern West Africans (Table 38, p. 136). However, more precisely, the Iron Age Senegalese (e.g. wide nose, wide orbit, tall face, broad mandible) overlap mainly the variation of a Ghanaian ethnic group (Ashanti). Even when considering sexes separately, the sample of Dioronboumak is particularly variable, especially in relation to vault size.

   According to various multivariate analyses, these Iron Age individuals have affinities with several groups but mostly from sub-Saharan Africa such as: Central Africans (e.g. Basuku), West Africans (e.g.

Tellem, Griots Lebu, Nokara, Ashanti), East Africans (e.g. Teita, Bahutu) and North Africans. However, the Ashanti (e.g. large nose and face) represent the modern group that is morphologically the closest to the Dioronboumak individuals. This morphological heterogeneity of the Dioronboumak population is in contradiction with the cultural homogeneity of the site observed by Descamps and Thilmans (2001). This fact however does not exclude the hypothesis of the arrival in Senegal of populations that are biologically different. Between the 4th century and 8th century AD, Senegalese coastal settlements could have been exposed to various biological influences despite a cultural continuity during the previous five centuries. They were in fact located at the cross-roads of various influential West African empires that were founded along or near the savanna belt (e.g. Nok culture, Ghanaian empire).

In conclusion, these results support both craniometric (Froment, 1998; Franklin *et al.*, 2010) and genetic (Biondi *et al.*, 1996; Tishkoff *et al.*, 2009) studies that show a marked geographical differentiation amongst modern Africans. However, in addition to data obtained from modern samples, these geographical differences are also observable up to the Iron Age. For example, the Iron Age Senegalese fit well within the variation of modern sub-Saharan Africans, originating especially from West Africa.

Archaeological specimens	Vault	Face	Nose	Orbit	Mandible	Affinities with modern populations
Middle to Late Iron Age (after 1st century AD)						
Dioronboumak	very variable length	very variable protrusion & malar length	high	high	-	West Africa (Ashanti) East Africa (Bahutu)
Faboura	moderately long, narrow frontal	moderately high, prognathic	moderately high & wide	-	narrow, small *ma*	West Africa (Ashanti, North West Cameroon)
Cap des Biches	moderately long & broad	moderately high & broad (palate too)	large	-	large, small *ma*	West Africa (Ashanti, North West Cameroon) South Africa (various)
Early Iron Age (before 1st century AD)						
Cayar	moderately long & broad	moderately broad (palate too)	-	-	-	West Africa (Ashanti)
Walalde	long & broad	relatively high, prognathic, long malars	moderately high & wide	moderately high	moderately large, high chin	West Africa (Ashanti, North West Cameroon) Central Africa (N'Komi)

Table 38. Summary of morphological trends observed from various factor analyses for Iron Age Senegalese crania.

CHAPTER 6

DISCUSSION AND CONCLUSION

In order to explore the biological effects on craniometry in terms of the Bantu-speakers expansion, the present research developed a "multi-layered" approach. It was important to focus on different aspects or "layers", which are considered as potential sources of biological diversity, such as: geography (and sex), ecology and time.

Although the dispersal of the Bantu-speakers was presented as the main historical event in all the microevolutionary models proposed previously for sub-Saharan Africa (Figures 11, 14 and 17, p. 19, 26, 49), its effects have probably been intermingled with and/or obscured by a series of confounding factors (e.g. geography, selection, other historical events). It is difficult to identify the biological effects of a single dispersal despite its scale and complexity. Therefore, several "layers", such as geography (and sexual dimorphism), ecology and time, that are probably responsible for variation to different degrees, will be examined separately in relative isolation.

Firstly, cranial variation in relation to geography is discussed on different levels of analysis. Morphological differences are re-evaluated especially within sub-Saharan Africa, as the latter represents the geographical background for the Bantu-speakers expansion. Geographical differentiation within each sex is similar for both males and females, except for slight variations. The latter however remains difficult to interpret because of some possible methodological biases.

Secondly, in order to explore the possible influence of equatorial climate on skull morphology, associations between morphological patterns and ecology are discussed. Although they are not directly related to the phenomenon of the Bantu-speakers expansion, their effects might be superimposed on those left behind by this large-scale migratory event.

Thirdly, the possible effects of time will be introduced when discussing African diversity in the Holocene. Despite the lack of human remains in West and Central Africa, the morphological trends could reflect continuity and/or differentiation of populations on a broad inter-regional level. As they occurred before (and during) the Bantu-speakers expansion, they might reflect indirectly this diaspora.

After having discussed these factors that represent the underlying key-features of the differentiation of sub-Saharan African populations, the morphological evidence for the Bantu-speakers phenomenon will be again addressed. Therefore, in light of the statistical results in **Chapters 4** and **5**, the microevolutionary models will be re-assessed in detail. For example, was the Bantu-speakers dispersal a simple founder effect phenomenon without any confounding factors? Or, was this phenomenon very complex, involving not only homogenization but also diversification due to various aspects (e.g. geography, ecology, population admixture)?

However, before discussing these microevolutionary processes, it is important to raise the following issues:

- the notion of "population" as a unit that was approached on different scales; and
- various methodological aspects that could have biased the results (e.g. size and nature of sample, methods, variables under study).

6.1. Population as the unit of biological processes

The notion of "population" as the unit of biological processes needs to be re-evaluated. In this work, the idea of a population was approached in a multi-layered manner through different scales of analysis (e.g. regional, ecological, and ethno-linguistic) that indirectly helped to better understand the impact of the Bantu-speakers expansion.

Although the definition of "population" involves basic biological principles (e.g. similar interbreeding, environment and/or culture), it varies according to the level of analysis. As previously mentioned, the major problem encountered (section 2.3.1.1., p. 27) is as follows: at what level can morphological differences be considered as an indicator of "population" differences? The definition of a population has therefore a variable hierarchy of clustering (e.g. continental, regional or ethnic scales). As suggested by Hiernaux (1974), the ideal anthropological definition of a "population" often corresponds to the ethnic group, sharing similar biological and cultural features, and whose cohesion is maintained by endogamy.

But, what are the best notions to delineate a "population" in craniometric terms? In order to see whether sub-Saharan Africans, especially the Bantu-speaking groups, follow the predictions of a theoretical definition, morphological diversity was explored on various levels of analysis (Table 39, p. 139).

Firstly, analyses of geographical (and sexual) and/or ecological differences were undertaken within Africa (and elsewhere) (**Chapter 4**). Secondly, on a more detailed scale (**Chapter 5**), other analyses were completed on a few ethnic groups and/or archaeological samples originating from different regions and/or mainly West and Central Africa.

In **Chapter 4**, as geography appears to be the main key-feature of "population" differentiation, different types of variables that were aimed at grouping individuals were tested in relation to several broad region subdivisions, namely geography and ecology (e.g. latitude, longitude, humidity). Predominant gene flow is well expressed inter-regionally with this type of "classification", as the latter was defined according to natural geographical barriers. But a "classification" that is not entirely geographical was also used here, as one attempted to consider hunter-gatherers separately (e.g. Pygmy, KhoiSan).

For sub-Saharan Africa, although South Africa was the least well represented region in the total sample (10%), marked morphological differences were often localized between the KhoiSan and the other groups. This fact also supported the relative homogeneity within sub-Saharan Africa, which is mainly composed of Bantu-speaking populations.

The ecological differences between groups reflected previous results, as ecology is closely related to geography. However, these differences reflecting mainly the importance of the equatorial climate (in sub-Saharan Africa) appeared to be less marked and more subtle than the geographical ones. This fact could be due to:

i) the morphological variation in relation to both geography and ecology is probably of a clinal nature, where differences between "populations" correspond to a long gradient of very small differences (Hiernaux, 1974); therefore the danger is to search for over-simplified clines, when trying to define a "population"; and
ii) as climate can be a very long-term process for the morphological differentiation of a "population", its effects might not always be clearly visible, especially on a relatively short-time scale.

In **Chapter 5**, other factors also related to geography, such as history, were explored. The questions of the effects of the Bantu-speakers dispersal and other microevolutionary processes occurring during the Holocene period were approached through various sample subdivisions (e.g. regional, ethno-linguistic or archaeological site).

Inter-regional differences were shown to be less marked when considering only Bantu-speaking populations. This fact may support the biological homogeneity within sub-Saharan Africa that might be due in part to the Bantu-speakers dispersal. However, grouping the latter into a single artificial unit remains an abstraction, as the underlying question about diversity is often related to a lower level such as the ethnic affiliation.

In fact, when using more detailed ethno-linguistic data for several Bantu- and non-Bantu-speaking groups (Murdock, 1959); morphological differences were more marked than on a purely geographical level. These results supported that the "classifications" based on ethno-linguistic data are of course closer to a more reliable definition of a "population". They are less misleading in biological terms than the geographical and ecological ones, as the cultural units resulting from varying but rather short-term phenomena are better localized geographically.

With this kind of approach, intra-regional diversity is also better identified, and therefore, the question about the relationship between morphology and "population" history can be addressed in a more accurate manner. For example, the observation of marked inter-ethnic differences between Central and East African Bantu-speakers could also support the "population" differentiation in relation not only to geography, but also to history. In fact, the pattern of the Bantu-speakers expansion was probably shaped on a broad level in (at least) two main streams, involving a high number of "populations" that should be considered as evolutionary units.

		On a broad scale		On a more detailed scale	
Chapter	Aspect of research analysed	Worldwide inter-regionally	Africans inter-regionally	Africans inter-ethnically	Africans intra-regionally
4	Geography (& sex dimorphism)	X	X		
	Ecology (e.g. climate)		X	X	
5	History		X	X	X

Table 39. Summary about the research aspect analysed, level of analysis and nature of sample.

An Iron Age Senegalese sample (e.g. Dioronboumak) illustrated quite well the following problem: when high levels of diversity are observed within a population originating from the same site and dated to the same period, the definition of a "population" is still questioned. In fact, the interpretations were confronted with the simultaneous presence of both cultural homogeneity and biological heterogeneity within a single "population". However, a common material culture does not necessarily exclude a biological heterogeneity.

In summary, the term "population" was used here in a broad sense, adapting its meaning to the questions addressed on different levels of analysis. It does not correspond to a single definition, and it reflects various interrelated factors, especially concerning the artificial notion of the Bantu-speaking populations.

6.2. Methodological aspects

Before discussing morphological trends, a few comments can be made about sampling procedures and methodology selected here. For example:

- nature and size of sample (e.g. heterogeneity of regional groupings);
- statistical approach (e.g. overwhelming effects in factor analyses);
- sexing methods (e.g. lack of standardization); and
- variables related to different parts of the skull (e.g. reliability of information).

6.2.1. Sample size and nature

Concerning the modern sample, geographical grouping on a worldwide scale (up to 10 groups) was achieved in a very broad manner (sections 3.1.1, 3.2., 3.2.4. and Table 4, p. 50, 52, 53, 55), as our main objective was addressing issues in relation to historical processes that occurred in Africa. The aim was to compare African diversity (subdivided into 4 regions) with 4 to 6 non-African groups, in order to evaluate broad differences and/or similarities between geographically distant populations.

For this purpose, a small non-African sample from Howells (1996) was used. The results of the present analysis for population affinities were not significantly different from his final analysis based on a larger "selected" data set.

Although similar geographical differences have been broadly observed before (Howells, 1973, 1989), they were less marked in this study. According to the multiple discriminant analyses (section 4.1.1.2, p. 72), all the rates of total correct classification based on either the vault (48.3-59.1%) or the face (48.6%-60.5%) were much lower than those previously obtained by Howells (1989) (up to 98%).

This was probably due to the heterogeneous nature of the sample under study. Two reasons are possibly responsible for the increased variation within the non-African groups. Firstly, the latter consisted of a small sample of "rejects" (Howells, 1989). Secondly, they were very broadly "classified" into several very artificial geographical areas, as sometimes Australians, Melanesians and Polynesians were pooled together.

Within Africa, the definition of a "population" on a regional level was not always very accurate. Regions were represented by very "heterogeneous" samples, especially in relation to ethno-linguistic background. For example, each artificial geographical area included a high number of ethnic groups. For example: for West Africa, 34; for Central Africa, 28; for East Africa, 10; and for South Africa, 7. Although the broad regional groupings might have underestimated intra-regional morphological diversity, they could also have inflated the inter-regional one. This was probably due to the fact that broad regional subdivisions included various genetically unrelated populations that sometimes adopted the same language in the same region (e.g. in East Africa, Bahutu and Tutsi).

In addition, the following question illustrates more precisely the problem of nature and sample size as a possible bias: why did the NOL variable provide much higher correlations ($r \geq 0.3$) with geographical and ecological variables than variable GOL (Appendix XVII, p. 212)? As the latter is, in metrical terms, very similar to NOL, this question is probably answered in relation to the heterogeneous nature of the sample and the number of groups compared.

In fact, variable NOL was analysed with a sample composed of mainly two groups (East Africans: 33%; Central Africans: 65%). Therefore, it reflected quite well climatic differences between two very different zones (e.g. wet heat, dry heat). In contrast, variable GOL was tested on a sample that represented more or less equally each region of sub-Saharan Africa. Therefore, the correlations might have been lowered, because of the broader geographical scale, reflecting less marked and more complex ecological differences.

Concerning the archaeological sample, interpretations about similarities and/or differences between various African populations dated from the Early, Mid- or Late Holocene had to err on the side of caution, due to the following: (i) sample size variability, (ii) intra-population variation (e.g. sexual dimorphism) and (iii) broad (e.g. inter-regional) level of analysis.

Firstly, the sample size is very variable, as there is a lack of fossil evidence between the Later Stone Age and Iron Age periods in West and Central Africa. Hypotheses about the presence of "pre-" or "proto-" Bantu-speaking populations (Brothwell and Shaw, 1971) were based on only one or two specimens (e.g. Iwo Eleru, Shum Laka).

Secondly, the levels of sexual dimorphism could have changed through time, as was suggested by Frayer (1980) for Europeans. The high variability of facial features observed amongst Late Pleistocene North Africans could be explained by marked sex differences (Chamla, 1978; Crevecoeur and Trinkaus 2004). The results obtained from the inter-regional comparisons through time, could have been slightly biased, when the sexes were not considered separately.

Finally, the regional subdivisions for archaeological populations were probably even more arbitrary than for modern ones. Ecological and climatic parameters that were used to define each modern region probably changed through time. For example, to some extent, the western zone of the savanna belt (e.g. Mali) could have been considered as a separate region.

In fact, according to archaeological data (MacDonald, 1999), the latter has probably been a meeting point for both northern and southern cultural influences during the Later Stone Age and Iron Age periods. This hypothesis is supported not only by paleoclimatological data (Muzzolini, 1993; Jousse, 2006), but also by present and previous craniometric studies that showed morphological affinities between Later Stone Age West Africans and North Africans (Sereno *et al.*, 2008).

6.2.2. Statistical approach

In terms of methodology, both previous and present work used a multivariate approach, using either multiple discriminant or factor analyses. Concerning the latter, the problems encountered were similar to those observed by Howells (1973, 1989).

For example, an "overwhelming" phenomenon noted for certain factor analyses was: geographical differences increased when highly correlated variables were used (analyses XII b & c; XIII b & c; Table 19, p. 91). In fact, the varimax procedure could not always minimize this phenomenon, nor could the preliminary selection of variables based on correlation matrices with lowest r value (<0.5) scores.

6.2.3. Sexing methods

The inter-regional differences observed between males and females may have been biased by methodological aspects, such as: the lack of standardization in the sex determinations, and the differential accuracy in sexing males and females.

As our data set was based on both compiled and personal data, the recording of sex determination was based on a large number of different methods and observers. Males and females could have been misclassified to some extent, affecting therefore the levels of variation within each sex separately. For example, the variation that appeared higher within males than females (Table 16, p. 85) is explained possibly by the fact that males could have been often misclassified as females. Therefore, interpretations about sexual dimorphism and geography have not been explored in detail here.

6.2.4. Cranium versus mandible: what is the best source of biological information?

In comparison to previous studies (Howells, 1989; Froment, 1992a, 1992b, 1998; Franklin *et al.*, 2010), new combinations of variables were tested in relation to not only the cranium but also the mandible (Table 40, p. 141). They allowed us to extend our analyses and interpretations to the whole skull.

Firstly, multiple discriminant analyses showed that mandibular variables were slightly better predictors of inter-regional differences within Africa (54.1%-79%) than cranial ones (33%-72.1%) (Table 18, p. 87).

In order to check whether sample size affected the results, a randomization procedure was carried out. However, as the results were very similar, they confirmed that the results were not biased. The best results were obtained from two variables (war and sh).

However, factor analyses showed that the variation of the mandible provide less significant geographical differences than the cranium (Table 21, p. 95). The modern sample did not cluster very clearly according to broad geographical subdivisions (analyses Xb & XIb - Figure 32, p. 98; Xc & XIc - Figure 33, p. 99). In comparison to the cranium, the position of the populations within the variation obtained from the mandible was similar, although the number of groups compared decreased. West Africans tended to present slightly broader and higher faces, and broader, larger and more robust mandibles than East Africans and/or Central Africans (Table 40, p. 141). These trends were well illustrated by the analyses XIb (Figure 32, p. 98) and XIc (Figure 33, p. 99).

In summary, is the cranium a better tool than the mandible for determining geographical differentiation of populations within Africa? It is premature to answer this question, as the number of variables analysed for the mandible was much smaller than those for the cranium. As the morphology of the face and the mandible are far more complex than the vault, it has been probably underexplored.

6.3. Geographical differentiation of modern Africans

The present section discusses the morphological differentiation of modern Africans in relation to geography (analysed in **Chapter 4**). It was necessary to re-evaluate this aspect, as a much larger sample of African populations was used here in comparison to other studies (e.g. Howells, 1989; Froment, 1992a, 1992b; Lahr, 1996; Franklin *et al.*, 2010). Therefore, past studies were compared with present results, in order to see whether they agree or disagree, when using different samples and/or different craniometric variables.

Despite its heterogeneity, the African sample analysed here was more representative of sub-Saharan African populations than previous studies. It had the advantage of being composed of many more ethnic groups (79 in total). The latter originated mainly from West and Central Africa (48%). In fact, Howells (1989) used only four (ethnic) groups (e.g. Teita, Dogons, Zulu, Bushman) for the whole African sub-continent. In contrast to Howells (1989), Froment (1992a, 1992b, 1998) used a high number of ethnic

Morphological feature (or variable)	Analysis n° (factor n°)	Most distant groups	
		Lowest values ⇐	**⇒ Highest values**
mandibular angle & ascending ramus width (war)	X b (fact. 1)	East Africans Central Africans	Bantu-speaking South Africans West Africans
mandibular breadth (bb)	XI b (fact. 1) XI c (fact. 2)	East Africans	Bantu-speaking South Africans West Africans
chin & face height	XI c (fact. 1)	Central Africans	West Africans

Table 40. Summary of the morphological trends (within Africa) in relation to the mandible.

groups originating from Central Africa. However, Franklin *et al.* (2010) used a high number of known ethnic groups (18 in total), 12 of which are similar to the present study.

6.3.1. On a worldwide scale: confirmation of the results of Howells

Firstly, the geographical differentiation of both African and non-African populations was re-evaluated on a broad inter-continental and/or inter-regional level. In general, amongst modern worldwide populations, both univariate and multivariate analyses showed a series of morphological trends similar to previous studies (Howells, 1989; Froment, 1992a, 1998; Hájek *et al.*, 2008) (Appendix VIII, p. 202).

Our results agreed quite well with Howells (1989) and Froment (1992a) especially when using the maximal cranial breadth (XCB) and nasal breadth (NLB). In fact, these two variables related to both vault and face, often appeared to be best predictors of population differences on a broad worldwide scale.

Despite high variability, sub-Saharan Africans often appeared to be at the extremity of variation with Europeans, Americans and/or Polynesians (Table 41, p. 142).

In evolutionary terms, this fact supported again that non-African groups are derived from African diversity, as suggested by Howells (1989), Lahr (1996), Manica *et al.* (2007) and Betti *et al.* (2009). On a worldwide scale, Africans tended to present the following series of features: a relatively narrow and short vault, a prognathic face, a large interorbital space, low orbits and wide nasal breadth. Most of the morphological traits observed here in sub-Saharan Africa appeared to be more variable (and relatively less gracile) than previously thought (Howells, 1989; Lahr, 1996) and this agrees with Franklin *et al.* (2010). Froment (1992a) and Hájek *et al.* (2008) also identified some West Africans (e.g. Cameroonians) as having the broadest vault. In addition, the East African sample often tended to be located amongst the highest values of the variation (on a worldwide scale). Despite high variation, they have larger and less gracile crania, in contrast to past studies with smaller samples (Lahr, 1996).

Our analyses confirmed that the highest levels of diversity are present within Africans as suggested by Howells (1989) and Relethford (2001b). On an inter-continental level, cranial morphologies vary in relation to a broad geographical "classification", as seen previously (Howells, 1989). On an intra-continental level within sub-Saharan Africa differentiation of populations is even more marked, as already suggested by Froment (1998). In contrast to the observations of Hiernaux (1974), this variation is not continuous (or "clinal"), as significant differences were observed between different regions.

Morphological feature (or variable)	Analysis n° (factor n°)	Most distant groups		Results agree with e.g.
		Lowest values ⇐	⇒ Highest values	
breadth of vault (XCB)	I b & c (2) VIb & XIIIc (2)	Africans various Africans	Europeans West Africans	Howells (1989) Froment (1992a, 1992b) Hájek *et al.* (2008)
length of vault (GOL) length of malar	III b & c (1)	Africans	Europeans & (Polynesians)	-
prognathism (BPL, BNL)	I b & c (1)	Americans or Europeans	Africans	Howells (1989)
interorbital breadth (DKB)	II b & c (1, 2)	Americans or Europeans	Africans	Howells (1989)
orbital height (OBH)	III b & c (2)	Africans	Polynesians or Europeans	Howells (1989)
nasal breadth (NLB)	IV,V b & c (2)	Americans	Africans	Howells (1989) Froment (1992a, 1992b)

Table 41. Summary of the morphological trends (on a worldwide scale) in comparison to previous studies.

6.3.2. Within sub-Saharan Africa: additional information

In comparison to previous studies (Howells, 1989; Froment, 1998), additional information has highlighted sub-Saharan African diversity. A larger sample subdivided into several regions allowed us to specify the affinities of the sub-Saharan Africans with various non-Africans. The use of mandibular variables that had been neglected in multivariate analyses, also showed other morphological trends within Africans.

Concerning most of the morphological trends observed on a worldwide scale, Africans (originating from both western and eastern regions) appeared to be relatively close to Australian-Melanesians (and various Asians). This morphological affinity between geographically distant groups was also observed previously by Howells (1989) and Froment (1998), but it was elaborated here. In particular, West Africans shared with Australian-Melanesians the following features: narrow vault; narrow, high and prognathic face; large malars; wide and low nasal aperture; low orbits; and large interorbital breadth.

Concerning the wide nasal aperture, it could be the result of an adaptive advantage to an equatorial climate. However, this hypothesis of Hiernaux and Froment (1976) cannot be applied to Australian-Melanesians who live in different climatic conditions. Although the origin of the latter could be related very hypothetically to equatorial populations (Lahr, 1996; Froment, 1998), they present other features (e.g. shape of malar, degree of facial protrusion) that are not comparable to Africans and Asians.

Therefore, the analyses showed only an "apparent" morphological affinity between West Africans and Australian-Melanesians, as this similarity could be due to a possible allometric effect of size. The correction for cranial size as well as the study of more detailed facial characters could perhaps reveal further information.

In addition, the presence of robusticity traits related to both supra-orbital ridges and zygomaxillary tuberosity in a few ethnically well-known populations (e.g. Bahutu, Basuku, Ashanti) showed that Africans are more robust than previously thought (Howells, 1989; Lahr, 1996). This observation is supported by Franklin *et al.* (2010). Although these features were not compared to other non-African groups, they suggested that some East Africans (e.g. Bahutu) are also relatively robust, although lesser than Australians.

Within sub-Saharan African populations, morphological differences on an inter-regional level were identified in a more subtle and complex manner than previously by Howells (1989) and Froment (1998), but in a similar way to a newer study by Franklin *et al.* (2010). However, they supported, only partially, morphological similarities which could result from a large-scale dispersal.

For most of the morphological trends (Table 42, p. 143), they showed a variation between extreme groups that often reflected size differences such as: the KhoiSan and/or various groups (in the lowest values); and the Bantu-speaking South Africans

Morphological feature (or variable)	Analysis n° (factor n°)	Most distant groups	
		Lowest values ⇐	⇒ Highest values
breadth of vault (XCB, ZYB)	VIb (fact. 2)	KhoiSan	West Africans Bantu-speaking South Africans
prognathism (BPL, BNL)	VIb & c (fact.1)	KhoiSan	West Africans Bantu-speaking South Africans
height of vault (BBH)	VIIIb (fact. 1)	East Africans	West Africans
orbital breadth (OBB)	VIIIc (fact. 2)	Central Africans	West Africans
palate breadth	XIIb & c (fact. 2)	KhoiSan (Bantu-speaking South Africans)	West Africans

Table 42. Summary of the morphological trends (within Africa) in relation to the cranium.

and/or West Africans (in the highest values) (analyses VIb & VIIIb, Figure 30, p. 96; analysis VIIIc, Figure 31, p. 97). Between these extremes, Central Africans and/or East Africans varied quite a lot, but they tended to have small crania.

Morphological trends in relation to facial prognathism within Africa (analysis VIb) showed similar geographical differences to those observed for vault breadth (Froment, 1992b; Hájek *et al.*, 2008). For example, there are differences:

- for the height of vault, between East Africans and West Africans (analysis VIIIb); and
- for the orbital breadth, between West Africans and Central Africans (analysis VIIIc).

Various parts of the cranium provided slightly different kinds of results about population affinities. For example, although broad regional differences in relation to palate breadth were similar to the vault and face in general, the Bantu-speaking South Africans appeared to be closer to KhoiSan. The latter have in fact much smaller palate than the West Africans. Except for this last observation, the above supports the hypothesis of homogenization due to the Bantu-speakers expansion.

However, inter-regional differences were sometimes difficult to identify because of lack of statistical significance and/or too small sample size. For example, as noted by Froment (1992b), nasal breadth did not show inter-regional differences within Africa (analyses VIIIb & c; factor 2; Figures 30-31, p. 96-97) as marked as those on a worldwide scale (analyses Vb & c; factor 2; Figures 27-28, p. 82-83).

In addition, although the (inter-ethnic) variation of the Pygmies has probably been under-estimated, most of the latter (despite their small sample size) were localized within the variation of the Central Africans. This fact also supports the hypothesis of a common origin for both Pygmies and Bantu-speaking Central Africans.

In summary, on a worldwide scale, the geographical differentiation appeared to be slightly less marked than on an intra-continental scale, as has been noted previously on living populations (Hiernaux, 1974). For example, the multiple discriminant analyses showed that, despite a high variability, the skull is a better predictor of regional differences within Africa (30%-79%) than between various continents (46.3%-60.5%). Factor analyses showed similar results, according to the degree of overlapping variability.

As distance increases, differences increase, as suggested by Betti *et al.* (2009). The present analyses illustrate these facts, as the two extremes of the variation correspond to the geographically most distant groups (e.g. KhoiSan and West Africans).

However, this does not apply to Bantu-speaking South Africans (who are often closest to West Africans). These morphological similarities within sub-Saharan Africa support a founder effect phenomenon in relation to the Bantu-speakers expansion.

Furthermore, according to the results obtained from the various correlations between the geographical coordinates and the regression factor scores (Table 21, p. 95), the variation within sub-Saharan Africa appeared to be directed mainly in a latitudinal (south-north or north-south) direction along the equator. This could reflect the predominant direction of gene flow that occurred within sub-Saharan Africa, especially in relation to the Bantu-speakers expansion. The latter was in fact mainly a southward dispersal despite variation. However, other factors not related to geography and history could also be responsible for this variation.

6.3.3. Geography and sexual dimorphism

On a broad scale, geographical differences appeared to be similar when using both pooled and separated sexes, as has been suggested by Howells (1989) and Lahr (1996). Although the variation due to sexual dimorphism is often significant, the localization of the geographical differences did not change (**Chapter 4**), either on a worldwide scale or within sub-Saharan Africa, when sexes were analysed together or separately.

However, a slightly different geographical pattern was observed for both sexes. Within sub-Saharan Africa, a slightly more marked geographical differentiation of female populations was observed when using both the vault and the face. For example, two analyses illustrated quite well this fact: analysis VIIc (using only females) (Table 19, p. 88) showed significant inter-regional differences, in contrast to analysis VIIb (using only males) (p. 88). Therefore, geographical differences are more visible within females than within males.

According to relatively high correlations between the regression factor scores and the geographical coordinates (Table 21, p. 95), it was also possible to show broadly the main direction of the variation. This was expressed for females, especially in a latitudinal way (north-south or south-north direction), in contrast to males whose regional differences were slightly more visible in a longitudinal (west-east or east-west) direction.

How does one explain all these results? Are inter-regional differences considered separately within both males and females an indicator of the intra-population variation? Although methodological biases could have affected these results (section 6.2., p. 139), one hypothesis could be proposed in relation to social patterns within the groups.

According to analyses of genetic systems characteristic of paternal or maternal inheritance, low levels of genetic polymorphisms within males or females could be an indicator of social features related to the mobility of the sexes (Mitchell, 1996). The genetic data (e.g. Seielstad *et al.,* 1998; Wood *et al.,* 2005) support the well-known anthropological models of exogamy/endogamy for each sex. In patrilineal societies, males are very local, but the few that disperse, tend to move far from their natal group. Therefore, intra-regionally, males will tend to be homogeneous, as genetic similarities increase with low mobility. While at large-scale comparisons, they will tend to be different from others. In contrast, females, who marry neighbouring groups, disperse much more locally, but remain close to their home. So, they will be quite different on an intra-regional scale and more homogeneous on a broader regional one (section 2.3.4.5., p. 42 and Table 1, p. 43).

Is morphological variation also an indicator of social structures or more precisely of differential mobility of sexes due to kinship systems? This question is related to the varying levels of diversity within each sex. But, its explanation in relation to kinship systems is difficult to attempt, as our sample is probably composed of populations with a high diversity of social systems. Although most of the African continent adopted a patrilineal system, numerous groups are also matrilineal in Central and East Africa (e.g. Basuku, N'Komi, Dschagga) as well as in Madagascar, and only a few in West Africa (e.g. Ashanti) and North Africa (e.g. Berbers) (Murdock, 1959; Hiernaux, 1974).

Nevertheless, it could be hypothetically assumed that, the levels of variation observed in both males and females are higher for whichever sex, being the most mobile, spreads its genes the most. Polygamy could favour this fact, and it often occurs in patrilineal societies. In sub-Saharan Africa despite its high cultural diversity, societies are mainly patrilineal, as it is suggested by the sample under study (approximately 74%).

Therefore, although information about polygamy could not be obtained here, the results could be interpreted as follows: the slightly higher inter-regional differences noted within females could mainly reflect the morphological outcome of male dispersal; and, as suggested by geneticists (e.g. Mitchell, 1996; Seielstad *et al.*, 1998), the results could support the high mobility of males on an inter-regional level.

However, once again, these hypotheses remain extremely speculative, as methodological aspects (e.g. lack of standardization in sexing methods) could have affected all the results.

6.4. Geography versus ecology: the influence of equatorial climate on cranial morphology?

Ecology as a major source of population differentiation is discussed in relation to geography. The question of the influence of climate on cranial morphology was raised in **Chapter 4**.

This issue with regard to ecology and population could not be answered simply, as it involves long-term microevolutionary processes that probably interacted with other historical factors (e.g. Bantu-speakers expansion).

In fact, cranial morphology in comparison with various ecological variables could not reveal a direct relationship with selective pressures. Nevertheless, it could suggest that morphological features could be potentially related to ecology (and/or geography).

Although different results have been obtained from different parts of the skull, a few main morphological trends have been observed, and they are summarised in Table 43 (p. 146). Do they result from the impact of equatorial climates which are characterized by extreme humidity and heat (especially in Central Africa)? Or, more precisely, are some of the facial features (especially nasal breadth) more affected by climate than other parts of the skull (Hiernaux and Froment, 1976; Roseman and Weaver, 2004)?

In the first instance, it was shown that most of the ecological variables (e.g. temperature, rainfall, vegetation, humidity) are interrelated to each other as well as to geography (e.g. latitude, longitude). They tend to reflect the importance of the equatorial area in sub-Saharan Africa that presents extreme climatic conditions (Appendix II, p.192-194).

For example, mean annual temperature was highly correlated with longitude, and therefore it reflected a pattern of variation mainly in a west-east (or east-west) direction (Table 23, p. 102).

Furthermore, most of the variables related to climate (e.g. rainfall, temperature) and geography (e.g. latitude, longitude) appeared to be dependent on vegetation and humidity (see Table 24, p. 102).

These preliminary observations underlined the fact that climatic parameters (and their effects on morphology) are difficult to study as a single phenomenon. They are part of the multiple aspects related to ecology and geography that most probably reflect long-term processes.

6.4.1. The observed morphological trends

Although it was not possible to determine directly the influence of selective pressures on cranial morphology, there is evidence that various morphological trends are related to climatic variables on a broad inter-regional level. In order to visualize the patterns in relation to both geography and ecology, it was also attempted to summarise all the results visually (Figure 49, p. 149).

Concerning the vault and face, most of the inter-regional differences follow a west-east direction in relation to increasing temperature. This fact was supported by several morphological features, such as an increase in vault length, a decrease in vault height and a decrease in palate breadth.

According to previous analyses (section 4.2.1., p. 101), temperature is highly correlated to humidity and rainfall, and it also reflects indirectly the marked climatic differences between two main zones: Central Africa (and West Africa) with wet heat and East Africa with dry heat.

Skull morphology		Direction of trend		Revealed by....[2]		Agree &/or disagree with...
features	trend	geography[1]	ecology	individual variables	regression factor scores	
VAULT						
length	↗	W⇐0°⇒E ↗	temperature ↗	NOL	-	
breadth	↗	W⇐0°⇒E ↙	temperature ↗	Q	XIIb (fact.1): EKB,WFB	not supported by Beals (1974)
height	↗	E⇒W	temperature ↗	-	XIIIb (fact.1): BBH	
CRANIAL BASE						
breadth	↙	N⇐0°⇒S ↙	pluviometry ↗	EAM/EAM	-	not supported by Hiernaux (1974)
FACE						
prognathism	↗	S⇒N	temperature ↗	-	VIc (fact.1): BPL, BNL	
nose length	↗	N⇐0°⇒S ↗	(humidity ↙)	NLH	VIIc (fact.2): OBH, NLH	= Froment & Hiernaux (1976)
nasal breadth	↗	W⇐0°⇒E ↗		NLB	VIIIc (fact.2): OBB, NPH, NLB	not entirely supported by Froment & Hiernaux (1976)
palate breadth	↗	E⇒W	temperature ↗	-	XIIb (fact.2): PABR	
skull breadth	↗	(N⇐0°⇒S ↙)	pluviometry ↗	-	XIc (fact.2): WFB, bb	not supported by Beals (1974)
total facial height	↗	N⇐0°⇒S ↗	temperature ↗	-	XIc (fact.1): NPH, sh	

1 Direction of morphological trends in relation to geographical coordinates: east to west (E⇒W), south to north (S⇒N), distance from 0° latitude or equator (N⇐0°⇒S), and distance from 0° longitude (W⇐0°⇒E);
2 Spearman Rank correlation coefficients (Table 23, p. 102).

Table 43. Morphological trends shown in relation to geography and climate.

These observations concerning the shape of the cranial vault were in fact supported by previous work (Hiernaux 1974; Ribot, 1998; Hájek *et al.*, 2008). In Ribot (1998), variation was observed in relation to climatic stress zones with two broad groupings. Firstly, in wet heat, Nigerians and Gabonese with broad and short vaults were grouped together. Secondly, in dry heat, Somalians and Kenyans with narrow and long vaults were grouped together. Tanzanians seemed to be a mixture of both wet and dry heat adaptations, as genetically East Africans are the result of admixture (e.g. Cavalli-Sforza *et al.*, 1994; Tishkoff *et al.*, 2009).

However, as vault (and whole skull) breadth appear to increase with increasing temperature (mainly towards the equator), this fact contradicts the hypothesis of Beals (1974), who supported the opposite trend. In this study, an increase in skull breadth follows a climatic trend towards increasing wet heat.

Concerning only the cranial base, a size reduction in breadth was noticed as one neared the equator. It could reflect a phenomenon of general size reduction of the vault between Central Africans and other groups, but an opposite trend was observed for other parts of the cranium such as the vault breadth. Similarly Hiernaux (1974) demonstrated no general cranial size reduction amongst populations adapting to forest environment. However this disagrees with Franklin *et al.* (2010).

Concerning only the facial features, the inter-regional differences were identified mainly starting from the equator in a south-north or north-south direction. This pattern showed an increasing prognathism, increasing nasal aperture height and decreasing skull breadth.

6.4.2. Do the results agree or disagree with previous literature?

Most of these morphological trends concur with previous work (Hiernaux, 1974; Hiernaux and Froment, 1976; Howells, 1989; Froment, 1993; Roseman and Weaver, 2004). Nasal height and facial height, which are both interrelated, reflect climatic trends in relation to dry heat (increasing in either northern or southern directions). As Hiernaux and Froment (1976) found a negative correlation between nasal height and humidity, it supported the fact that the nose works well as a humidifier in dry conditions, where it increases its surface in length. This general phenomenon also seemed related to the whole cranial height (both cranium and mandible).

However, as nasal breadth tends to increase in a west-east direction, this fact contradicted previous observations showing an opposite trend (Froment and Hiernaux, 1976; Vecchi *et al.*, 1987). The latter found a high correlation between nasal breadth and average rainfall, as well as a cline of the nasal index showing a west-east gradient along the equator.

Is therefore the face particularly "plastic" to environmental stresses, as was suggested by Hiernaux and Froment (1976)? It is difficult to answer this question, as several trends were observed here, not only for the face but also for the whole skull.

Although adaptation to air moisture could be a possible explanation for the observed inter-regional differences, they can also often be questionable when they are analysed on a too small scale. In fact, because of the size and nature of the sample, the differences noted within sub-Saharan Africa might not be as marked as those observed by Hiernaux and Froment (1976) on living populations.

6.5. The biological effects of the Bantu-speakers expansion: re-assessing the microevolutionary models on the basis of the results

After having discussed each potential source of biological diversity such as geography, sex, and ecology, it is now necessary to raise again the question about the biological effects due to the Bantu-speakers expansion. A series of microevolutionary models presented previously (Figure 50, p. 154) are systematically re-evaluated here. Starting from a null-hypothesis where a minimal amount of factors (e.g. founder effect) were acting on diversity, other parameters (e.g. geography, ecology and admixture) were gradually added in the following models.

Firstly, what does the literature review on linguistic, archaeological (see section 2.2., p. 13-27) and biological (see section 2.3.2., p. 32-34) data, tell us about the Bantu-speakers expansion?

The latter appeared to be essentially a linguistic phenomenon (Heine, 1980; Bastin *et al.*, 1983). Bantu languages have a wide distribution that probably originated from West-Central Africa. Due to linguistic affinities observed throughout all of Bantu-speaking sub-Saharan area, this phenomenon could result from a large-scale dispersal. If languages spread simultaneously via a movement of people, this expansion also corresponds to a biological reality. However, if languages spread only by cultural diffusion, there is no migratory event and therefore no gene flow, as supported by Robertson and Bradley (2000).

Archaeological evidence indirectly supports the first option where there is an exchange of both cultural and biological aspects. In addition to language expansion, (semi-) sedentary populations seemed to have spread throughout all sub-Saharan Africa (de Maret, 1977; Phillipson, 1985; Clist, 1991; Vansina, 1995). In fact, an increasing number of settlements providing evidence of a (semi-) sedentary way of life have been discovered. They date to the beginning of this dispersal (*c.* 3,000-1,000 BC) and they also seem to agree with linguistic data. This cultural diffusion appeared however to be a complex process of sedentism associated with different kinds of food productions (e.g. horticulture, agriculture and pastoralism), which probably emerged subsequently to various demographic pressures (Clark, 1980; David, 1980; Van Bakel, 1981).

In addition, a high frequency of some genetic markers was observed throughout sub-Saharan Africa. They support the dispersal of Bantu-speakers and the founder effect phenomenon resulting from the latter (Cavalli *et al.*, 1994; Pereira *et al.*, 2001; Takasaka *et al.*, 2006; Tishkoff *et al.*, 2009; Bryc *et al.*, 2010).

Finally, what has this present work found about the possible biological effects left by the Bantu-speakers expansion?

The results showed closer craniometric affinities of populations that speak Bantu languages even geographically far apart (e.g. South Africa, West Africa) than between neighbouring populations who speak different languages (e.g. KhoiSan, Bantu-speaking South Africans). This supports the hypothesis of a founder effect phenomenon due to the Bantu-speakers dispersal. However, this thesis has also shown much complexity due to the influence of confounding factors (e.g. geography, ecology, and admixture).

Assuming that these key results together with the archaeological, linguistic and genetic data support the historical event of the Bantu-speakers expansion, the following question is raised. How can we interpret the morphological patterns observed in the context of that event?

In attempting to answer this problem, different microevolutionary models (1 to 4) are proposed, as they progressively integrate various factors to the historical event. ***Model 5*** (Figure 50, p. 154), which was related to one hypothesis tested by geneticists (Seielstad *et al.*, 1994), is not discussed here in relation to the Bantu-speakers expansion. In fact, the interpretations of the results in relation to social patterns appeared at present to be too speculative (section 6.3.3., p. 144).

6.5.1. Model 1: a founder effect phenomenon (without any confounding factors) as a too simple phenomenon

The ***model 1*** (Figure 50, p. 154) corresponds to the simplest solution (one sudden and continuous historical event without any confounding factors) showing a founder effect phenomenon, in other words, a morphological homogeneity throughout the whole Bantu-speaking world. The first small populations that originally spread over a large area are responsible for the decrease in diversity that is often observed in the next generations.

Homogenization effects due to a founder effect phenomenon and related to the Bantu-speakers expansion have been observed here within sub-Saharan Africa, through several statistical tests. For example, analyses of variance on individual variables showed that, the inter-regional differences tended to decrease when considering only Bantu-speaking populations and excluding all other linguistic groups (Appendices XII & XVIII, p. 207 & 213). These facts suggest that there is slightly less diversity within only Bantu-speakers.

However, despite this general trend, inter-regional differences related to the vault (e.g. CC, ASB, Q, Q1, FAR, OAR, TOTARCH) and mandible (e.g. phar, ma) were increasing slightly. This suggests therefore that the homogenization processes must have been only partial.

Post hoc comparisons tests also supported indirectly these lower differences within sub-Saharan African Bantu-speakers (section 5.1.2.2, p. 109-111). For example, the most marked morphological differences were often localized between non-Bantu-speakers from South Africa

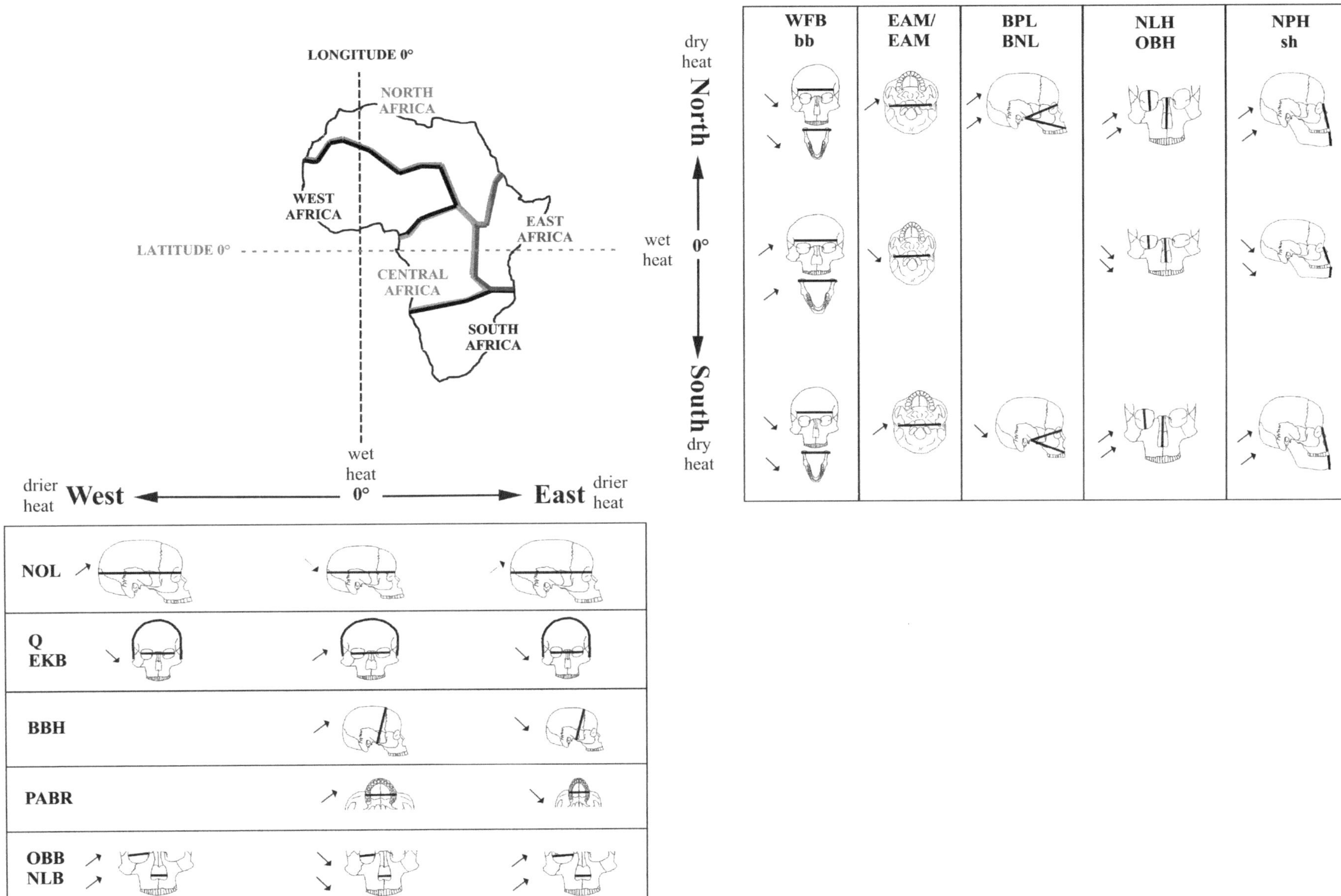

Figure 49. Redefining model 3: morphological trends of the skull according to their geographical and climatic information.

(KhoiSan) and other regions (predominantly composed of Bantu-speaking groups; up to 48%).

These observations support previous work (Hiernaux, 1976a; Froment, 1992a, 1992b) that suggested morphological homogeneity within Bantu-speakers. In particular, Froment (1995; *pers.comm.*) also noted marked differences between KhoiSan (and Afro-Asiatic), and all other linguistic groups (e.g. Niger-Khordofanian, Nilo-Saharan). Although the present work did not test the differences between all these linguistic families, it supported indirectly the results obtained by the previous author. However, Froment (1995, 1998) also showed that the Niger-Khordofanian group composed mainly of Bantu-speakers was difficult to separate from the Nilo-Saharan one.

In addition, did this migration of a small group of "proto-" Bantu-speakers over a large area, produce homogenization of cranial morphologies in the same way as for genetics? In fact, present results broadly agree with genetics and dental anthropology that both observed high levels of homogeneity within Bantu-speaking populations (e.g. Cavalli-Sforza *et al.*, 1994; Irish, 1997; Pereira *et al.*, 2001; Berniell-Lee *et al.*, 2009; Bryc *et al.*, 2010). However, it was not possible to compare the different degrees of homogenization that were observed on genes, teeth and skulls. Although all of them were probably affected by the same founder effect phenomenon, they might reflect other microevolutionary processes that were of variable intensity (e.g. selection).

As the simplest model can only be partially supported, other factors (e.g. geography influencing routes of dispersal) must be added.

6.5.2. Model 2: geography as the key-element in directing gene flow and shaping differences on a broad regional level

Model 2 (Figure 50, p. 154) corresponds to a bow wave pattern with two main streams of dispersal (e.g. West-Central, East). This dichotomous pattern supported by linguistics (Greenberg, 1964; Bastin *et al.*, 1983) and archaeology (Phillipson, 1977b; de Maret, 1989; Vansina, 1995), involves a diversification in relation to geography.

In this model, the geographical background is strongly responsible for the diversification of the populations, as its natural topographical barriers affect population movements and therefore gene flow direction. In addition to this phenomenon, an underlying homogenization is still occurring due to not only a relative "permeability" of the natural barriers but also to the Bantu-speakers dispersal.

Is geography really the key-element in directing gene flow and shaping the differences on a broad regional level? To test this question in **Chapter 5**, a broad regional "classification" was used. Although the data pooled broadly within each region in a rather heterogeneous sample, it reflected approximately the geographical (e.g. continental, inter-regional) differences.

Within Africa, increasing or decreasing geographical proximity broadly seemed to reflect inter-regional differences and/or similarities. For example, most marked differences were noted between the most distant groups (West Africa and KhoiSan), and the least marked ones between the closest groups (East Africa and Central Africa). However, the similarities between distant groups such as West (and Central) Africans and Bantu-speaking South Africans suggested the impact historically of the Bantu-speakers expansion.

Furthermore, ***model 2*** was supported by other results that showed not only homogenization but also regional diversification in relation to the dichotomous pattern of the Bantu-speakers expansion. Although slightly fewer morphological differences were observed for most of the skull, the latter still did not appear as homogeneous (based on analyses of variance), both on inter-regional and inter-ethno-linguistic levels.

For example, Central Africans showed systematically a size reduction of the skull in comparison to Eastern and Southern Bantu-speakers. According to analyses of variance and post hoc multiple comparisons tests (section 5.1.2.2, p. 109-111), differences were localized between all groups, but always present between Eastern Bantu (e.g. Bahutu) and another linguistic group (e.g. West Atlantic-speaking Ashanti). The latter were closer to the Central Bantu-speakers than the Eastern ones, as they were often misclassified with the Basuku. In this case, it appeared that similarities reflected quite well geographical proximity but also probably history.

Therefore, it appears that homogeneity within Bantu-speaking populations is not systematically present, as suggested by other morphological and genetic studies (Hiernaux, 1980; Cavalli-Sforza *et al.*, 1994; Franklin *et al.*, 2010). This regional diversification increased even more, when it was analysed on an inter-ethnic level. In fact, ethnic units were relatively more homogeneous (because of endogamy) than the broad regional groupings, and therefore increased the geographical differences. These facts verified the work of Relethford (2001b), who observed an underestimation of the diversity when analysing populations on a too broad regional level.

For example, multiple discriminant analyses showed that the craniometric variables are better predictors of population differences on an inter-ethnic level (Table 28, p. 111) than on an inter-regional one (Table 18, p. 87). In fact, their rates of total correct classification (Table 44, p. 152) were slightly higher than those previously obtained. Factor analyses performed previously also supported this higher inter-ethnic differentiation (section 5.1.2.2.3, p. 111).

In summary, ***model 2*** could explain better the results obtained for the Bantu-speakers expansion than the previous simplest hypothesis. Traces of founder effects are still visible (as they are less easily lost for macro-scale phenomena than for micro-scale ones), but a regional diversification occurred. The latter is clearly shown on an inter-ethnic level, and it could reflect two main streams for the dispersal.

6.5.3. Model 3: problems in identifying selective pressures

Model 3 (Figure 50, p. 154) is similar to the previous model, but it adds the impact of selection, drift and irregular advancements and retreats of the Bantu-speakers dispersal/expansion. In fact, it is assumed that the biological effects of the expansion are slightly obscured by other factors (e.g. climate, other historical events). These selective pressures (mainly related to equatorial climate) are long-term processes. Therefore, it is improbable that the time scale of the Bantu-speakers expansion is sufficient to really reflect a regional differentiation due only to selection, although a few traits might be under strong selective pressures.

However, the impact of selective pressures remains unknown in terms of time, although according to Hiernaux (1963), craniometric change is of a longer term process (more than a few generations) than postcranial morphology.

As we saw previously, within Bantu-speaking populations, equatorial climate could have had a major impact, as most morphological changes were localized between Central Africa (around 0° latitude) and the other regions (Figure 49, p. 149). Most of the variation in skull morphology reflected differences in size and shape between two main parts, e.g. the vault and the face. In addition, most of the variables related to the mandible were often highly correlated with facial features. However, previous studies (Hiernaux and Froment, 1976; Howells, 1989; Friess *et al.,* 2002; Roseman and Weaver, 2004; Smith, 2009) underlined the fact that facial features are under greater selective pressures than the vault. This was also supported by the fact that, the present morphological trends obtained agreed more often with previous ecological hypotheses for the face than the vault. As suggested by Hiernaux and Froment (1976), the height of the nose (related to total surface area) increased with increasing dryness, and therefore supported the role of the nose as an air humidifier. But, the increasing breadth of the vault within Africa did not seem to follow trends related to decreasing temperature as was previously observed on a worldwide scale by Beals (1974).

However, the main problem encountered in ***model 3*** was the identification of a "causal" relationship with selective pressures. Morphological trends shown in relation to ecology could not provide any evidence of selection, but only suggestions. It is important to be aware of the fact that, the statistical results obtained did not necessarily reflect a direct "causal" association. The Spearman Rank correlation coefficients (especially when r was greater than 0.3) showed a possible relationship between morphological and ecological variables. When the r values were rather low (never above 0.5), it suggested the presence of several confounding factors that might have attenuated the correlations

The present model raised another methodological question: were the selected environmental parameters here sufficiently reliable for a study in relation to ecology? In terms of "reliability", geographical coordinates appeared to be more accurate and meaningful than the categorical variables that subdivided sub-Saharan Africa in broad regional units. In particular, they revealed more information on the direction of morphological trends. Furthermore, most of the geographical and ecological variables appeared to be interrelated to each other. This phenomenon of redundancy is also related to the problem of studying the effects of each environmental parameter in isolation.

Variables entered	Inter-regional level on a worldwide scale	Inter-regional level within sub-Saharan Africa	Inter-ethnic level within sub-Saharan Africa
GOL, XCB, PAC, OCC, BBH, WCB, BNL	48.3% (N=904)	42.6% (N=1389)	77.1% (N=201)
BIMAS.BR, EAM/EAM, FOL, FOB	-	64% (N=347)	68.5% (N=201)
BPL, ZYB, JUB, STB, XFB, FMB, ZMB	53.8% (N=847)	44.5 (N=59%)	64.2% (N=201)
BBH, OBB, GOL, DKB, EKB, ZYB, FMB	-	72.1% (N=401)	65.2% (N=201)
FMB, OBH, Q1, DKB, NLB	-	60.3% (N=453)	66.2% (N=201)

Table 44. Comparison of the total rates of correct classification obtained from the various multiple discriminant analyses performed on various levels of analysis in Chapters 4 and 5.

As suggested by Hiernaux (1974), population differentiation does not reflect only climate. In fact, his biological distance matrix for sub-Saharan Africa showed that many cases had large distances between populations in the same climatic zone, as well as low distances between populations of different climatic zones.

In summary, the results obtained here fit ***model 3***, which however could not separate the impact of the three following factors:

- the *in situ* evolution related to the importance of natural selection (e.g. climate influence);
- gene flow resulting from the dispersal of Bantu-speaking populations; and
- genetic drift (whose impact on variation was however not that high, according to RMET analyses) (section 5.1.3, p. 114-119).

Consequently, is selection a bias when studying historical factors? This question depends on the level of analysis, as selection could affect morphology on long-term and broad geographical scale, rather than "shorter" historical episodes. Nevertheless, within sub-Saharan Africa, traces of the large-scale and long-term migratory event of Bantu-speakers could be observed, although they could have been confused with other factors.

6.5.4. Model 4: synthesis and impact of population admixture

Model 4 (Figure 50, p. 154) summarised the main factors involved in the diversification of sub-Saharan Africans during the Bantu-speakers expansion. Geography, climate and history might have been both long- and short-term processes, and it is not possible to ignore them, as they could be potential influences on morphology. Signatures of the Bantu-speakers dispersal (e.g. founder effect, diversification in two streams) have probably been obscured by other factors (climate, other historical events.

This model also underlines the importance of population admixture, as the Bantu-speakers expansion could not have been a simple process of replacement, but more a progressive assimilation of local hunter-gatherers (e.g. Pygmies, KhoiSan) and pastoralists. Therefore, one would expect to see: morphological diversity within sub-Saharan Africa should reflect decreasing homogeneity, especially from the "proto-" Bantu-speaking source of the dispersal.

In fact, the results obtained here confirmed that population admixture with indigenous populations might have contributed to additional diversification, although to a minor degree in comparison to the large-scale homogenization. For example, although the broad definition of regional groupings probably led us to underestimate population admixture within each region, the categories of both Pygmy and Bantu-speaking Central Africans, were difficult morphologically to distinguish from each other.

This hypothesis is supported by Vansina (1990), who suggested that, in West-Central Africa, "proto-" Bantu-speaking immigrants could have been not very different from the indigenous populations such as the Pygmy, as they probably lived in symbiosis for a long time.

However, the more marked differences observed between the Bantu-speaking South Africans and the KhoiSan suggested that, in the southern areas of the continent, the phenomenon of assimilation involving gene flow was less intense and continuous through time than in the equatorial areas. But, this observation also does not exclude some recent population admixture as a source of diversification in both Bantu- and non-Bantu-speaking South Africans, as the present study did not explore in detail the intra-regional diversity.

In summary, during the Bantu-speakers expansion, agriculturalists admixed to some extent with hunter-gatherers and pastoralists within each region of sub-Saharan Africa. This phenomenon occurred at various degrees and between different and/or similar groups, as is supported by genetic data (e.g. Cavalli-Sforza *et al.*, 1994; Destro-Bisol *et al.*, 2000; Henn *et al.*, 2008; Tishkoff *et al.*, 2009; Bryc *et al.*, 2010).

For example, population admixture between Central Africans and Pygmies who appeared to share many features for cranial morphology, probably produced homogenization and lower levels of diversity. However, it is unknown whether the same phenomenon of homogenization (or an increase in levels of diversity) occurred between groups that initially were more different morphologically, such as the Bantu-speaking South Africans and the KhoiSan. Recent studies (Morris and Ribot, 2006) showed in fact that it was most probably a very variable process within each region of Southern Africa.

Therefore, ***model 4*** as a synthetic hypothesis fits best the results obtained here on cranial morphology to explain the complex biological effects of the Bantu-speakers expansion. Many questions could not be answered relating to confounding factors, as probably, the effects of various microevolutionary processes (e.g. selection, admixture, random processes) were superimposed on those during the whole period of the dispersal. Furthermore, simultaneously to the Bantu-speakers expansion, other numerous historical events occurred on a smaller scale, which could have obscured the founder effect phenomenon.

6.5.5. Additional factors: origins and temporal continuity or differentiation of "proto-" Bantu-speakers?

In addition to previous models, the notion of time is introduced here, in order to see whether the results provided some indication on the origins and temporal continuity (or differentiation) of the "proto-" Bantu-speaking populations.

Firstly, was it still possible to observe an increasing morphological similarity towards the geographical source of the dispersal? This question could not be answered, as the variation of most of the Bantu-speaking sub-Saharan Africans often overlapped. In contrast to Hiernaux (1974) whose study was based on a wide range of biological data, a "clinal" variation with West-Central Africans in the middle of the global diversity was not revealed here.

Furthermore, are there any traces of survivorship of the "proto-" Bantu-speakers dispersal in the morphology of modern Bantu-speaking populations? The "proto-" Bantu-speaking populations may have had a more distinguishable biological and cultural "identity" before their expansion. Afterwards, especially in the later phases of the expansion, the definition of the word "bantu" is broadened to the whole African sub-continent, and it concerned a wide range of populations.

However, in this study using craniometrics, the inter-regional levels of variation were difficult to interpret. It could correspond to modern populations that had admixed even more in the last few centuries. For example, in Cameroon, an increased rate of admixture between groups belonging to various linguistic families (e.g. Sudanic, Afro-Asiatic) has been noted both ethnographically and genetically (Spedini *et al.*, 1999). In sub-Saharan Africa, the populations who retreated southwards because of desertification, contributed to increased levels of diversity by admixing with existing Bantu-speaking populations.

In fact, since "neolithic" times, along the savanna belt, periodical climatic phenomena probably triggered advances and retreats of both pastoralists and agriculturalists. As a result of these population movements on a micro-scale, the original morphological diversity of the "proto-" Bantu-speakers who expanded 5,000 years ago is difficult to trace. However, is this lost diversity only a hypothetical concept? Or, is there any evidence of this phenomenon through the paleoanthropological records?

6.6. Morphological trends through time in Africa

Our main objective here is to discuss the results obtained in **Chapter 5** about past sub-Saharan African Holocene populations. In order to explore the direction and the intensity of gene flow between and within regions before and during the Bantu-speakers expansion, the issues are raised concerning the relationship of Later Stone Age and/or Iron Age human groups with modern populations, on both macro- and micro-scales.

Firstly, geographical differentiation was explored through time on a broad inter-regional level. As geography is a key-aspect in understanding diversity, this study provided a preliminary re-evaluation of the affinities of several archaeological specimens with a larger modern sample.

Secondly, some of the modern groups were historically well documented and therefore, it was possible to address more specific questions in relation to history. Iron Age Senegalese populations that have been recently discovered were the focus of the present work, as their sample size enabled us to carry out several comparisons.

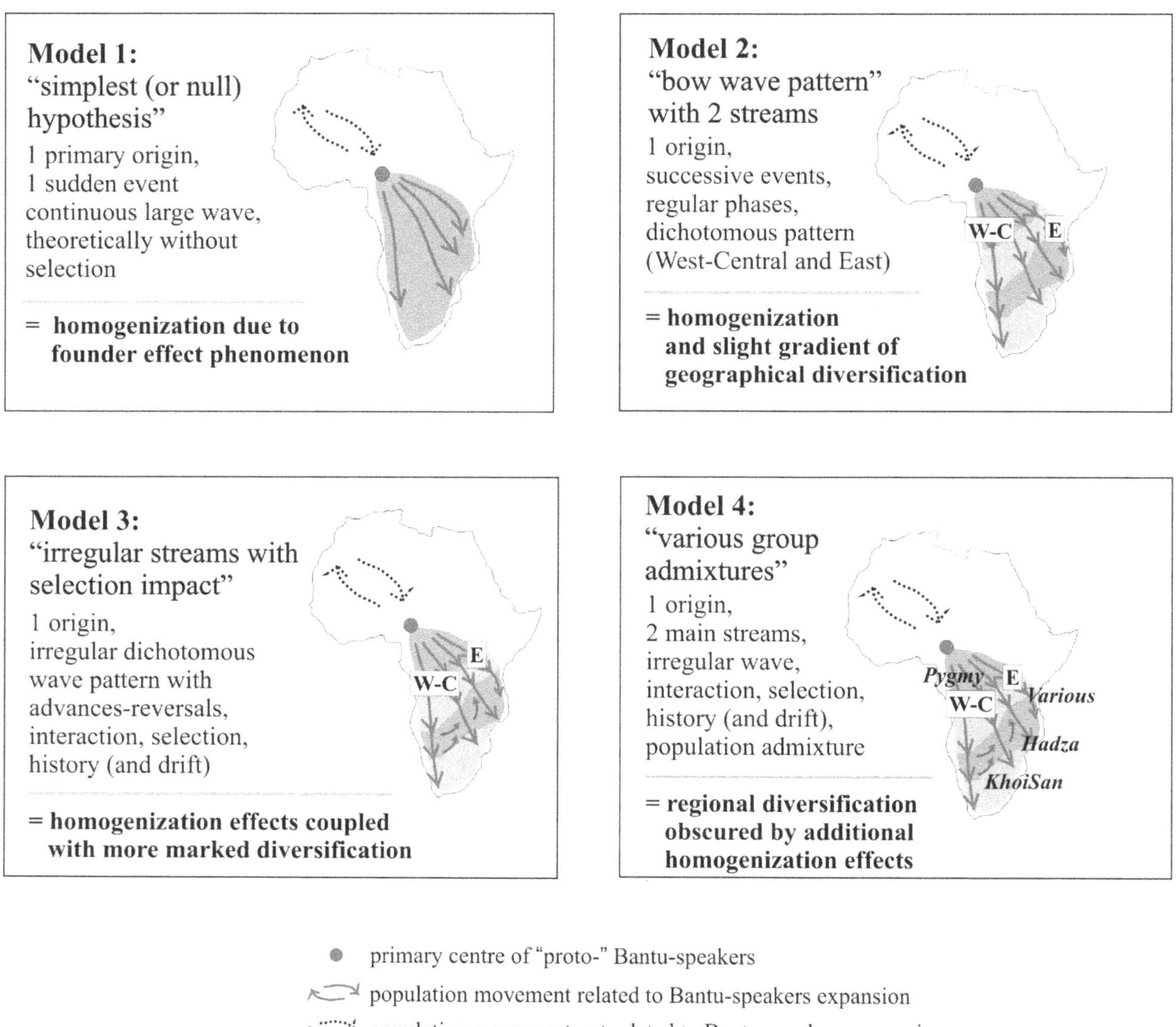

Figure 50. Four microevolutionary models for attempting to explain the biological effects of the Bantu-speakers expansion, and diversity in a wider sense within sub-Saharan Africa.

6.6.1. Increasingly varying geographical differences through time before (and during) the Bantu-speakers expansion

Various archaeological specimens dated to both "pre-" Bantu-speakers (>4,000 BC) and Bantu-speakers expansion (≤4,000 BC) periods, were analysed. The results are summarised in Table 45 (p. 160). They show that many archaeological specimens within each region fall within or outside one standard deviation of African modern variation.

The geographical differences that were present between modern populations on a broad regional level appeared to vary through time. Although, after 4,000 BC, they tend to be similar to present populations, they are increasingly different before 4,000 BC. For example, 53% of the archaeological specimens dated to the "pre-" Bantu-speakers period fall outside one standard deviation of modern variation. In contrast, when they are dated to later periods, only a small percentage of them (23%) were localized outside modern variation.

However, the archaeological sample size of each region was very diverse: "pre-" Bantu-speakers specimens originated mainly from North Africa (81%), and the later ones from West Africa (71%). Therefore, the interpretations about population affinities through time could refer to a few regions only. Nevertheless, when focusing on one area, past West Africans, despite high levels of morphological diversity, appeared to be often the closest to modern populations of their region of origin. This fact suggests a broad morphological continuity through time.

Concerning the Later Stone Age period, the results in general agreed with previous observations. West Africans (e.g. Hassi-el-Abiod) had often the highest values of variation for vault size, prognathism and nasal breadth, (e.g. Dutour, 1992). They are closer morphologically to Late Pleistocene North Africans (e.g. Afalou, Taforalt) and this may support the fact that they are related. Similar morphological affinities have been observed for mid-Holocene specimens in Niger (Senero *et al.*, 2008).

Later Stone Age sub-Saharan Africans originating from Central (e.g. Ishango) and Eastern (e.g. Gamble, Bromhead) Africa appeared to be very different from those originating from other Northern areas. This supports that LSA East Africans, had unique morphological traits (Rightmire, 1975; Protsch, 1978). Although the latter were also localized very often within the highest values of the modern variation (e.g. Bantu-speaking South Africans), they have slightly smaller vault, less prognathic face (especially Bromhead) and narrower mandible than (Later Stone Age-) Iron Age West Africans.

Concerning the Later Stone Age-Iron Age period (3,000-1,000 BC) that could correspond to the early phase of the Bantu-speakers dispersal in West-Central Africa, very few morphological observations could be made. Only one mandible originating from Shum Laka (Cameroon) provided some information about the presence of features that appeared to be very different from modern populations (see factor analysis XIXc, Figure 39, p. 125). However, it was shown previously that other parts of the skull (e.g. palate breadth, mandibular breadth) from the same specimen were very similar to the variation of modern West Africans (Ribot *et al.*, 2001; Pinhasi and Semal, 2000). In fact, the affinities of the archaeological sample were easier to determine when using the cranium as geographical differentiation of the comparative modern sample was more marked than with the mandible.

Most of the Iron Age (after 1,000 BC) samples were very similar to modern West Africans (Table 45, p. 160). However, more than 20% of samples from seven sites fall outside one standard deviation of the modern variation, for example at Attiri (North Africa) and Walalde, Cap des Biches, Dioronboumak and Late Tellem (West Africa). Nevertheless, in contrast to previous Later Stone Age specimens, all these Iron Age individuals are still localized within two to three standard deviations (of the modern variation).

6.6.2. Trends in relation to both robusticity and gracilization

As a series of Later Stone Age and/or Iron Age specimens were compared to modern African populations, they showed a gracilization trend of the skull from about the first millennium BC (Dutour, 1992; Lahr, 1996; Orban *et al.*, 2001). This phenomenon was already observed within Africa since the Late Pleistocene (Grine *et al.*, 2007; Crevecoeur *et al.*, 2009). According to paleoanthropological records (section 2.3.3.2., p. 33-38), it was possible to show marked morphological differences between populations from the Early to the Late Holocene.

Concerning the cranium, breadth and length of the vault increased markedly in Late Pleistocene (e.g. Afalou, Taforalt) specimens, in comparison to more recent samples. However, this morphological trend reflected mainly differences related to size rather than shape.

Concerning the mandible, Later Stone Age (-Iron Age) specimens (e.g. Shum Laka, Ishango, Gamble Cave, Bromhead) often present common features related mainly to robusticity (e.g. very wide ascending ramus, small mandibular angle) (Figure 39, p. 125). For example (Figure 39b), although the Shum Laka individual has a very narrow mandible, it also has a very wide ascending ramus. This observation supported previous comments made by several authors about the presence of common mandibular features amongst Later Stone Age West-Central Africans such as Iwo Eleru (Brothwell and Shaw, 1971), Mbi Crater (Asombang, 1988) and Ishango (Orban *et al.*, 2001). To some extent, despite marked size differences, similar features have also been observed on earlier North African specimens (e.g. Nazlet Khater) (Thoma, 1984; Pinhasi and Semal, 2000; Crevecoeur and Trinkaus, 2004).

However, through time and space, this gracilization trend did not occur systematically as a continuous process, especially amongst some of the Iron Age West Africans (e.g. Cap des Biches, Dioronboumak, Walalde). The Senegalese specimens tended to be, despite variability, in the highest values of the modern variation (e.g. West Africans, Bantu-speaking South Africans) for both vault size and/or various facial features (prognathism degree, nasal width, orbital breadth).

Various authors agree about the trend towards reduced robusticity and increased gracilization (Howells, 1989, 1995; Lahr, 1994, 1996; Lahr and Foley, 1994; Grine *et al.*, 2007). The loss of skull robustness is a worldwide phenomenon that occurred in parallel to several regional diversifications. However, the origin of robust features is still not fully understood: it could reflect for example plesiomorphies inherited from an early modern population, or a biomechanical adaptation to a coarse diet (Armelagos *et al.*, 1984).

Numerous experimental observations in biomechanics showed that both the mandible and other facial features (e.g. interorbital region), are subject to high loadings during mastication (Arbel and Herskovitz, 2000; Daegling and Hylander, 2000; Ross, 2001). Therefore, "environmental" hypotheses have often been proposed to explain the reduction in size of craniofacial features, a phenomenon that occurred especially during the last 12,000 years.

Concerning North East Africa, Armelagos *et al.* (1984) interpreted the general decrease of skeletal robusticity from Early to Late Nubians as a result of nutritional change induced by the advent of agriculture. This question remains controversial, as morphological changes could also be due to gene flow or other random factors (Froment, 2002).

However, is it possible to extrapolate this hypothesis as an explanation about the phenomena observed on a very broad inter-regional level within sub-Saharan Africa? In fact, it is easier to identify a continuous morphological trend on an intra-regional level such as in the case of the Nubians, than on a broad inter-regional one. Therefore, although the results obtained here reflected broadly a reduction in size through time, they did not provide in detail for each region the changes related to skull shape.

Furthermore, in contrast to the Nubians (Carlson and Van Gerven, 1977; Van Gerven, 1982; Armelagos *et al.*, 1984), a decrease of the cranial length, an increase in cranial height and a less protruding face, are not systematically observed within West Africans. In fact, the "stresses" responsible for ontogenic changes, leading to a morphological reduction, were probably of different types and intensities from one region to another. Furthermore gene flow could also have been responsible for these changes too.

The consequences of both changes in food production and increasing sedentism (e.g. softer diet, nutritional deficiencies and infectious diseases) might have been more diversified and less intense in West and Central Africa than in North East Africa. This hypothesis could be supported by the following facts: a greater diversity of environments and ways of subsistence, and a possible persistence of hunting-gathering in some areas during the Iron Age period (e.g. MacDonald, 1999). However, as there is less skeletal evidence in equatorial areas than elsewhere, the morphological (and pathological) consequences of these phenomena are more difficult to verify, especially in Central Africa.

In summary, as populations were diverse, it is therefore difficult to identify a single morphological trend for the whole African continent. Although a decrease in cranial capacity or "microcranialization" has been observed since the early Holocene on a continental level (Henneberg and Steyn, 1993), intra-regional diversity still needs to be evaluated in detail.

However, from the Later Stone Age to the Iron Age period, morphological changes in relation to size and robusticity are more marked when using only mandibular variables: this fact could reflect changes in relation to subsistence strategies, because the mandibular morphology may be more biomechanically affected by external "stresses" (e.g. diet) than the cranium. In comparison, the mandible reflected less geographical differentiation (within modern populations), and this observation could also support indirectly the "biomechanical" hypothesis. However, the importance of gene flow as a source of variation cannot be excluded.

6.6.3. Focusing on Iron Age West Africa

The question was raised whether, on a more detailed regional level, there is continuity and/or differentiation between various past and modern West Africans. This is also related to the question of biological contribution of sub-Saharan and/or North Africans to West Africans. The high variability of an Iron Age Senegalese population (e.g. Dioronboumak) is finally discussed, taking into account both previous results obtained on a broad regional level and archaeological hypotheses about coastal settlements.

6.6.3.1. "Apparent" morphological continuity up to present times?

According to various analyses (section 5.2.1.2., p. 121-133), it was observed that there is "apparently", on a broad inter-regional level, morphological continuity between Iron Age and modern West Africans. This fact was also observed on an inter-ethnic level, as the latter reflected more or less the same geographical differences.

More precisely, the Iron Age Senegalese, despite high variability (e.g. Walalde, Dioronboumak, Faboura), often overlapped the variation of a Ghanaian ethnic group (Ashanti). The latter tend to present the following morphological features: wide nose, wide orbit, tall and prognathic face, and broad mandible.

However, the Late Iron Age Malian populations (e.g. Early and Late Tellem) are closer to modern Central Africans, especially in terms of the face that is relatively smaller and less prognathic (Figure 37, p. 123), as espoused before (Huizinga *et al.*, 1967, 1979).

In addition, it was noted that the early Iron Age individual from Walalde was located within the variation of the largest and most prognathic skulls. However, in comparison with another Senegalese specimen (Cayar) dated to the same time frame, it was different in both size and shape. Within the same region, slight differences could have been present between contemporaneous populations inhabiting either inland (e.g. Walalde) or coastal (e.g. Cayar) areas. Therefore, despite the presence of a broad morphological continuity in West Africa, additional analyses revealed in more detail the affinities between various regions and ethnic groups.

6.6.3.2. Biological input originating mainly from sub-Saharan area?

For West Africa, the question of the possible presence of various biological influences through time was already suggested by Dutour (1992), Huizinga *et al.* (1967, 1979) and Sereno *et al.*, (2008). However, these authors focused only on Malian populations of different periods, and they did not extend their approach to neighbouring countries. Therefore, by adding Iron Age Senegalese samples, it was an opportunity to explore whether both sub-Saharan and North Africans contributed biologically to a diversification of West African populations.

The analysis focused on two Iron Age Senegalese coastal populations, Dioronboumak and Faboura dated to the same period (1st-8th century AD) and located in the same area (Figure 21, p. 59). Initially, it was suggested that these coastal settlements were at the crossroads of various influences, although they belonged to a similar cultural backgrounds (Descamps and Thilmans, 1997). In fact, multiple discriminant analyses showed that Dioronboumak had marked affinities with Central Africans (e.g. Basuku), as they were often misclassified with the latter (section 5.2.2., p. 133).

This observation also supported by Huizinga *et al.* (1967, 1979) who observed an important biological input originating from sub-Saharan Africa in Late Iron Age West Africa (e.g. Mali). For example, concerning vault breadth and nasal width, the authors noted morphological affinities between Tellem populations and Central Africans (e.g.

Batetela). In fact, according to ethno-historical data, the Tellem might be associated with the southward and eastward population movements caused by the decline of the Ghana Empire. Since the 12th century AD, Mali became more important and Tellem populations possibly took refuge in less accessible places such as in the high Bandiagara cliffs of the Middle Niger Valley.

Although there is no ethno-historical information available for the earlier coastal populations of Senegal, it is assumed that, in West Africa, key-areas that were located between desert and forest probably attracted many populations through time. Most of the Iron Age groups originated from sub-Saharan Africa, as, the desert expanded and became increasingly a natural geographical barrier between both northern and southern parts of the continent.

As a consequence, was the biological impact from North Africa greater during the Later Stone Age than during later periods which were characterized by increasing desertification?

Previous work showing morphological affinities between Later Stone Age Malian specimens (e.g. Asselar) and Late Pleistocene North Africans (e.g. Afalou) supported this hypothesis (Dutour, 1992). Present results also noted a series of mandibular morphological features (e.g. wide mandibular ascending ramus) that are frequent amongst many Later Stone Age (-Iron Age) Africans (e.g. Gamble Cave, Shum Laka). However, it is unknown whether these features were the result of population admixture and/or adaptation to similar selective pressures.

6.6.3.3. Affinities of Senegalese reflecting various influences: the case of the Dioronboumak population

Even when considering sexes separately, Dioronboumak individuals are still variable in relation to facial features, especially amongst males. This fact is also well illustrated by factor analyses XXIIb (Figure 44, p. 130) and XXIIc (Figure 45, p. 130). In order to explain this morphological heterogeneity within the Dioronboumak sample, two hypotheses could be proposed in relation to history and some social features.

Firstly, as seen in **Chapters 4** and **5**, morphological affinities often reflect geographical proximity and sometimes indirectly historical aspects. Therefore, does the morphological heterogeneity of Dioronboumak population reflect a high diversity of influences?

In fact, this heterogeneity is in contradiction with the cultural homogeneity noted by Descamps and Thilmans (1997, 2001) at the site of Dioronboumak. However, cultural homogeneity does not exclude the hypothesis of the arrival of populations that are biologically different. Between the 4th and 8th centuries AD, Senegalese coastal settlements could have been exposed to various biological influences despite a cultural continuity during five centuries. They were located at the crossroads of various influential West African empires that originated below or along the savanna belt (e.g. Nok culture, Ghana Empire).

Secondly, as the heterogeneity of the Dioronboumak population is mainly due to males, another explanation could be proposed here in relation to social features. As seen above (section 6.3.3., p. 144), marriage patterns could affect the levels of diversity within both males and females. Is therefore exogamy responsible for this morphological heterogeneity? In addition, why should males more affected than females at Dioronboumak?

These facts could simply be the result of methodological biases (e.g. sample size), but it is questioned here whether, the kinship system practised by a society could also have had an influence on morphology. Although it is not possible to extrapolate ethno-historical data of present populations on archaeological ones, it is interesting to know that probably all Senegambians were initially matrilineal (Arcin, 1907; Murdock, 1959).

Therefore, other questions could be raised: along the Senegalese coast, were the Iron Age societies matrilineal, and did they encourage exogamy or endogamy within the "dominant" sex?

6.6.3.4. Traces of the Bantu-speakers expansion through the paleoanthropological record?

When analysing only the cranium, a broad morphological continuity was observed on an inter-regional level through the Holocene. In West Africa, it suggested the impact of gene flow originating increasingly from sub-Saharan Africa during the Iron Age, in contrast to the Later Stone Age. This fact also supported indirectly that, demographic pressures encouraged population expansions along the equator not only southwards (Bantu-speakers expansion) but also northwards.

When analysing only the mandible, differences within modern populations were not always as significant as those observed for the crania. But trends through time suggested the possible impact of selection: the loss of robusticity and a decrease in size of certain mandibular features might be due to several biomechanical stresses that occurred during periods of subsistence shifts. However other potential factors (e.g. gene flow, genetic drift) cannot be excluded as well. This phenomenon was however supported by only a very small number of West(-Central) African specimens (e.g. Shum Laka) dated to the transitional phase of the Later Stone Age-Iron Age period (*c.* 1,000 BC).

In summary, so far, it is difficult to find any direct evidence for the "proto-" Bantu-speakers expansion through the paleoanthropological record. Nevertheless, as there is a broad inter-regional morphological continuity through time between West and Central Africans, the origins of the "proto-" Bantu-speakers were probably localized on a broad scale (e.g. region) rather than on a smaller one (e.g. ethnic group).

6.7. Conclusions: suggestions of some microevolutionary mechanisms

As the present craniometric approach was "multi-layered", it reviewed the main factors considered as potential sources of morphological diversity (e.g. geography, sex, ecology and history), both on macro- and micro-scales, within sub-Saharan Africa. The question of the definition of a "population" in geographical and biological terms was often raised during the analyses and interpretations of the results. It also appeared that the ethnic unit often provided more "reliable" information (in relation to the nature of the sample) than the broad regional groupings, as was seen on an intra-population level before (Hiernaux, 1974).

However, the main question remains: What is the best model to explain the Bantu-speakers expansion and therefore the diversity of skull morphologies in sub-Saharan Africa? Or, in other words, were the proposed microevolutionary models of any use for the interpretation of the results obtained about modern diversity? In fact, they helped us to direct the main lines of the research, by analysing each factor in relative isolation.

Model 1 was a kind of null-hypothesis that anyway could not be verified. Most probably, the effects of several confounding factors were superimposed in addition to the Bantu-speakers expansion. But, the latter appeared to illustrate the founder effect phenomenon that left some signatures on both phenetic and genetic levels (Hiernaux, 1974; Pereira *et al.*, 2001; Manica *et al.*, 2007; Bryc *et al.*, 2010).

Models 2 and 3 were more realistic and complex hypotheses were proposed by adding successively the effects of geography (e.g. routes of dispersal), selection (e.g. climate) and other micro-scale historical events (e.g. population admixture).

Traces of two main streams dispersal might have been observed in relation to the Bantu-speakers expansion, although they could be due to other confounding factors (e.g. geography, ecology, micro-scale dispersals). Throughout the interpretations, the effects of climate on skull morphology could not be excluded. However, the relationship between morphology and climate appeared to be not strictly causal.

Model 4 corresponded to the most synthetic hypothesis including the highest amount of confounding factors. However, it could not be explored fully especially in relation to population admixture, which has probably been very diverse within each region of sub- Saharan Africa.

Model 5 does not correlate with the Bantu-speakers expansion, as the interpretations remained too speculative. The study of the impact of, for example, kinship systems (e.g. patrilineal, matrilineal) and marriage patterns (e.g. endogamy, exogamy) on skull morphology is also another crucial research perspective to develop. In fact, it was difficult to interpret the levels of variation within both males and females in relation to those patterns (e.g. differential mobility of sexes). Furthermore morphological analyses on extreme cases of marriage patterns, could help to define more accurately, whether the social features can have an influence on the diversity of a population substructure on a short- or long-term.

In summary, the present work supported craniometrically the following observations: a phenomenon such as the Bantu-speakers expansion in sub-Saharan Africa probably left traces on a large-scale and long-term. Although the earliest phases of the "proto-" Bantu-speakers diversification could have been lost through later

Archaeological populations	In comparison to modern variation[1]		Multivariate analysis comments: closest to....
	within 1 SD	outside 1 SD	
"Pre-" Bantu-speakers expansion period: Later Stone Age period (before 4,000 BC)			
North Africa			
Afalou	15	22	Bantu-speaking South Africans
Taforalt	8	9	
total	**23**	**31**	
West Africa			
Hassi-el-Abiod	0	1	
total	**0**	**1**	
East Africa			
Gamble	6	0	various groups
Bromhead	1	2	various groups
total	**7**	**2**	
Central Africa			
Ishango	1	0	various groups
total	**1**	**0**	
All pooled	**30**	**34**	various groups
During and/or after the Bantu-speakers expansion: Later Stone Age and/or Iron Age periods (after 4,000 BC)			
North Africa			
Attiri	33	10	East Africans
Duweishat	30	3	Bantu-speaking South Africans
total	**66**	**13**	
West Africa			
Shum Laka	0	1	
Cayar	1	0	East Africans
Walalde	1	1	Bantu-speaking South Africans
Cap des Biches	6	4	Bantu-speaking South Africans
Faboura	8	1	Bantu-speaking South Africans
Dioronboumak	16	4	
Early Tellem	26	0	
Late Tellem	86	21	
Kainji	3	0	West & East Africans
total	**147**	**32**	
East Africa			
Ngorongoro Crater	5	0	Bantu-speaking South Africans
total	**5**	**0**	
Central Africa			
Koro Toro	1	0	Central Africans
Sanga	1	0	Central Africans
total	**2**	**0**	
All pooled	**220**	**45**	various groups

[1] Observations corresponding to a maximum number of individuals falling within or outside one standard deviation of the pan-African modern variation on an inter-regional level.

Table 45. Number of archaeological specimens (by region) falling within or outside the modern pan-African variation.

developments, a general process of homogenization increased the similarities between all regions of sub-Saharan Africa. Slight differences in west-central and eastern regions could be the result of the nature of the dispersal itself (which occurred in two main streams).

Morphological diversity in sub-Saharan Africa reflects complex processes in which selection, population replacement and assimilation of foraging groups by early agriculturalists all took place. However, according to this and past work (Hiernaux, 1974; Froment, 1998), the concept of a single "bantu" identity which could have been supported by a series of common morphological features in addition to genetic data, remains poorly defined. Furthermore, there is no "bantu" entity in "cultural" or archaeological terms, but in Linguistics, the concept "bantu" exists.

In fact, Bantu-speakers did not appear systematically homogeneous in morphology, as it should be expected if they represented the sole descendants of a small group recently expanded. In addition, it was not possible to determine craniometrically a founder effect phenomenon that should have shown an increasing similarity towards the source of the expansion, as the "proto-" Bantu-speaking source was probably obscured by later admixtures. This agrees in particular with Franklin *et al.* (2010).

Nevertheless, for a better understanding of microevolutionary processes, the effects of population admixture have to be explored in more detail, taking into account the initial variation in the groups under study. They should help to identify traces of homogenization left by not only the Bantu-speakers dispersal but also other micro-scale historical events of a shorter time span.

Furthermore, is there any preferential model for the differentiation of West and Central African populations through time? In fact, throughout the Holocene, two complementary evolutionary mechanisms co-existed; e.g. geographical isolation (or *in situ* evolution) and gene flow (or migratory events). Paleoclimatological data can provide evidence about periods during which population expansions were possible (Muzzolini, 1993; Jousse 2006).

The inter-regional differences that were observed here within Africa have been supported by craniometric (Froment, 1998; Franklin *et al.*, 2010), genetic (Biondi *et al.*, 1996; Bryc *et al.*, 2010) and both craniometric and genetic (Hájek *et al.*, 2008) studies that showed a marked geographical differentiation in modern Africans.

In addition to data obtained from modern samples, these geographical differences were also broadly observable until the Iron Age with an important biological input originating from sub-Saharan Africans in the west. For example, the Senegalese dated to this period fitted well within the variation of modern sub-Saharan Africans originating especially from both central and western regions. However, this morphological continuity was also apparent throughout time, as when looking in detail at the intra-regional variation, the latter appeared to be higher especially for the site of Dioronboumak in Senegal.

Therefore, it was not possible to exclude some northern influences, as has been suggested by genetics (Dios *et al.*, 2001). Further comparisons with well-documented modern samples, originating from North Africa could help to clarify the high levels of variation in some of these Iron Age Senegalese coastal populations.

In conclusion, the present approach was mainly based on a broad inter-regional level for identifying large-scale phenomena such as the Bantu-speakers expansion, but not ignoring an intra-regional approach. Further intra-regional studies could help to decrease the possible influence of methodological biases, especially in relation to the nature of the sample. They might provide more detailed results about the craniometric variation within each region of sub-Saharan Africa, the historical context of which is extremely complex.

In particular, this study needs further crania from West and Central Africans of known "ethnic" origin. Further ethnographical work could be done about the origin of various West-Central Africans (e.g. Cameroonians) who show the highest levels of cultural and linguistic diversity.

In order to explore the diversity of temporal trends, intra-regional studies of archaeological populations are also needed. In the last few thousand years, the magnitude of cultural processes affecting population subsistence and language in relation to the Bantu-speakers expansion has been extremely large throughout sub-Saharan Africa. Their effects on the biological relationships between recent and past regional populations remain unclear, especially on a micro-scale. However, preliminary results here show past population diversity especially for the (Later Stone Age-) Iron Age periods in West Africa.

As skeletal morphology probably reflects both historical and environmental parameters, hypotheses in relation to the latter should also be explored further. For example, the question was raised, whether changes in both shape and size of mandible can be the result of biomechanical stresses induced by subsistence shifts. Therefore, it was not possible to fully appreciate the apparent "outlier" position of some Later Stone Age-Iron Age specimens (Shum Laka) that could belong to "proto-" Bantu-speaking populations.

In addition, analyses using three-dimensional methods continue to provide more detailed information on morphological patterns (e.g. Hennessy and Stringer, 2002; Franklin *et al.*, 2010), although similar to two-dimensional methods (Franklin *et al.*, 2005). This type of research still needs to be expanded on both past and present sub-Saharan African populations, especially in West and Central Africa. Therefore, current research is focusing not only on expanding the data set and methods, but also on comparing even more closely African craniometric variation to recent genetic data (Tishkoff *et al.*, 2009).

LITERATURE CITED

Adiwas Kouerey, G., Ekarga Mba, E., Mombo, J.-B., Mouleingui Boukoussou, V., Mvele Ndango'o, P.-C., Simon, I. and Walter, P. (1989) Unité et diversité du monde bantu. *In:* T. Obenga (ed.) *Actes du Colloque international: les peuples Bantu, migration, expansion et identité culturelle, Vol. I,* Libreville, C.I.CI.B.A., April 1985. Paris, L'Harmattan. p. 167-186.

Alexander, J. (2001) Islam, archaeology and slavery in Africa. *World Archaeology,* 33, 44-72.

Andah, B. and Bagodo, O. (1993) Research and theory in Archaeology since the 1960s: an assessment of the African especially West African scene. *West African Journal of Archaeology,* 22, 1-23.

Anderson, J.E. (1968) Late paleolithic skeletal remains from Nubia. *In:* F. Wendorf (ed.) *The Prehistory of Nubia.* Dallas, Southern Methodist University Press. p. 996-1040.

Anderson, S., Bankier, A., Barrell, B., de Bruijn, M., Coulson, A., Drouin, J., Eperon, I., Nierlich, D., Roe, B. and Young, I. (1981) Sequence and organization of the human mitochondrial genome. *Nature,* 290, 457-465.

Andrea, A.C. and Casey, J. (2002) Pearl millet and Kintampo subsistence. *African Archaeological Review,* 19(3), 147-173.

Arbel, G. and Herskovitz, R.I. (2000) Strain distribution on the skull due to occlusal loadings: an anthropological perspective. *Homo,* 51(1), 30-55.

Arcin, A. (1907) *La Guinée Française*. Paris: A. Challamel. 752p.

Armelagos, G.J., Van Gerven, D.P., Martin, D.L. and Huss-Ashmore, R. (1984) Effects of nutritional change on the skeletal biology of Northeast African (Sudanese Nubian) populations. *In:* J.D. Clark and S.A. Brandt (eds) *The Causes and Consequences of Food Production in Africa.* California, University of California Press. p. 132-146.

Asombang, R.N. (1988) *Bamenda in Prehistory. The Evidence from Fiye Nkwi, Mbi Crater and Shum Laka Rockshelters.* Unpublished Ph.D. dissertation, University College of London. 496p.

Auger, F. (1967) *Analyse de la variation morphologique de quelques populations du Sud-Cameroun.* Unpublished Ph.D. dissertation, Faculté des Sciences de Paris. 145p.

Ballouche, A. and Neumann, K. (1995) A new contribution to the Holocene vegetation history of the West African Sahel: pollen from Oursi, Burkina Faso and charcoal from three sites in northeast Nigeria. *Vegetation History and Archaeobotany,* 4, 31-39.

Bandelt, H.-J., Forster, P., Sykes, B.C. and Richards, M.B. (1995) Mitochondrial portraits of human populations using median networks. *Genetics,* 141, 743-753.

Barbujani, G., Sokal, R. and Oden, N.L. (1995) Indo-European origins: a computer-simulation test of five hypotheses. *American Journal of Physical Anthropology,* 96, 109-132.

Bastin, Y. (1980) Statistique grammaticale et innovations en Bantou. *In:* L.H. Hyman and J. Voorhoeve (eds) *L'expansion bantoue*. Viviers, 4-16th April 1977. Paris, SELAF. p. 387-400.

Bastin, Y., Coupez, A. and De Halleux, B. (1979) Statistiques lexicale et grammaticale pour la classification historique des langues bantoues. *Bulletin des Séances de l'Académie Royale des Sciences d'Outremer*, 3, 375-387.

Bastin, Y., Coupez, A. and De Halleux, B. (1983) Classification lexicostatistique des langues bantoues (214 relevés). *Bulletin des Séances de l'Académie Royale des Sciences d'Outremer,* 27(2), 173-199.

Beals, K.L. (1974) Head form and climatic stress. *American Journal of Physical Anthropology,* 37, 85-92.

Beleza, S., Gusmaõ, L., Amorim, A., Carracedo, A., Salas, A. (2005) The genetic legacy of western Bantu migrations. *Human Genetics*, 117, 366–375.

Benington, R.C. (1912) A study of the negro skull with special reference to the Congo and Gaboon crania. *Biometrika*, VIII, 292-339.

Berniell-Lee, B., Calafell, F., Bosch, E., Heyer, E., Sica, L., Mouguiama-Daouda, P., van der Veen, L., Hombert, J.-M., Quintana-Murci, L. and Comas, D. (2009) Genetic and demographic implications of the Bantu Expansion: insights from human paternal lineages. *Molecular Biology and Evolution*, 26(7), 1581-1589.

Betti, L., Balloux, F., Amos, W., Hanihara, T. and Manica, A. (2009) Distance from Africa, not climate, explains within-population phenotypic diversity in humans. *Proceedings of the Royal Society of Biology*, 276, 809–814.

Biondi, G., Rickards, O., Martinez-Labarga, C., Taraborelli, T., Ciminelli, B. and Gruppioni, G. (1996) Biodemography and genetics of the Berba of Benin. *American Journal of Physical Anthropology*, 99, 519-535.

Blench, R. (1993) Recent developments in African language classification and their implications for prehistory. *In:* T. Shaw, P. Sinclair, B. Andah and A. Okpoke (eds) *The Archaeology of Africa*. London, Routledge. p. 126-138.

Blench, R. (1997) Language studies in Africa. *In:* J.O. Vogel (ed.) *Encyclopedia of Precolonial Africa*. London, Altamira Press. p. 90-100.

Blench, R. (1999) The languages of Africa: macrophyla proposals and implications for archaeological implications. *In:* R. Blench and M. Spriggs (eds) *Archaeology and Language (vol.) IV: language change and cultural transformation.* London, Routledge. p. 29-47.

Boaz, N.T., Pavlakis, P.P. and Brooks, A. (1990) Late Pleistocene-Holocene human remains from the Upper Semliki, Zaire. *Virginia Museum of Natural History Memoir*, 1, 273-299.

Bocoum, H., McIntosh, S.K. and McIntosh, R.J. (1999) L'âge du fer dans la moyenne vallée du fleuve Sénégal. *Notes Africaines*, 196, 9-18.

Boule, M. and Vallois, H.V. (1932) *L'homme fossile d'Asselar*. Archives de l'Institut de Paléontologie Humaine, mémoire n°9. Paris, Masson. 90p.

Bower, J. (1995) Early food production in Africa. *Evolutionary Anthropology*, 4 (4), 130-139.

Brabant, H. (1963) *Observations anthropologiques et odontologiques sur les dents des Hutu du Rwanda.* Annales du Musée Royal de l'Afrique Centrale, Tervuren, series in-8° (Human Sciences), 47. 30p.

Brauer, G. (1980) Human skeletal remains from Mumba rock shelter, Northern Tanzania. *American Journal of Physical Anthropology*, 52, 71-84.

Brauer, G. and Rimbach, K.W. (1990) Late archaic and modern *Homo sapiens* from Europe, Africa, and Southwest Asia: craniometric comparisons and phylogenetic implications. *Journal of Human Evolution,* 19, 789-807.

Brooks, A. and Smith, C. (1987) Ishango revisited: new age determinations and cultural interpretations. *African Archaeological Review,* 5, 65-78.

Brooks, A.N. (2006) Cultural responses to aridity in the Middle Holocene and increased social complexity. *Holocene*, 151, 29-49.

Brothwell, D.R. (1963) Evidence of early population change in Central and Southern Africa: doubts and problems. *Man,* 63(132), 101-104.

Brothwell, D.R. and Shaw, T. (1971) A late Upper-Pleistocene proto-West African negro from Nigeria. *Man,* new series, 6(2), 221-227.

Bruner, E. and Manzi, G. (2004) Variability in facial size and shape among North and East African human populations. *Italian Journal of Zoology*, 71(1), 51-56.

Bryc, K., Auton, A., Nelson, M.R., Oksenberg, J.R., Hauser, S.L., Williams, S., Froment, A., Bodo, J.-M., Wambebe, C., Tishkoff, S.A., and Bustamante, C.D. (2010) Genome-wide patterns of population structure and admixture in West Africans and African Americans. *Proceedings of the National Academy of Sciences, USA,* 107(2), 786–791.

Burton, M.L., Moore, C.C., Whiting, J.W.M. and Romney, A.K. (1996) Regions based on social structure. *Current Anthropology,* 37, 87-123.

Cahen, D. (1978) Vers une révision de la nomenclature des industries préhistoriques de l'Afrique Centrale. *L'Anthropologie,* 82(1), 5-36.

Caldararo, N. and Guthrie, M. (1998) Mitochondrial DNA, the Y chromosome and the origins of modern humans. *Homo,* 49(2), 225-240.

Campbell, M.C. and Tishkoff, S.A. (2008) African genetic diversity: implications for human demographic history, modern human origins, and complex disease mapping. *Annual Review of Genomics and Human Genetics*, 9, 403–433.

Campbell, M.C. and Tishkoff, S.A. (2010) The evolution of human genetic and phenotypic variation in Africa (review). *Current Biology*, 20, 166-173.

Cann, R.L., Stoneking, M. and Wilson, A.C. (1987) Mitochondrial DNA and human evolution. *Nature,* 325, 31-36.

Carlson, D.S. and Van Gerven, D.P. (1977) Masticatory function and post-Pleistocene evolution in Nubia. *American Journal of Physical Anthropology,* 46, 495-506.

Castri, L., Tofanelli, S, Garagnani, P., Bini, C., Fosella, X., Pelotti, S., Paoli, G., Pettener, D. and Luiselli, D. (2009) mtDNA variability in two Bantu-Speaking populations (Shona and Hutu) from Eastern Africa: implications for peopling and migration patterns in Sub-Saharan Africa. *American Journal of Physical Anthropology*, 140, 302-311.

Cavalli-Sforza, L.L. (1986) *African Pygmies.* New York, Academic Press. 461p.

Cavalli-Sforza, L.L., Piazza, A., Menozzi, P. and Mountain, J. (1988) Reconstruction of human evolution: bringing together genetic, archaeological and linguistic data. *Proceedings of the National Academy of Sciences, USA,* 85, 6002-6006.

Cavalli-Sforza, L.L., Menozzi, P. and Piazza, A. (1994) *The History and Geography of Human Genes*. New Jersey, Princeton University Press. 518p.

Chabeuf, M. (1959) Anthropologie du Moyen-Congo et du Gabon méridional. *Bulletins et Mémoires de la Société d'Anthropologie de Paris,* 10, series X, 97-185.

Chamla, M.-C. (1978) Le peuplement de l'Afrique du nord de l'épipaléolithique à l'époque actuelle. *L'Anthropologie,* 82(3), 385-430.

Chen, Y.-S., Torroni, A., Excoffier, L., Santachiara-Benerecetti, A.S. and Wallace, D.C. (1995) Analysis of mtDNA variation in African populations reveals the most ancient of all human continent-specific haplogroups. *American Journal of Human Genetics,* 57, 133-149.

Clark, J.D. (1980) Prehistoric populations and pressures favoring plant domestication in Africa. *In:* B.K. Savartz, J. Raymond and A. Dumett (eds) *West African Culture Dynamics.* Bristol, Mouton Publishers. p. 67-105.

Clist, B. (1987) Early bantu settlements in West-Central Africa: a review of recent research. *Current Anthropology,* 28, 380-382.

Clist, B. (1991) Synthèse régionale du Néolithique. *In:* R. Lanfranchi and B. Clist (eds) *Aux Origines de l'Afrique Centrale.* Libreville, Sepia editions. p. 180-189.

Coetzee, J.A. and Van Zinderen B.E.M. (1984) Preliminary results from sediment cores from Lake Bosumtwi, Ghana. *Palaeoecology of Africa and the Surrounding Islands,* 16, 173-192.

Coia, V., Sansonnetti, B., Paoli, G., Tofanelli, S., Spedini, G. and Destro-Bisol, G. (2002) Variation at 10 protein coding loci in the Mbenzele Pygmies from the Central African Republic and a comparison with microsatellite data. *American Journal of Human Biology*, 14, 9-14.

Comrie, B., Mathews, S. and Polinsky, M. (1997) *The Atlas of Languages.* London, Bloomsbury. 224p.

Connah, G. (1976) The Daima sequence and the prehistoric chronology of the Lake Chad region of Nigeria. *Journal of African History,* 17, 321-52.

Coon, C.S. (1971) A fossilised human mandibular fragment from Kangatotha, Kenya, East Africa. *American Journal of Physical Anthropology,* 34, 157-163.

Cornelissen, E. (1997) Central African transitional cultures. *In:* J.O. Vogel (ed.) *Encyclopedia of Precolonial Africa.* London, Altamira Press. p. 312-320.

Cornelissen, E. (2002) Human responses to changing environments in Central Africa between 40,000 and 12,000 B.P. *Journal of World Prehistory*, 16(3), 197-235.

Cornevin, M. (1993) *Archéologie Africaine: à lumière des découvertes récentes.* Paris: Maisonneuve et Larose. 270p.

Corruccini, R.S. (1974) An examination of the meaning of cranial discrete traits for human skeletal biological studies. *American Journal of Physical Anthropology,* 40, 425-446.

Corruccini, R.S. (1976) The interaction between non-metric and metric cranial variation. *American Journal of Physical Anthropology,* 44, 285-293.

Coupez, A. (1977) L'oeuvre de H. Johnston et la linguistique moderne. *Bulletins des Séances de l'Académie Royale des Sciences d'outre-mer,* 3, 224-239.

Coupez, A. (1989) Lexicostatistique bantu, état de la question. *In:* T. Obenga (ed.) *Actes du Colloque international: les peuples Bantu, migration, expansion et identité culturelle, Vol. I,* Libreville, C.I.CI.B.A., April 1985. Paris, L'Harmattan. p. 43-49.

Creighton, G. (1966) Prehistoric populations of Africa. *In:* J. Butler (ed.) *Boston Papers on Africa. African History, Vol. II.* Boston, Boston University Press. p. 3-37.

Crevecoeur, I. (2008) Étude anthropologique du squelette du Paléolithique supérieur de Nazlet Khater 2 (Egypte). Apport à la compréhension de la variabilité passée des hommes modernes. Leuven, Presses Universitaires de Louvain. *Egyptian Prehistory Monographs, 8*. 318p.

Crevecoeur, I. and Trinkaus, E. (2004) From the Nile to the Danube: a comparison of the Nazlet Khater 2 and Oase 1 early modern human mandibles. *Anthropologie (Brno)*, XLII(3), 229-239.

Crevecoeur, I., Rougier, H., Grine, F. and Froment, A. (2009) Modern human cranial diversity in the Late Pleistocene of Africa and Eurasia: evidence from Nazlet Khater, Peştera cu Oase, and Hofmeyr. *American Journal of Physical Anthropology*, 140,: 347–358.

Crevecoeur, I., Semal, P., Cornelissen, E. and Brooks, A. (2010) The Late Stone Age human remains from Ishango (Democratic Republic of Congo). Contribution to the study of the African Late Pleistocene modern human diversity. *American Journal of Physical Anthropology (Annual Meeting Issue)*, 48(Suppl.), 60.

Crognier, E. (1972) *Adaptation morphologique d'une population africaine au biotope tropical: les Sara du Tchad.* Unpublished Ph.D. dissertation in Natural Sciences, University of Paris VII. 240p.

Cruciani, F., Trombetta, B., Sellitto, D., Massaia, A., Destro-Bisol, G., Watson, E., Beraud Colomb, E., Dugoujon, J.-M., Moral, P. and Scozzari, R. (2010) Human Y chromosome haplogroup R-V88: a paternal genetic record of early mid Holocene trans-Saharan connections and the spread of Chadic languages. *European Journal of Human Genetics*, 18, 800–807.

Curtin, P., Feierman, S., Thompson, L. and Vansina, J. (1995) *African History from Earliest Times to Independence.* London, Longman. 546p.

Daegling, D.J. and Hylander, W.L. (2000) Experimental observation, theoretical models, and biomechanical inference in the study of mandibular form. *American Journal of Physical Anthropology,* 112, 541-551.

David, N. (1980) History of crops in North Cameroon to A.D. 1900. *In:* B.K. Swartz, J. Raymond and E. Dumett (eds) *West African Culture Dynamics*. 9th International Congress of Anthropology and Ethnological Sciences (Chicago, 1973). The Hague, Mouton Publishers. p. 139-182.

David, N. (1982) Early Bantu expansion in the context of central African prehistory. *In:* L. Bouquiaux (ed.) *Expansion bantou.* Paris, CNRS Editions. p. 609-644.

Davies, A. (1932) A re-survey of the morphology of the nose in relation to climate. *Journal of the Royal Anthropological Institute* (London), 62, 337-359.

De Heinzelin, J. (1962) Ishango. *Scientific American,* 206 (6), 105-116.

De Maret, P. (1977) Sanga: new excavations, more data and some related problems. *The Journal of African History,* 18, 321-337.

De Maret, P. (1980) Bribes, débris et bricolage. *In:* L.H. Hyman and J. Voorhoeve (eds) *L'expansion bantoue*, Viviers, 4-16th April 1977. Paris, SELAF. p. 715-730.

De Maret, P. (1982) New survey of archaeological research and dates for west-central and north-central Africa. *Journal of African History*, 23, 1-15.

De Maret, P. (1985) Recent archaeological research and dates from Central Africa. *Journal of African History*, 26, 129-148.

De Maret, P. (1989) Le contexte archéologique de l'expansion bantu en Afrique Centrale. *In:* T. Obenga (ed.) *Actes du Colloque international: les peuples Bantu, migration, expansion et identité culturelle, Vol. I,* Libreville, C.I.CI.B.A., April 1985. Paris, L'Harmattan. p. 118-138.

De Maret, P. (1990) Phases and facies in the archaeology of central Africa. *In:* P. Robertshaw (ed.) *A History of African Archaeology*. London, James Currey Ltd & Heinemann. p. 109-134.

De Maret, P. (1996) Shum Laka (Cameroon): human burials and general perspectives. *In:* G. Pwiti and R. Soper (eds) *Aspects of African Archaeology,* Papers from 10th Congress of the Panafrican Association for Prehistory and Related Studies, Harare, June 1995. Harare, University of Zimbabwe Publications. p. 274-279.

De Maret, P. and Nsuka, F. (1977) History of Bantu metallurgy: some linguistics aspects. *History in Africa,* 4, 43-65.

De Maret, P., Clist, B. and Van Neer, W. (1987) Résultats des premières fouilles dans les abris de Shum Laka et d'Abeke au nord-ouest du Cameroun. *L'Anthropologie,* 91, 559-584.

De Maret, P., Asombang, R., Cornelissen, E., Lavachery, P., Moeyersons, J. and Van Neer, W. (1993) Preliminary results of the 1991-1992 field season at Shum Laka, Northwestern Province, Cameroon. *Nyame Akuma,* 39, 13-15.

De Quatrefages, A. and Hamy, E.T. (1882) *Les crânes des races humaines. Crania Ethnica.* Paris, Baillière & fils. 528p.

De Villiers, H. and Fatti, L.P. (1982) The antiquity of the Negro. *South African Journal of Sciences,* 78, 321-332.

Defrise-Gussenhoven, E. (1955) Éloignement équiprobables et taux d'éloignement en biométrie. *Bulletin de l'Institut Royal des Sciences naturelles de Belgique,* XXXI (26), 1-31.

Descamps, C. and Thilmans, G. (1997) Fouille d'un tumulus à Ndiamon-Badat, îles du Saloum, Sénégal. *Saint Louis-Lille-Liège,* 3, 1-14.

Descamps, C. and Thilmans, G. (2001) Fouille de tumulus coquilliers à Djouta (îles du Saloum, Sénégal). *XIV International Congress of Prehistoric and Protohistoric Sciences*, Liège (2-8 Sept.), abstract.

Destro-Bisol, G. (1993) Migrations, genetic variability, and DNA polymorphisms. *Current Anthropology,* 34, 765-775.

Destro-Bisol, G. and Spedini, G. (1989) Anthropological survey on red cell glutathione peroxidase (GPX1) polymorphism in Central Western Africa: a tentative hypothesis on the interaction between GPX1*2 and Hbβ*S allelic products. *American Journal of Physical Anthropology,* 79, 217-224.

Destro-Bisol, G. and Spedini, G. (1991) The Nigeria-Cameroon plateau as the homeland of the Bantu spread: archaeological, linguistic and anthropological evidence. *Animal & Human Biology,* II, 107-138.

Destro-Bisol, G., Battaggia, C., Macchiarelli, R., Bailly, C., Scozzari, M.R. and Spedini, G. (1992) A bio-anthropological study of the Bakakas of Cameroon. *Annals of Human Biology,* 19(2), 185-195.

Destro-Bisol, G., Presciuttini, S., d'Ajola, E., Dobosz, M. and Spedini, G. (1994) Genetic variation at the ApoB3'HVR, D2S44, and D7S21 Loci in the Ewondo ethnic group of Cameroon. *American Journal of Human Genetics,* 55, 168-174.

Destro-Bisol, G., Giardina, B., Sansonetti, B. and Spedini, G. (1996) Interaction between oxidized hemoglobin and the cell membrane: a common basis for several *falciparum* malaria-linked genetic traits. *Yearbook of Physical Anthropology,* 39, 137-159.

Destro-Bisol, G., D'Aloja, E., Spedini, G., Scatena, R., Giardina, B. and Pascali, V. (1999) Brief communication: resistance to *falciparum* malaria in α-thalassemia, oxidative stress, and hemoglobin oxidation. *American Journal of Physical Anthropology,* 109, 269-273.

Destro-Bisol, G., Boschi, I., Caglià, A., Tofanelli, S., Pascali, V., Paoli, G. and Spedini, G. (2000) Microsatellite variation in Central Africa: an analysis of intrapopulational and interpopulational genetic diversity. *American Journal of Physical Anthropology,* 112, 319-337.

Destro-Bisol, G., Coia, V., Boschi, I., Verginelli, F., Caglia, A., Pascali, V., Spedini, G. and Calafell, F. (2004a) The analysis of variation of mtDNA hypervariable region 1 suggests that Eastern and Western Pygmies diverged before the Bantu expansion. *American Naturalist,* 163, 212–226.

Destro-Bisol, G., Donati, F., Coia, V., Boschi I., Verginelli, F., Caglia, A., Tofanelli, S., Spedini, G. and Capelli, C. (2004b) Variation of female and male lineages in sub-Saharan populations: the importance of sociocultural factors. *Molecular Biology and Evolution,* 21, 1673–1682.

Diamond, J. (1997) *Guns, germs, and steel: the fates of human societies.* New York, W.W. Norton & Company. 480p.

Diamond, J. and Bellwood, P. (2003) Farmers and their languages: the first expansions. *Science,* 300, 597-603.

Dios, S., Luis, J.R., Carril, J.C. and Caeiro, B. (2001) Sub-Saharan genetic contribution in Morocco: microsatellite DNA analysis. *Human Biology,* 73(5), 675-688.

Drontschilow, K. (1913) Metrische studien an Schädeln aus Kamerun. *Archives für Anthropologie,* XL, 161-183.

Dutour, O. (1989) *Hommes fossiles du Sahara, peuplements holocènes du Mali septentrional.* Marseille, Editions of Centre National de la Recherche Scientifique. 285p.

Dutour, O. (1992) Palimpseste paleoanthropologique sur l'homme fossile d'Asselar (Sahara). *Préhistoire et Anthropologie Mediterranéennes,* 1, 73-83.

Eastman, J.R. (1997) *Idrisi for Windows User's Guide, Version 2.0 (January 1997).* Worcester, Clark University, Clark Labs for Cartographic Technology and Geography Analysis.

Eggert, M.K.H. (1993) Central Africa and the archaeology of the equatorial rainforest: reflections on some major topics. *In:* T. Shaw, P. Sinclair, B. Andah and A. Okpoko (eds) *The Archaeology of Africa: Food, Metals and Towns.* London, Routledge. p. 289-329.

Eggert, M.K.H. (1997) Equatorial African Iron Age. *In:* J.O. Vogel (ed.) *Encyclopedia of Precolonial Africa.* London, Altamira Press. p. 429-435.

Ehret, C. (1972) Bantu origins and history: critique and interpretation. *Transafrican Journal of History,* 2(1), 1–9.

Ehret, C. (1973) Patterns of Bantu and Central Sudanic Settlement in Central and Southern Africa (ca. 1000 B.C. – 500 A.D.). *Transafrican Journal of History,* 3(1/2), 1–71.

Ehret, C. (1984) Historical/linguistic evidence for early African food production. *In:* J.D. Clark and S.A. Brandt (eds) *From Hunters to Farmers. The Causes and Consequences of Food Production in Africa.* Berkeley, University of California Press. p. 26-35.

Ehret, C. (1997) African languages: a historical survey. *In:* J.O. Vogel (ed.) *Encyclopedia of Precolonial Africa.* London, Altamira Press. p. 159-166.

Ehret, C. (1998) *An African Classical Age: eastern and southern Africa in world history 1000 B.C. to A.D. 400.* Oxford, James Currey. 354p.

Ehret, C. (2001) Bantu Expansions: re-envisioning a central problem of early African history. *The International Journal of African Historical Studies,* 34(1), 5–27.

Ehret, C. (2002a) Populations and languages in the expansion of the Afroasiatic language family. *American Journal of Physical Anthropology (Annual Meeting Issue),* 34(Suppl.), 67.

Ehret, C. (2002b) *The Civilizations of Africa: A history to 1800.* Charlottesville, University of

Virginia Press. 480p.

Eller, E. (1999) Population substructure and isolation by distance in three continental regions. *American Journal of Physical Anthropology,* 108, 147-159.

Eller, E. (2001) Estimating relative population sizes from simulated data sets and the question of greater African effective size. *American Journal of Physical Anthropology,* 116, 1-12.

Eltis, D. (2001) The volume and structure of the Transatlantic slave trade: A reassessment source. *The William and Mary Quarterly*, 58(1), 17-46.

Essomba, J.-M. (1992) *L'Archéologie au Cameroun.* Paris, Karthala. 383p.

Excoffier, L., Pellegrini, B., Sanchez-Mazas, A., Simon, C. and Langaney, A. (1987) Genetics and history of sub-saharan Africa. *Yearbook of Physical Anthropology*, 30, 151-194.

Fagg, B. (1972) Rop rock shelter excavations. *West African Journal of Archaeology,* 2, 1-12.

Ferembach, D., Schwidetzki, I. and Stloukal, M. (1980) Recommendations for age and sex diagnoses of skeletons. *Journal of Human Evolution,* 9 (7), 517-549.

Fix, A.G. (1999) *Migration and Colonization in Human Microevolution.* Cambridge, Cambridge University Press. 236p.

Fleagle, J.G. and Gilbert, C.C. (2008) Modern human origins in Africa. *Evolutionary Anthropology,* 17, 1-2.

Foley, R. (1998) The context of human genetic evolution. *Genome Research*, 8, 339-347.

Foley, R. and Lahr, M.M. (1998) Mode 3 technologies and the evolution of modern humans. *Cambridge Archaeology Journal*, 7, 3-36.

Franklin, D., Freedman, L. and Milne, N. (2005) Three-dimensional technology for linear morphological studies: a re-examination of cranial variation in four southern African indigenous populations. *Homo*, 56, 17–34.

Franklin, D., Cardini, A. and Oxnard, C.E. (2010) A geometric morphometric approach to the quantification of population variation in sub-Saharan African crania. *American Journal of Human Biology*, 22, 23–35.

Frayer, D.W. (1980) Sexual dimorphism and cultural evolution in the Late Pleistocene and Holocene of Europe. *American Journal of Physical Anthropology,* 9, 399-415.

Friess, M., Marcus, L.F., Reddy, D.P. and Delson, E. (2002) The use of 3D laser scanning techniques for the morphometric analysis of facial shape variation. *In:* B. Mafart, H. Delingette and G. Subsol (eds) *Three-Dimensional Imaging in Paleoanthropology and Prehistoric Archaeology,* Acts of the XIVth UISPP Congress, University of Liège, Belgium, 2-8 September 2001. British Archaeological Reports International Series 1049. p. 31-35.

Frisancho, A.R. (1996) *Human Adaptation and Accommodation.* Michigan, University of Michigan Press. 209p.

Froment, A. (1989) Body morphology and the savanna-forest transition: a West African example. *International Journal of Anthropology,* 4, 61-74.

Froment, A. (1992a) La différenciation morphologique de l'homme moderne: congruence entre forme du crâne et répartition géographique du peuplement. *Comptes Rendus de l'Académie des Sciences de Paris,* 315, series III, 323-329.

Froment, A. (1992b) Origines du peuplement de l'Egypte ancienne: l'apport de l'anthropobiologie. *Archéo-Nil,* 2, 79-98.

Froment, A. (1993) Adaptation biologique et variation dans l'espèce humaine: le cas des Pygmées d'Afrique. *Bulletins et Mémoires de la Société d'Anthropologie de Paris,* 5, 417-448.

Froment, A. (1995) Diversification culturelle et différenciation physique dans l'espèce humaine: une contribution de la biologie à la linguistique. *Cahiers ORSTOM* (Human Sciences series), 31, 239-251.

Froment, A. (1998) Le peuplement de l'Afrique Centrale: contribution de la paléoanthropologie. *In:* M. Delneuf, J.M. Essomba and A. Froment (eds) *Paléo-anthropologie en Afrique centrale: un bilan de l'Archéologie au Cameroun.* Paris, L'Harmattan. p. 13-90.

Froment, A. (2002) Biological evolution of populations during the early Holocene transitions. *Biennial Books of European Anthropological Association*, 2, 41-60.

Froment, A. and Hiernaux, J. (1984) Climate associated variation between populations of the Niger Bend. *Annals of Human Biology,* 11, 189-200.

Garcea, E.A. (2004) An alternative way towards food production: the perspective from the Libyan Sahara. *Journal of World Prehistory*, 18(2), 107-154.

Gautier, A. (1987) The archaeozoological sequence of the Acacus. *In:* B. Barich (ed.) *Archaeology and Environment in the Libyan Sahara: the Excavations in the Tadrat Acacus, 1978-1983.* British Archaeological Reports International Series 368. p. 283-308.

Gazetters (1954-1982) *Official standard names approved by the United States board on geographic names,* n°10 (South Africa), n°20 (Angola), n°45 (Egypt), n°59 (Gabon), n°60 (Cameroon), n°61 (Republic of Congo), n°64 (CAR), n°65 (Chad), n°68 (Sudan), n°78 (Kenya), n°80 (Republic of Congo), n°82 (Uganda), n°84 (Burundi), n°85 (Rwanda), n°89 (Ivory Coast), n°90 (Guinea), n°91 (Dahomey), n°92 (Tanzania), n°93 (Mali), n°98 (Togo), n°102 (Ghana), n°106 (Liberia), n°109 (Mozambique), n°138 (Kenya), n°140 (Somalia) and n°141 (Ethiopia). Washington, Washington Press.

Gebara, I. (1947) Sur quelques indices de longueur et de largeur des mandibules humaines. *Bulletins et Mémoires de la Société d'Anthropologie de Paris,* 8, series 9, 55-62.

Georgeon, E., Dutour, O. and Raimbault, M. (1992) Paléoanthropologie du gisement lacustre néolithique de Kobadi (Mali). *Préhistoire et Anthropologie Méditerranéennes*, 1, 85-97.

Giles, B., Doyle, M., Johnston, M., Wood, I. and Harding, D. (1997a) *Peoples of Southern Africa.* New York, The Diagram Group, Facts on file Inc. 112p.

Giles, B., Doyle, M., Johnston, M., Wood, I. and Harding, D. (1997b) *Peoples of Northern Africa.* New York: The Diagram Group, Facts on file Inc. 112p.

Giles, B., Doyle, M., Johnston, M., Wood, I. and Harding, D. (1997c) *Peoples of Eastern Africa.* New York, The Diagram Group, Facts on file Inc. 112p.

Giles, B., Doyle, M., Johnston, M., Wood, I. and Harding, D. (1997d) *Peoples of Central Africa.* New York, The Diagram Group, Facts on file Inc. 112p.

Giles, B., Doyle, M., Johnston, M., Wood, I. and Harding, D. (1997e) *Peoples of Western Africa.* New York, The Diagram Group, Facts on file Inc. 112p.

Giles, E. and Elliot, O. (1963) Sex determination by discriminant function analysis of crania. *American Journal of Physical Anthropology,* 21, 53-58.

Giresse, P., Maley, J. and Kossoni, A. (2005) Sedimentary environmental changes and millennial climatic variability in a tropical shallow lake (Lake Ossa, Cameroon) during the Holocene. *Palaeogeography, Palaeoclimatology, Palaeoecology*, 218, 257-285.

Goebel, T. (2007) The Missing years for modern humans. *Science,* 315, 194-196.

Gomila, J. (1969) *Anthropométrie des Bédik (Sénégal Oriental): contribution à l'étude de certains effets de l'endogamie*. Unpublished Ph.D. dissertation, Faculté des Sciences de Paris. 242p.

Gomila, J. (1971) *Les Bédik (Sénégal oriental). Barrières culturelles et hétérogénéité biologique.* Montréal, Montréal University Press. 273p.

Górnay, S. (1957) Crania Africana: Uganda. *Polska Akademia Nauk zaklad Anthropologii Materialy I prace Antropologiczne*, 14, 10-399.

Goudie, A.S. (1996) Climate: past and present. *In:* W.M. Adams, A.S. Goudie and A.R. Orme (eds) *The Physical Geography of Africa.* Oxford, Oxford University Press. p. 34-59.

Graw, M. (1999) Metric sex determination of the skull base. *Homo*, 50(2), 101-106.

Greenberg, J.H. (1964) Historical inferences from linguistic research in sub-Saharan Africa. *Boston*

University Papers in African History, 1, 1-15.

Grine, F.E., Bailey, R.M., Harvati, K., Nathan, R.P., Morris, A.G., Henderson, G.M., Ribot, I. and Pike, A.W.G. (2007) Late Pleistocene human skull from Hofmeyr, South Africa, and modern human origins. *Science*, 315, 226–229.

Grove, A.T. (1993) Africa's climate in the Holocene. *In:* R. Blench and M. Spriggs (eds) *Archaeology and Language, Vol. IV: language change and cultural transformation.* London: Routledge. p. 32-42.

Groves, C.P. and Thorne, A. (1999) The terminal Pleistocene and early Holocene populations of northern Africa. *Homo,* 50(3), 249-262.

Guglielmino-Matessi, C.R., Gluckman, P. and Cavalli-Sforza, L.L. (1979) Climate and the evolution of skull metrics in man. *American Journal of Physical Anthropology,* 50, 549-564.

Guthrie, M. (1962) Some developments in the prehistory of the Bantu languages. *Journal of African History,* III(2), 273-282.

Guthrie, M. (1967-1971) *Comparative Bantu. An introduction to the comparative linguistics and prehistory of the Bantu languages: the comparative linguistics of the Bantu languages,* Vols 1-4. Farnborough, Gregg International.

Gutierrez, M. and Valentin, F. (1995) Archéologie et anthropologie des tumulus de Kapanda (Angola). *Journal des africanistes,* 65(2), 145-169.

Hájek, M., Čeṃṇ V. and Brůžek, J. (2008) Mitochondrial DNA and craniofacial covariability of Chad basin females indicate past populations events. American *Journal of Human Biology,* 20, 465-474.

Hammer, M.F. and Zegura, S. (1996) The role of the Y chromosome in human evolutionary studies. *Evolutionary Anthropology,* 5(4), 116-134.

Hammer, M.F., Spurdle, A.B., Karafet, T., Bonner, M.R., Wood, E.T., Novelletto, A., Malaspina, P., Mitchel, R.J., Horai, S., Jenkins, T. and Zegura, S.L. (1997) The geographic distribution of human Y chromosomes. *Genetics,* 145, 787-805.

Hamy, E.T. (1882) Quelques observations sur l'anthropologie des Somalis. *Bulletins et Mémoires de la Société d'Anthropologie de Paris,* series III, 5, 697-706.

Hanihara, T. (1996) Comparison of craniofacial features of major human groups. *American Journal of Physical Anthropology,* 99, 389-412.

Harlan, J.R. (1971) Agricultural origins: centers and non-centers. *Science,* 174, 468-474.

Harlan, J.R. (1993) The tropical African cereals. *In:* T. Shaw, P. Sinclair, B. Andah and A. Okpoko (eds) *The Archaeology of Africa. Food, Metals and Towns.* London, Routledge. p. 53-60.

Harpending, H., Batzer, M., Gurven, M., Jorde, L., Rogers, A. and Sherry, S. (1998) Genetic traces of ancient demography. *Proceedings of the National Academy of Sciences, USA,* 95, 1961-1967.

Harris, M. (1997) *Culture, People, Nature: an introduction to general anthropology.* New York, Longman. 491p.

Hartle, D.D. (1970) Preliminary report of the University of Ibadan's Kainji rescue archaeology project, 1968. *West African Archaeological Newsletters*, 12, 7-19.

Harvati, K. and Weaver, T.D. (2006) Human cranial anatomy and the differential preservation of population history and climate signatures. *The Anatomical Records,* Part A, 288, 1225-1233.

Hassan, F.A. (1997) Holocene palaeoclimates of Africa. *African Archaeological Review,* 14(4), 213-230.

Hassan, H.Y., Underhill, P.A., Cavalli-Sforza, L.L. and Ibrahim, M.E. (2008) Y-Chromosome variation among Sudanese: restricted gene flow, concordance with language, geography, and history. *American Journal of Physical Anthropology*, 137, 316–323.

Haviland, W.A. (2000) *Anthropology.* Orlando, Harcourt College Publishers. 823p.

Heine, B. (1980) Methods in comparative Bantu linguistics. The problem of Bantu linguistic classification. *In:* L.H. Hyman and J. Voorhoeve (eds) *L'expansion bantoue*, Viviers, 4-16th April 1977. Paris: SELAF. p. 295-320.

Henn, B.M., Gignoux, C., Lin, A.A., Oefner, P.J.,

Shen, P., Scozzari, R., Cruciani, F., Tishkoff, S.A., Mountain, J.L. and Underhill, P.A. (2008) Y-chromosomal evidence of a pastoralist migration through Tanzania to southern Africa. *Proceedings of the National Academy of Sciences, USA,* 105(31), 10693–10698.

Henneberg, M. and Steyn, M. (1993) Trends in cranial capacity and cranial index in Subsaharan Africa during the Holocene. *American Journal of Human Biology,* 5, 473-479.

Henneberg, M. and Steyn, M. (1994) Preliminary report on the paleodemography of the K2 and Mapungubwe populations (South Africa). *Human Biology,* 66(1), 105-120.

Hennessy, R.J. and Stringer, C.B. (2002) Geometric morphometric study of the regional variation of modern human craniofacial form. *American Journal of Physical Anthropology,* 117, 37-48.

Herskovits, M.J. (1937) Physical types of West African negroes. *Human Biology,* 9(4), 483-497.

Hewes, G.W., Irwin, H., Papworth, M. and Saxe, A. (1964) A new fossil human population from the Wadi Halfa area, Sudan. *Nature,* 4943, 341-343.

Hiernaux, J (1964a) Les dermatoglyphes digitaux des Tutsi du Rwanda et des Shi du Congo. *Bulletins et Mémoires de la Société d'Anthropologie de Paris,* 6 (series XI), 369-385.

Hiernaux, J. (1953) Les caractères physiques des Bashi. *Mémoire de l'Institut Royal Colonial Belge,* series 8, 23(5). 47p.

Hiernaux, J. (1955a) Problèmes d'anthropologie physique en Afrique Centrale. *Folia Scientifica Africae Centralis,* I(4), 6-8.

Hiernaux, J. (1955b) Physical anthropology and the frequency of genes with a selective value: the sickle cell gene. *American Journal of Physical Anthropology,* 13, 455-472.

Hiernaux, J. (1963) Heredity and environment: their influence on human morphology. A comparison of two independent lines of study. *American Journal of Physical Anthropology,* 21, 575-589.

Hiernaux, J. (1964b) Luba du Katanga et Luba du Kasai (Congo): comparaison de deux populations de même origine. *Bulletins et Mémoires de la Société d'Anthropologie de Paris,* 6 (series XI), 611-622.

Hiernaux, J. (1965a) Hérédité, milieu et morphologie. *Biotypologie*, XXVI(1-2), 1-36.

Hiernaux, J. (1965b) Note sur les Tutsi de l'Itombwe (République du Congo): la position anthropologique d'une population émigrée. *Bulletins et Mémoires de la Société d'Anthropologie de Paris,* 7 (series XI), 361-379.

Hiernaux, J. (1966a) Peoples of Africa from 22° N to the Equator. *In:* Baker, P.T. and Weiner, J.S. (eds), *The Biology of Human Adaptability*. Oxford, University Press. p. 91-110.

Hiernaux, J. (1966b) Human biological diversity in Central Africa. *Man*, I(3), 287-306.

Hiernaux, J. (1966c) Les Bushong et les Cwa du royaime Kuba (Congo-Kinshasa): pygmées, pygmoïdes et pygméisation: anthropologie, linguistique et expansion bantoue. *Bulletins et Mémoires de la Société d'Anthropologie de Paris,* 9 (series XI), 299-336.

Hiernaux, J. (1968) *La diversité humaine en Afrique subsaharienne. Recherches biologiques.* Bruxelles, Editions de l'Institut de Sociologie de l'Université Libre de Bruxelles. 261p.

Hiernaux, J. (1970) La diversité biologique des groupes ethniques. *In:* J. Ki-Zerbo (ed.) *Histoire générale de l'Afrique,* Vol. I. Paris, UNESCO, Jeune Afrique. p. 53-72.

Hiernaux, J. (1972) Notes sur les restes osseux trouvés dans la tombe de Cyirima II, (Annex I). *In:* F.L. Van Noten (ed.) *Les tombes du roi Cyirima Rujugira et de la reine-mère Nyirayuhi Kanjogera. Description archéologique,* Vol. 77. Tervuren, Annales du Musée Royal de l'Afrique Centrale. p. 63-73.

Hiernaux, J. (1974) *The People of Africa.* London, Weidenfeld and Nicolson, Scribner's Sons. 217p.

Hiernaux, J. (1976a) Physical anthropology of the living populations of Sub-saharan Africa. *Annual Review of Anthropology,* 5, 149-168.

Hiernaux, J. (1976b) Blood polymorphisms in the Sara Majingay of Chad. *Annals of Human Biology*, 3, 127-140.

Hiernaux, J. (1977) Long-term biological effects of human migration from the African savanna to the equatorial forest: a case study of human adaptation to a hot and wet climate. *In:* G.A. Harrison (ed.) *Population Structure and Human Variation.* IBP Monograph 11. Cambridge, Cambridge University Press. p. 187-217.

Hiernaux, J., de Longrée, E. and de Buyst, J. (1971) Fouilles archéologiques dans la vallée du Haut-Lualaba – I. Sanga, 1958. *Annales du Musée Royal de l'Afrique Centrale,* Tervuren (Human Sciences series), 73.

Hiernaux, J. and Froment, A. (1976) The correlation between anthropobiological and climate variables in sub-saharan Africa: revised estimates. *Human Biology,* 48, 757-767.

Hiernaux, J. Vincke, E. and Commelin, D. (1976) Les Oto et les Twa des Konda (zone de l'équateur, Zaïre). *L'Anthropologie,* 80(3), 449-464.

Hiernaux, J. and Gauthier, A-M. (1977) Comparaison des affinités linguistiques et biologiques de douze populations de langue bantu. *Cahiers d'Histoire Africaine,* 66-67, 241-254.

Hiernaux, J. and Basu, A. (1980) *La diversité biologique humaine*. Paris, Masson. 420p.

Hiernaux, J., Plantier, M. and De Buyst, J. (1992) Etude ostéométrique des restes humains de Sanga et Katoto (Age du Fer, Zaïre). *Anthropologie et Préhistoire* (Bruxelles), 103, 9-44.

Hitzeroth, H.W. (1986) On the genetic interrelationships of South African negroes. *American Journal of Physical Anthropology,* 69, 389-401.

Hodder, I. (1981) Society, economy and culture: an ethnographic case study amongst the Lozi. *In:* I.I.G. Hodder and N. Hammond (eds) *Patterns of the Past.* Cambridge, Cambridge University Press. p. 67-95.

Hoelzmann, P., Keding, B., Berke, H, Kroepelin, S. and Kruse, H.-J. (2001) Environmental change and archaeology: lake evolution and human occupation in the Eastern Sahara during the Holocene. *Palaeogeography, Palaeoclimatology, Palaeoecology,* 169, 193–217.

Holden, C.J. (2001) Bantu language trees and cultural evolution. *Conference on Population History: Genes, Language and Culture* (University College of London, London, 16th June), abstract.

Holden, C.J. (2002) Bantu language trees reflect the spread of farming across sub-Saharan Africa: a maximum-parsimony analysis. *Proceedings of The Royal Society of London,* 269, 793-799.

Holl, A. (1993) Transition from Late Stone Age to Iron Age in the Sudano-Sahelian zone: a case study from the Perichadian plain. *In:* T. Shaw, P. Sinclair, B. Andah and A. Okpoko (eds) *The Archaeology of Africa: Food, Metals and Towns.* London, Routledge. p. 330-343.

Holl, A. (1997a) Pan-Africanism, diffusionism, and the Afrocentric idea. *In:* J.O. Vogel (ed.) *Encyclopedia of Precolonial Africa.* London, Altamira Press. p. 58-65.

Holl, A. (1997b) Western Africa: the prehistoric sequence. *In:* J.O. Vogel (ed.) *Encyclopedia of Precolonial Africa.* London, Altamira Press. p. 305-312.

Howells, W.W. (1973) Cranial variation in man: a study by multivariate analysis of pattern of difference among recent human populations. *Papers of the Peabody Museum of Archaeology and Ethnology n° 67.* Peabody Museum, Harvard University. 259p.

Howells, W.W. (1989) Skull shapes and the map: craniometric analyses in the dispersion of modern homo. *Papers of the Peabody Museum of Archaeology and Ethnology n° 79.* Peabody Museum, Harvard University. 189p.

Howells, W.W. (1995) Who's who in skulls. Ethnic identification of crania from measurements. *Papers of the Peabody Museum of Archaeology and Ethnology n° 82.* Peabody Museum, Harvard University. 108p.

Howells, W.W. (1996) Howell's craniometric data on the internet. *American Journal of Physical Anthropology*, 101, 441-442.

Huffman, T. (2007) *Handbook to the Iron Age: The archaeology of pre-colonial farming societies in southern Africa.* Scottsville, University of

KwaZulu-Natal Press. 504p.

Huizinga, J. (1968a) Human biological observations on some African populations of the thorn savanna belt. *Koninklijke Nederlandse Akademie van Wetenschappen (Proceedings of the Section of Sciences),* series C, 71(4), 356-390.

Huizinga, J. (1968b) New physical anthropological evidence bearing on the relationships between Dogon, Kurumba, and the extinct West African Tellem populations. *Koninklijke Nederlandse Akademie van Wetenschappen (Proceedings of the Section of Sciences),* series C, 71(1), 16-30.

Huizinga, J. (1974) Comparative survey of African people living in the hot-dry-savanna belt (1964-1972). *The Netherlands Contribution to the International Biological Programme, Final Report 1966-1971.* Amsterdam, North-Holland Publication. p. 113-117.

Huizinga, J. (1977) A comparative survey of African people living in the northern semi-arid zone: a search for a base line. *In:* G.A. Harrison (ed.) *Population Structure and Human Variation.* Cambridge, Cambridge University Press. p. 241-271.

Huizinga, J., Birnie-Tellier, N.F. and Glanville, E.V. (1967) Description and carbon-14 dating of Tellem cave skulls from the Mali republic: a comparison with other Negroid groups. *Koninklijke Nederlandse Akademie van Wetenschappen (Proceedings of the Section of Sciences),* series C, 70, 338-367.

Huizinga, J., Bedaux, R.M.A. and van der Waals, J.D. (1979) Anthropological research in Mali. *National Geographic Society Research Reports Projects*, 11, 281-307.

Hunter, J.S. (1980) The national system of scientific measurement. *Science,* 210, 869-874.

Huysecom, E., Rasse, M., Lespez, L., Neumann, K., Fahmy, Ballouche, A., Ozainne1, S., Maggetti, M., Tribolo, C. and Soriano, S. (2009) The emergence of pottery in Africa during the tenth millennium cal BC: new evidence from Ounjougou (Mali). *Antiquity,* 83, 905–917

Irish, J. (1995) High frequency archaic dental traits in modern sub-Saharan African populations. *American Journal of Physical Anthropology*, 20 (Suppl.), 117.

Irish, J. (1997) Characteristic high- and low-frequency dental traits in recent Sub-saharan Africans and the origins of modern humans. *American Journal of Physical Anthropology*, 102, 455-467.

Irish, J. (1998) Ancestral dental traits in recent sub-Saharan Africans and the origins of modern humans. *Journal of Human Evolution*, 34, 81-98.

Jablonski, N.G. and Chaplin, G. (2000) The evolution of human skin coloration. *Journal of Human Evolution,* 39, 57-106.

Jackson, S.P. (1961) *Climatological Atlas of Africa,* Plates 47 and 50. Pretoria: Commission for Technical Co-operation in Africa South of the Sahara.

Jamison, P.L. and Zegura, S.L. (1974) A univariate and multivariate examination of measurement error in Anthropometry. *American Journal of Physical Anthropology*, 40, 197-204.

Johnston, H.H. (1913) A survey of the ethnography of Africa and the former racial and tribal migrations in that continent. *Journal of the Royal Anthropological Institute,* 43, 375-421.

Johnston, H.H. (1919/1922) *A comparative study of the Bantu and semi-Bantu languages.* Vols 1-2. Oxford: Clarendon Press.

Jorde, L., Rogers, A., Bamshad, M., Watkins, A., Krakowiak, P. Sung, S., Kere, J. and Harpending, H. (1997) Microsatellite diversity and the demographic history of modern humans. *Proceedings of the National Academy of Sciences, USA,* 94, 3100-3103.

Jorde, L.B., Watkins, W.S., Bamshad, M.J., Dixon, M.E., Ricker, C.E. and Seielstad, M.T. (2000) The distribution of human genetic diversity: a comparison of mitochondrial, autosomal, and Y-chromosome data. *American Journal of Human Genetics,* 66, 979-988.

Jousse, H. (2006) What is the impact of Holocene climatic changes on human societies? Analysis of West African Neolithic populations dietary customs. *Quaternary International*, 151, 63-73.

Kappers, A. (1936) The character and spread of associated cephalic and cranial index peaks in Africa. *Koninklijke Nederlandse Akademie van Wetenschappen (Proceedings of the Royal Acadamie, Amsterdam)*, 39, 32-49.

Kasule, S. (1998) *The History Atlas of Africa*. New York, Macmillan. 160p.

Katzmarzyk, P.T. and Leonard, W.R. (1998) Climatic influences on human body size and proportions: ecological adaptations and secular trends. *American Journal of Physical Anthropology*, 106, 483–503.

Keita, S.O.Y. (1988) An analysis of crania from Tell-Duweir using multiple discriminant functions. *American Journal of Physical Anthropology*, 75, 375-390.

Keita, S.O.Y. (1990) Studies of ancient crania from Northern Africa. *American Journal of Physical Anthropology*, 83, 35-48.

Keita, S.O.Y. (2005) History in the interpretation of the pattern of p49a,f TaqI RFLP Y-Chromosome variation in Egypt: a consideration of multiple lines of evidence. *American Journal of Human Biology*, 17, 559–567.

Keita, S.O.Y. (2007) On Meroitic Nubian crania, Fordisc 2.0, and Human Biological History. *Current Anthropology*, 48(3), 425.

Keith, A. (1911) On certain physical characters of the Negroes of the Congo Free State and Nigeria. *Journal of the Royal Anthropological Institute (of Great Britain and Ireland)*, 41, 40-70.

Kim, N.C. and Kusimba, C.M. (2008) Pathways to social complexity and state Formation in the southern Zambezian region. *African Archaeological Review*, 25, 131-152.

Kimmel, M., Chakraborty, R., King, J., Bamshad, M., Watkins, W. and Jorde, L. (1998) Signatures of population expansion in microsatellite repeat data. *Genetics*, 148, 1921-1930.

Kitson, E. (1931) A study of the negro skull with special reference to the crania from Kenya colony. *Biometrika*, 23, 271-314.

Kittles, R. and Keita, S.O.Y. (1999) Interpreting African genetic diversity. *African Archaeological Review*, 16(2), 87-91.

Knight, A., Underhill, P.A., Mortensen, H.M., Zhivotosky, L.A., Lin, A.A., Henn, B.M., Louis, D., Ruhlen, M. and Mountain, J.L. (2003) African Y chromosome and mtDNA divergence provides insight into the history of click languages. *Current Biology*, 13(1), 1-20.

Knip, A. (1970) Metrical and non-metrical measurements on the skeletal remains of Christian populations from two sites in Sudanese Nubia: I & II. *Koninklijke Nederlandse Akademie van Wetenschappen (Proceedings of the Section of Sciences)*, series C, 73(5), 433-468.

Knip, A. (1971) The frequencies of non-metrical variants in Tellem and Nokara skulls from the Mali Republic. I. & II., *Koninklijke Nederlandse Akademie van Wetenschappen (Proceedings of the Section of Sciences)*, series C, 74, 422-443.

Krings, M., Salem, A.H., Bauer, K., Geisert, H., Malek, A.K., Chaix, L., Simon, C., Welsby, D., Di Rienzo, A., Utermann, G., Sajantila, A., Pääbo, S. and Stoneking, M. (1999) mtDNA analysis of Nile river valley populations: a genetic corridor or a barrier to migration? *American Journal of Human Genetics*, 64: 1166-1176.

Krogman, W.M. and Iscan, M.Y. (1986) *The Human Skeleton in Forensic Medicine*, Illinois: Springfield C.C.Thomas. 551p.

Kurki, H.K., Ginter, J.K., Stock, J.T. and Pfeiffer, S. (2008) Adult proportionality in small-bodied foragers: a test of ecogeographic expectations. *American Journal of Physical Anthropology*, 136, 28-38.

Lahr, M.M. (1992) *The Origins of Modern Humans: a test of the multiregional hypothesis*. Unpublished Ph.D. dissertation, University of Cambridge. 399p.

Lahr, M.M. (1994) The multiregional model of modern human origins: a re-assessment of its morphological basis. *Journal of Human Evolution*, 26, 23-56.

Lahr, M.M. (1995) Patterns of modern human diversification: implications for Amerindian origins. *Yearbook of Physical Anthropology*, 38, 163-198.

Lahr, M.M. (1996) *The Evolution of Modern Human Diversity: a study of cranial variation.*

Cambridge, Cambridge University Press. 416p.

Lahr, M.M. and Foley, R. (1994) Multiple dispersals and modern human origins. *Evolutionary Anthropology,* 3(2), 48-60.

Lasker, G.W. and Mascie-Taylor, C.G.N. (1988) The framework of migration studies. *In:* C.G.N. Mascie-Taylor and G.W. Lasker (eds) *Biological Aspects of Human Migration*. Cambridge Studies in Biological Anthropology. Cambridge, Cambridge University Press. p. 1-13.

Lavachery, P. (1996). Shum Laka Rockshelter Holocene deposits: From stone to metal (Northwestern Cameroon). *In:* G. Pwiti and R. Soper (eds.) *Aspects of African Archaeology.* Papers from the 10th Congress of the Panafrican Association for Prehistory and Related Studies. Harare, University of Zimbabwe Publications. p. 265–274.

Lavachery, P., Cornelissen, E., Moeyersons, J., and de Maret, P. (1996) 30.000 ans d'occupation, 6 mois de fouilles: Shum Laka, un site exceptionnel en Afrique Centrale. *Anthropologie et Préhistoire* 107, 197–211.

Leakey, L.S.B. (1931) *Stone Age Cultures of Kenya Colony.* Cambridge, Cambridge University Press. 288p.

Leakey, L.S.B. (1935) *The Stone Age Races of Kenya*. London, Oxford University Press. 150p.

Leschi, J. (1961) Etude des crânes de Tellem recueillis en pays Dogon (République du Mali). *Bulletin de l'Institut Français d'Afrique Noire,* XXIII, series B (3-4), 413-423.

Lester, P. (1927) Contribution à l'anthropologie des Somalis. *Bulletins et Mémoires de la Société d'Anthropologie de Paris,* 8, 175-187.

Lester, P. (1928) Etude anthropologique des populations de l'Ethiopie. *L'Anthropologie*, 38, 289-315.

Lieberman, D.E. (2008) Speculations about the selective basis for modern human craniofacial form. *Evolutionary Anthropology*, 17, 55–68.

Lieberman, D.E., McBratney, B.M. and Krovitz, G. (2002) The evolution and development of cranial form in *Homo sapiens*. *Proceedings of the National Academy of Sciences*, 99(3), 1134-1139.

Lovejoy, P. (1983) *Transformations in Slavery: a history of slavery in Africa.* Cambridge, Cambridge University Press. 349p.

MacDonald, K.C. (1994) *Socio-economic Diversity and the Origins of Cultural Complexity along the Middle Niger (2000 BC to AD 300).* Unpublished Ph.D. dissertation, Cambridge University. 298p.

MacDonald, K.C. (1997) Korounkorokale revisited: the *Pays* Mande and the West African microlithic technocomplex. *African Archaeological Review*, 14(3), 161-200.

MacDonald, K.C. (1999) Archaeology, language and the peopling of West Africa: a consideration of the evidence. *In:* R. Blench and M. Spriggs (eds) *Archaeology and Language, Vol. II: correlating archaeological and linguistic hypotheses.* London: Routledge. p. 33-66.

MacEachern, S. (2000) Genes, tribes, and African history. *Current Anthropology*, 41(3), 357-384.

MacEachern, S. (2006) Africanist archaeology and ancient IQ: racial science and cultural evolution in the twenty-first century. *World Archaeology*, 38(1), 72-92.

MacEachern, S. (2008) Where in Africa does Africa starts? Identity, genetics and African studies from the Sahara to Darfur. *Journal of Social Archaeology*, 7(3), 393-412.

Magory, C.C. and Day, M.H. (1994) The human skeletal remains excavated from burial mounds at Ngorongoro Crater; Northern Tanzania. *Homo*, 45(2), 103-118.

Maho, J. (2001) The Bantu area: (towards clearing up) a mess. *Africa and Asia*, 1, 40-49.

Malaspina, P., Cruciani, F., Ciminelli, B.-M., Terrenato, L., Santolamazza, P., Alonso, A., Banyko, J., Brdicka, R., Garcia, O., Gaudiano, C., Guanti, G., Kidd, K.K., Lavinha, J., Avila, M., Mandich, P., Moral, P., Qamar, R., Mehdi, S.Q., Ragusa, A., Stefanescu, G., Caraghin, M., Tyler-Smith, C. and Scozzari, R. (1998) Network analyses of Y chromosome types in Europe, Northern Africa, and Western Asia reveal specific patterns of geographic distribution. *American Journal of Human Genetics,* 63, 847-860.

Maley, J. (1993) The climatic and vegetational history of the equatorial regions of Africa during the upper Quaternary. *In:* T. Shaw, P. Sinclair, B. Andah and A. Okpoko (eds) *The Archaeology of Africa: Food, Metals and Towns.* London, Routledge. p. 43-52.

Maley, J. (1998) Vegetation dynamics, palaeoenvironments and climatic changes in the forests of western Cameroon during the last 28,000 years B.P. *Review of Palaeobotany and Palynology*, 6667(97), 157-187.

Mandela, N. (1996) *Long Walk to Freedom.* London, Abacus. 768p.

Manica, A., Amos, W., Balloux, F. and Hanihara, T. (2007) The effect of ancient population bottlenecks on human phenotypic variation. *Nature,* 448, 346–349.

Marquer, P. (1972) Nouvelle contribution à l'étude du squelette des pygmées occidentaux du centre africain comparé à celui des pygmées orientaux. *Mémoires du Museum National d'Histoire Naturelle,* new series A (Zoology), LXXII. 122p.

Marshall, F. and Hildebrand, E. (2002) Cattle before crops: the beginnings of food production in Africa. *Journal of World Prehistory*, 16(2), 99-143.

Martin, R. and Saller, K. (1959) *Lehrbuch der Anthropologie.* Stuttgart, Gustav Fischer Verlag. 1477p.

Martorana, M.C., Battaggia, C., Destro-Bisol, G., Sansonetti, B., Aebischer, M.L. and Spedini, G. (1996) A seven year anthropobiological survey on the distribution of abnormal hemoglobins in Cameroon. *Anthropologischer Anzeiger,* 54, 229-237.

Mauny, R. (1955) Baobabs-cimetières à griots. *Notes Africaines*, 67, 72-76.

Mayor, A., Huysecom, E., Gallay, A., Rasseb, M. and Ballouche, A. (2005) Population dynamics and paleoclimate over the past 3000 years in the Dogon Country, Mali. *Journal of Anthropological Archaeology*, 24, 25–61.

Mayr, E. (1974) *Populations, espèces et évolution.* Paris, Hermann Publications. 496p.

Mbida, C.M., Van Neer, W., Doutrelepont, H. and Vrydaghs, L. (2000) Evidence for banana cultivation and animal husbandry during the first millennium BC in the forest of Southern Cameroon. *Journal of Archaeological Science,* 27, 151–162.

McIntosh, S.K. and McIntosh, R.J. (1992) The Middle Senegal Valley project: preliminary results from the 1990-91 field season. *Nyame Akuma,* 38, 47-61.

Mellars, P. (2006) Why did modern human populations disperse from Africa *ca.* 60,000 years ago? A new model. *Proceedings of the National Academy of Sciences, USA*, 103(25), 9381–9386.

Melnick, D.J. and Hoezler, G.A. (1993) What is mtDNA good for in the study of primate evolution? *Evolutionary Anthropology,* 2(1), 2-10.

Menard, J. (1983) Les restes humains exhumés dans l'île de Toala. *L'Anthropologie,* 87(1), 127-133.

Mercader, J. (2002) Forest people: the role of African rainforests in human evolution and dispersal. *Evolutionary Anthropology*, 11, 117-124.

Mercader, J., Garralda, M.D., Pearson, O.M. and Bailey, R.C. (2001) Eight hundred-year-old human remains from the Ituri tropical forest, Democratic Republic of Congo: the rock shelter site of Matangai Turu Northwest. *American Journal of Physical Anthropology,* 115, 24-37.

Mitchell, R.J. (1996) Y-chromosome-specific restriction fragment length polymorphisms (RFLPs): relevance to human evolution and human variation. *American Journal of Human Biology,* 8, 573-586.

Montandon, G. (1928) *L'ologénèse humaine.* Paris, Félix Alcan. 478p.

Morales, J.C. and Melnick, D.J. (1998) Phylogenetic relationships of the macaques (*Cercopithecidae*: a *Macaca*), as revealed by high resolution restriction site mapping of the mitochondrial ribosomal genes. *Journal of Human Evolution,* 34, 1-23.

Morris, A.G. (1992) *A Master Catalogue:*

Holocene Human Skeletons from South Africa. Johannesburg, Witwatersrand University Press. 160p.

Morris, A.G. (2002) Isolation and the origin of the Khoisan: Late Pleistocene and Early Holocene human evolution at the southern end of Africa. *Human Evolution,* 17(3-4), 231-240.

Morris, A.G. and Ribot, I. (2006) Morphometric cranial identity of prehistoric Malawians in the light of sub-Saharan African diversity. *American Journal of Physical Anthropology*, 130, 10-25.

Murdock, G.P. (1959) *Its peoples and their culture history.* Toronto, McGraw-Hill. 456p.

Murdock, G.P. (1967) *Ethnographic Atlas.* Pittsburgh, University of Pittsburgh. 128p.

Murdock, G.P. (1981) *Atlas of World Cultures.* Pittsburgh, University of Pittsburgh. 151p.

Murray, J. (1998) *Cultural Atlas of Africa.* New York, Checkmark Books, Facts on file Inc. 240p.

Muzzolini, A. (1992) Le 'Bovidien' dans l'art rupestre saharien: un réexamen critique. *L'Anthropologie,* 96, 737-758.

Muzzolini, A. (1993) The emergence of a food-producing economy in the Sahara. *In:* T. Shaw, P. Sinclair, B. Andah and A. Okpoko (eds) *The Archaeology of Africa: Food, Metals and Towns.* London, Routledge. p. 227-239.

N'Goie-N'Galla, D. (1989) L'expansion bantoue à la lumière de l'onomastique. *In:* T. Obenga (ed.) *Actes du Colloque international: les peuples Bantu, migration, expansion et identité culturelle, Vol. II*, Libreville, C.I.C.I.B.A., April 1985. Paris: L'Harmattan. p. 301-335.

Nenquin, J. (1963) Excavations at Sanga, 1957. *Annales du Musée Royal de l'Afrique Centrale, Tervuren,* Human Sciences series, n°45.

Newman, J.L. (1995) *The Peopling of Africa, a Geographic Interpretation.* London, Yale University Press. 235p.

Nurse, G.T., Santos, D.J.H., Steinberg, A.G. and Jenkins, T. (1979) Serogenetic studies on the Njinga of Angola. *Annals of Human Biology,* 6, 335-344.

Ogoti Kiriama, O. (1993) The iron-using communities in Kenya. *In:* T. Shaw, P. Sinclair, B. Andah and A. Okpoko (eds) *The Archaeology of Africa: Food, Metals and Towns.* London: Routledge. p. 484-498.

Ojikutu, R.O., Nurse, G.T. and Jenkins, T. (1977) Red cell enzyme polymorphisms in the Yoruba. *Human Heredity,* 27, 444-453.

Oliver, R. (1991) *The African Experience.* New York, Icon Editions. 284p.

Olson, J.S. (1996) *The Peoples of Africa. An Ethnohistorical Dictionary.* London, Greenwood Press. 681p.

Orban, R., Ribot, I., Fenaux, S. and de Maret, P. (1996) Les restes humains de Shum Laka (Cameroun, LSA - Age du fer). *Anthropologie et Préhistoire*, 107, 213-225.

Orban, R., Twiesselmann, F. and Semal, P. (2001) Sur la biométrie des mandibules et des dents humaines d'Ishango (LSA, République démocratique du Congo). *Bulletin et Mémoires de la Société d'Anthropologie de Paris*, 13(1-2), 97-110.

Oumar, I., Mariac, C., Pham, J.-L. and Vigouroux, Y. (2008) Phylogeny and origin of pearl millet (*Pennisetum glaucum* [L.] R, Br) as revealed by microsatellite loci. *Theoretical and Applied Genetics,* 117, 489-497.

Ousley, S., Jantz, R. and Freid, D. (2009) Understanding Race and Human Variation: Why Forensic Anthropologists are Good at Identifying Race. *American Journal of Physical Anthropology,* 76, 68-76.

Page, J.W. (1976) A note on interobserver error in multivariate analyses of populations. *American Journal of Physical Anthropology*, 44, 521-526.

Pales, L. (1953) Raciologie comparative des populations de l'Afrique occidentale. *Bulletins et Mémoires de la Société d'Anthropologie de Paris,* 4, 187-293.

Papastergiadis, N. (2000) *The Turbulence of Migration, Globalization, Deterritorialization and Hybridity.* Cambridge, Polity Press. 256p.

Paris, F. (1990) Les sépultures monumentales d'Iwelen (Niger). *Journal des Africanistes,* 60, 42-

74.

Passarino, G., Semino, O., Quintana-Murci, L., Excoffier, L., Hammer, M. and Santachiara-Benerecetti, S. (1998) Different genetic components in the Ethiopian population, identified by mtDNA and Y-chromosome polymorphisms. *American Journal of Human Genetics,* 62, 420-434.

Pereira, L., Macaulay, V., Torroni, A., Scozzari, R., Prata, M.-J. and Amorim, A. (2001) Prehistoric and historic traces in the mtDNA of Mozambique: insights into the Bantu expansions and the slave trade. *Annals of Human Genetics,* 65, 439-458.

Petit-Maire, N. and Dutour, O. (1987) Holocene populations of the Western and Southern Sahara: Mechtoids and paleoclimates. *In:* A. Close (ed.) *Prehistory of arid North Africa.* Dallas, Southern Methodist University. p. 259-285.

Philip, G. (1998) *Modern College Atlas for Africa.* London: Philip's. 104p.

Phillipson, D.W. (1977a) Lowasera. *Azania,* 12, 1-32.

Phillipson, D.W. (1977b) The spread of the Bantu language. *Scientific American,* 236(4), 106-114.

Phillipson, D.W. (1985) An archaeological reconsideration of Bantu expansion. *Muntu,* 2, 69-85.

Phillipson, D.W. (1994) *African Archaeology.* Cambridge, Cambridge University Press. 268p.

Pietrusewsky, M. (1990) Craniofacial variation in Australasian and Pacific populations. *American Journal of Physical Anthropology,* 82, 319-340.

Pietrusewsky, M. (2008) Metric analysis of skeletal remains: methods and applications. *In:* M.A. Katzenberg and S.R. Saunders (eds) *Biological Anthropology of the Human Skeleton.* New York, John Wiley & Sons Inc. p. 487-532.

Pinhasi, R. and Semal, P. (2000) The position of the Nazlet Khater specimen among prehistoric and modern African and Levantine populations. *Journal of Human Evolution,* 39, 269–288.

Poutrin, (1910) Contribution à l'étude des pygmées d'Afrique: les négrilles du Centre Africain. *L'Anthropologie*, XXI, 435-504.

Protsch, R. (1975) The absolute dating of Upper Pleistocene sub-Saharan fossil hominids and their place in human evolution. *Journal of Human Evolution,* 4, 297-322.

Protsch, R. (1978) The chronological position of Gamble's Cave II and Bromhead's site (Elmenteita) of the Rift Valley, Kenya. *Journal of Human Evolution,* 7, 101-109.

Prowse, T.L. and Lovell, N.C. (1996) Concordance of cranial and dental morphological traits and evidence for endogamy in ancient Egypt. *American Journal of Physical Anthropology*, 101, 237-246.

Quintana-Murci, L., Quach, H., Harmant, C., Luca, F., Massonnet, B., Patin, E., Sica, L., Mouquiama-Daouda, P., Comas, D. and Tzur, S. (2008) Maternal traces of deep common ancestry and asymmetric gene flow between Pygmy hunter-gatherers and Bantu-speaking farmers. *Proceedings of the National Academy of Sciences, USA*, 105, 1596–1601.

Ramachandran S., Deshpande, O., Roseman, C.C., Rosenberg, N.A., Feldman, M.W. and Cavalli-Sforza, L.L. (2005) Support from the relationship of genetic and geographic distance in human populations for a serial founder effect originating in Africa. *Proceedings of the National Academy of Sciences, USA*, 102(44), 15942-15947.

Rapacz, J., Chen, L., Bütler-Brenner, E., Wu, M.J., Hasler-Rapcz, J.O., Bütler, R. and Schumaker, V.N. (1991) Identification of the ancestral haplotype for apolipoprotein B suggests an African origin of *Homo sapiens sapiens* and traces their subsequent migration to Europe and the Pacific. *Proceedings of the National Academy of Sciences, USA,* 88, 1403-1406.

Reed, F.A. and Tishkoff, S.A. (2006) African human diversity, origins and migrations. *Current Opinion in Genetics & Development*, 16, 597–605.

Reich, D. and Goldstein, D. (1998) Genetic evidence for a Paleolithic human population expansion in Africa. *Proceedings of the National Academy of Sciences, USA,* 95, 8119-8123.

Reinecke, P. (1896) Beschreibung einiger

Rassenskelette aus Afrika. *Archiv für Anthropologie,* XXV, 185-227.

Relethford, J.H. (1994) Craniometric variation among modern human populations. *American Journal of Physical Anthropology,* 95, 53-62.

Relethford, J.H. (1998) Mitochondrial DNA and ancient population growth. *American Journal of Physical Anthropology,* 105, 1-7.

Relethford, J.H. (2001a) *Genetics and the search for modern human origins.* New York, Wiley-Liss. 264p.

Relethford, J.H. (2001b) Global analysis of regional differences in craniometric diversity and population substructure. *Human Biology,* 73(5), 629-636.

Relethford, J.H. (2002) Apportionment of global human genetic diversity based on craniometrics and skin color. *American Journal of Physical Anthropology*, 118, 393–398.

Relethford, J.H. and Harpending, H.C. (1994) Craniometric variation, genetic theory, and modern human origins. *American Journal of Physical Anthropology,* 95, 249-270.

Relethford, J.H., Crawford, M.H. and Blangero, J. (1997) Genetic drift and gene flow in post-famine Ireland. *Human Biology*, 69, 443–465.

Relethford, J.H. and Jorde, L.B. (1999) Genetic evidence for larger African population size during recent human evolution. *American Journal of Physical Anthropology,* 108, 251-260.

Renfrew, C. (1992) Archaeology, genetics and linguistic diversity. *Man*, 27, 445-478.

Renfrew, C. (1999) World linguistic diversity and farming dispersals. *In:* R. Blench and M. Spriggs (eds) *Archaeology and Language: Theoretical and Methodological Orientations,* Vol. I. London, Routledge. p. 82-90.

Rexova, K., Bastin, Y. and Frynta, D. (2006) Cladistic analysis of Bantu languages: a new tree based on combined lexical and grammatical data. *Naturwissenschaften*, 93, 189-194.

Rhine, R.J. (1986) Ten years of cooperative research at Mikumi National Park. *In:* J.G. Else and P.C. Lee (eds) *Ontogeny, Cognition and Social Behavior of Primates.* Cambridge, Cambridge University Press. p. 13-22.

Ribot, I. (1998) *Cranial Variation in Equatorial Africa.* M.Phil. dissertation, University of Cambridge. 59p.

Ribot, I. (2002) *Craniomandibular variation in sub-Saharan Africa: sexual dimorphism, geography, ecology and history.* Ph.D. dissertation, University of Cambridge, 392p.

Ribot, I. (2003) Craniometrical analysis of Central and East Africans in relation to history. A case study based on unique collections of known ethnic affiliation. *Anthropologica et Praehistorica,* 114, 25-50.

Ribot, I. (2004) Differentiation of modern sub-Saharan African populations: craniometric interpretations in relation to geography and history. *Bulletins et Mémoires de la Société d'Anthropologie de Paris,* 16(3-4), 143-165.

Ribot, I., Orban, R. and de Maret, P. (2001) The prehistoric burials of Shum Laka rockshelter (North-West Cameroon). *Annales du Musée Royal de l'Afrique Centrale,* Tervuren (Sciences Humaines), n° 164. 218p.

Ribot, I., Froment, A. and Lahr, M. (2002) Une approche craniométrique sur la diversité morphologique des populations Holocènes originaires de l'Afrique occidentale (et centrale): essai d'interpretations en relation avec la géographie et l'histoire. *Bulletins et Mémoires de la Société d'Anthropologie de Paris*, 14(1-2), 225-226.

Ribot, I., Lahr, M. and Storto, C. (2006a) Iron Age Senegalese populations: a craniometric approach in a historical perspective. Acts of the XIVth UISPP Congress, University of Liège, Belgium, 2-8 September 2001. *British Archaeological Reports, International Series* 1522, 199-206.

Ribot, I., Lahr, M. and Storto, C. (2006b) Populations sénégalaises de l'âge du fer: une approche craniométrique dans une perspective historique. *In:* A. Camara and C. Descamps (eds) *Sénégalia: travaux sur le patrimoine ouest-africain.* Paris, Sépia. p. 183-192.

Rightmire, G.P. (1972) Cranial measurements and discrete traits compared in distance studies of

African negro skulls. *Human Biology,* 44(3), 263-276.

Rightmire, G.P. (1975) New studies of Post-Pleistocene human skeletal remains from the rift Valley, Kenya. *American Journal of Physical Anthropology,* 42, 351-370.

Rightmire, G.P. (1976a) Metric versus discrete traits in African skulls. *In:* E. Giles and J.S. Friedlander (eds) *The Measures of Man, Methodologies in Biological Anthropology.* Cambridge, Peabody Museum Press. p. 383-407.

Rightmire, G.P. (1976b) Multidimensional scaling and the analysis of human biological diversity in sub-Saharan Africa. *American Journal of Physical Anthropology,* 44, 445-452.

Rightmire, G.P. (1977) Appendix. Notes on the human burials from Lowasera. *Azania,* XII, 30-32.

Rightmire, G.P. (1990) *The Evolution of Homo erectus.* Cambridge, Cambridge University Press. 260p.

Robertson, J.H. and Bradley, R. (2000) A New Paradigm: The African Early Iron Age without Bantu Migrations. *History in Africa*, 27, 287-323.

Rogers, A. and Harpending, H. (1992) Population growth makes waves in the distribution of pairwise genetic differences. *Molecular Biology and Evolution,* 9(3), 552-569.

Rogers, J. and Kidd, K.K. (1993) Nuclear DNA polymorphisms in a wild population of yellow baboons (*Papio hamadryas cynocephalus)* from Mikumi National Park, Tanzania. *American Journal of Physical Anthropology,* 90, 477-486.

Roseman, C.C. (2004) Detecting interregionally diversifying natural selection on modern human cranial form by using matched molecular and morphometric data. *Proceedings of the National Academy of Sciences, USA*, 101(35), 12824-12829.

Roseman, C.C. and Weaver, T.D. (2004) Multivariate apportionment of global human craniometric diversity. *American Journal of Physical Anthropology,* 125, 257-263.

Roset, J.-P. (1987) Paleoclimatic and cultural conditions of neolithic development in the Early Holocene of Northern Niger (Air and Ténéré). *In:* A. Close (ed.) *Prehistory of Arid North Africa: Essays in Honor of Fred Wendorf,* Dallas, Southern Methodist University Press. p. 163-187.

Ross, C.F. (2001) In vivo function of the craniofacial haft: the interorbital "pillar". *American Journal of Physical Anthropology,* 116, 108-139.

Rossi, P.F., Battagia, C., Sansonetti, B., Destro-Bisol, G., Capucci, E., Aebischer, M.L., Kaptué, L. and Spedini, G. (1991) Les Foulbé du Nord-Cameroun: origine, migrations, peuplement et variabilité génétique. *Rivista di Antropologia* (Rome), LXIX, 49-72.

Rouger, P., Salmon, D., Ruffie, J. and Salmon, C. (1983) Studies of blood group antigens in Aka Pygmies. *Journal of Human Evolution,* 12, 361-366.

Ruff, C.B. (1994) Morphological adaptation to climate in modern and fossil hominids. *Yearbook of Physical Anthropology,* 37, 65-107.

Ruhlen, M. (1991) *A Guide the World's Languages. Classification* (I). London, Edward Arnold. 463p.

Salas, A., Richards, M., De la Fe, T., Lareu, M.-V., Sobrino, B., Sanchez-Diz, P., Macaulay, V. and Carracedo, A. (2002) The making of the African mtDNA landscape. *American Journal of Human Genetics*, 71, 1082-1111.

Salmons, J.E. (1925) The Bush and the Bantu mandibles. *The South African Journal of Science,* 22, 470-479.

Salzmann, U. and Hoelzmann, P. (2005) The Dahomey Gap: an abrupt climatically induced rain forest fragmentation in West Africa during the late Holocene. *The Holocene*, 15, 190–199.

Sanchez-Mazas, A., Graven, L. and Pellegrini, B. (1992) Génétique, linguistique et préhistoire du peuplement sub-saharien. *Bulletin du Centre Genevois d'Anthropologie*, 3, 3-21.

Schepartz, L.A. (1987) *From Hunters to Herders: Subsistence Pattern and Morphological Change in Eastern Africa.* Unpublished Ph.D. dissertation,

Michigan University. 141p.

Schuster, S.C., Miller, W., Ratan, A., Tomsho, L.P., Giardine, B., Kasson, L.R., Harris, R.S., Petersen, D.C., Zhao, F., Qi, J., Alkan, C., Kidd, J.M., Sun, Y., Drautz, D.I., Bouffard, P., Muzny, D.M., Reid, J.G., Nazareth, L.V., Wang, Q., Burhans, R., Riemer, C., Wittekindt, N.E., Moorjani, P., Tindall, E.A., Danko, C.G., Siang Teo, W., Buboltz, A.M., Zhenhai Zhang1, Qianyi Ma1, Oosthuysen, A., Steenkamp, A.W., Oostuisen, H., Venter, P., Gajewski, J., Zhang, Y., Pugh, B.F., Makova, K.D., Nekrutenko, A., Mardis, E.R., Patterson, N., Pringle, T.H., Chiaromonte, F., Mullikin, J.C., Eichler, E.E., Hardison, R.C., Gibbs, R.A., Harkins, T.T. and Hayes, V.M. (2010) Complete Khoisan and Bantu genomes from southern Africa. *Nature*, 463(18), 943-947.

Scozzari, R., Torroni, A., Semino, O., Cruciani, F., Spedini, G. and Santachiara-Benerecetti, S.A. (1994) Genetics studies in Cameroon: mitochondrial DNA polymorphisms in Bamileke. *Human Biology,* 66, 1-12.

Seielstad, M.T., Herbert, J.M., Lin, A.A., Underhill, P.A., Ibrahim, M.D., Vollrath, D. and Cavalli-Sforza, L.L. (1994) Construction of the human Y-chromosomal haplotypes using a new polymorphic A to G transition. *Human Molecular Genetics,* 3, 2159-2161.

Seielstad, M.T., Minch, E. and Cavalli-Sforza, L.L. (1998) Genetic evidence for a higher female migration rate in humans. *Nature Genetics,* 20, 278-280.

Semino, O., Passarino, G., Brega, A., Fellous, M. and Santachiara-Benerecetti, A.S. (1996) A view of the Neolithic demic diffusion in Europe through two Y chromosome-specific markers. *American Journal of Human Genetics,* 59, 964-968.

Semino, O., Santachiara-Benerecetti, A.S., Falaschi, F., Cavalli-Sforza, L.L. and Underhill, P.A. (2002) Ethiopians and Khoisan share the deepest clades of the human Y-Chromosome phylogeny. *American Journal of Human Genetics,* 70, 265-268.

Sereno, P.C., Garcea, E.A.A., Jousse, H., Stojanowski, C.M., Saliège, J.-F., Abdoulaye Maga, A., Ide, O.A., Knudson, K.J., Mercuri, A.M., Stafford, T.W., Kaye, T.G., Giraudi, C., Massamba N'siala, I., Cocca, E., Moots, H.M., Dutheil, D.B. and Stivers, J.P. (2008) Lakeside Cemeteries in the Sahara: 5000 years of Holocene population and environmental change. *PLoS ONE (Public Library of Science)* 3(8), 1-22.

Sergi, S. (1912) *Crania Habessinica: contributo all'antropologia dell'Africa orientale*. Berlin, Virchov-Stiftung publication. 519p.

Shaw, T. (1980) Early crops in Africa: a review of the evidence. *In:* B.K. Swartz, J. Raymond and E. Dumett (eds) *West African Culture Dynamics*. The Hague, Mouton Publishers. p. 107-153.

Shea, B.T. and Bailey, R.C. (1996) Allometry and adaptation of body proportions and stature in African pygmies. *American Journal of Physical Anthropology,* 100, 311-340.

Shinnie, P.L. and Kense, F.J. (1989) *Archaeology of Gonja, Ghana. Excavations at Daboya.* Calgary, The University Calgary Press. 369p.

Shriver, M., Jin, L., Ferrell, R. and Deka, R. (1997) Microsatellite data support an early population expansion in Africa. *Genome Research,* 7, 586-591.

Shrubsall, F. (1897) Crania of African Bush races. *Journal of the Royal Anthropological Institute (of Great Britain and Ireland),* 27, 263-290.

Shrubsall, F. (1898) A study of A-Bantu skulls and crania. *Journal of the Royal Anthropological Institute (of Great Britain and Ireland)*, 28, 55-94.

Shrubsall, F. (1899) Notes on Ashanti skulls and crania. *Journal of the Royal Anthropological Institute (of Great Britain and Ireland)*, 28, 95-103.

Shrubsall, F. (1907) Notes on some Bushman crania and bones from the South African Museum, Cape Town. *Annals of the South African Museum (Cape Town),* V(5), 227-270.

Smith, A.B. (1974) Preliminary report of excavations at Karkarichinkat nord and Karkarinchikat Sud, Tilemsi valley, Republic of Mali, Spring 1972. *West African Journal of Archaeology,* 4, 33-55.

Smith, A.B. (1993) New approaches to Saharan rock art of the Bovidien period. *In:* L. Krzyzaniak, M. Kobusiewicz and J. Alexander (eds)

Environmental Change and Human Culture in the Nile Basin and Northern Africa until the second millennium BC. Poznan, Archaeological Museum. p. 77-89.

Smith, A.B. (2005) The concepts of 'Neolithic' and 'Neolithisation' for Africa? *Before Farming*, 1, 1-6.

Smith, H.F. (2009) Which Cranial Regions Reflect Molecular Distances Reliably in Humans? Evidence from Three-Dimensional Morphology. *American Journal of Human Biology*, 47, 36-47.

Soodyall, H. (1993) *Mitochondrial DNA polymorphisms in southern African populations*. Unpublished Ph.D. dissertation, University of the Witwatersrand, Johannesburg.

Soodyall, H., Vigilant, L., Hill, A.V., Stoneking, M. and Jenkins, T. (1996) mtDNA control-region sequence variation suggests multiple independent origins of an "Asian-specific" 9-bp deletion in sub-Saharan Africans. *American Journal of Human Genetics*, 58, 595-608.

Spedini, G., Walter, H., Capucci, E., Fuciarelli, M., Rickards, O., Aebischer, M.L. and Crosti, N. (1983) An anthropobiological study in Basse Kotto (Central Africa). I. Erythrocyte and sero-genetic markers: an analysis of the genetic differentiation. *American Journal of Physical Anthropology*, 60, 39-47.

Spedini, G. and Bailly, C. (1987) Etude anthropobiologique au Cameroun. Les Bamiléké du plateau occidental. *Rivista di Antropologia* (Rome), LXV, 193-212.

Spedini, G., Destro-Bisol, G., Mondovi, S., Kaptue, L., Taglioli, L. and Paoli, G. (1999) The peopling of Sub-Saharan Africa: the case study of Cameroon. *American Journal of Physical Anthropology*, 110, 143-162.

SPSS Base 9.0 (1999a) *User's Guide*. Chicago: SPSS Inc. 740p.

SPSS Base 9.0 (1999b) *Applications Guide*. Chicago: SPSS Inc. 412p.

Stahl, A. (1985) Reinvestigations of Kintampo 6 rock shelter, Ghana: implications for the nature of culture change. *African Archaeological Review*, 3, 117-150.

Stahl, A. (1993) Intensification in the West African Late Stone Age: a view from Central Ghana. *In:* T. Shaw, P. Sinclair, B. Andah and A. Okpoko (eds) *The Archaeology of Africa: Food, Metals and Towns*. London, Routledge. p. 261-273.

Steegman, A.T. (1975) Human adaptation to cold. *In:* A. Damon (ed.) *Physiological Anthropology*. New York, Oxford University Press. p. 130-166.

Stefan, V.H. (1999) Craniometric variation and homogeneity in Prehistoric/Protohistoric Rapa Nui (Easter Island) regional populations. *American Journal of Physical Anthropology*, 110, 407-419.

Steyn, M.P. and Henneberg, M.A. (1997) Cranial growth in the prehistoric sample from K2 at Mapungubwe (South Africa) is population specific. *Homo*, 48(1), 62-71.

Stringer, C.B. (1979) A re-evaluation of the fossil human calvaria from Singa, Sudan. *Bulletin of the British Museum of Natural History, Geology*, 32, 77-83.

Stringer, C.B. (1989) Documenting the origin of modern humans. *In:* E. Trinkaus (ed.) *The Emergence of Modern Humans*. Cambridge, Cambridge University Press. p. 67-96.

Stringer, C.B., Harvati, K., Allsworth-Jones, P., Grün, R. and Folorunso, C.A. (2010) New research on the Iwo Eleru cranium from Nigeria. *American Journal of Physical Anthropology (Annual Meeting Issue)*, 48 (Suppl.), 292.

Stynder, D.D. (2009) Craniometric evidence for South African Later Stone Age herders and hunter–gatherers being a single biological population. *Journal of Archaeological Science*, 36, 798–806.

Stynder D.D., Ackermann R.R. and Sealy J.C. (2007a) Early to mid-Holocene South African Later Stone Age human crania exhibit a distinctly Khoesan morphological pattern. *South African Journal of Science*, 103, 349–352.

Stynder, D.D., Rogers Ackermann, R. and Sealy, J.C. (2007b) Craniofacial variation and population continuity during the South African Holocene. *American Journal of Physical Anthropology*, 134, 489–500.

Takasaka, T., Kitamura, T., Sugimoto, C., Guo, J., Zheng, H.-Y. and Yogo, Y. (2006) Phylogenetic analysis of major African genotype (Af2) of JC virus: implications for origin and dispersals of modern Africans. *American Journal of Physical Anthropology*, 129, 465–472.

Thilmans, G. (1968) Étude de quelques crânes lébou (Sénégal). *Bulletin de l'I.F.A.N.*, XXX (series B), 4, 1291-1297.

Thilmans, G. and Descamps, C. (1982) Recherches scientifiques dans les parcs nationaux du Sénégal. *Mémoires de l'Institut Fondamental d'Afrique Noire,* IFAN, Dakar, 92, 33-50.

Thilmans, G., Descamps, C. and Khayat, B. (1980) Protohistoire du Sénégal, recherches archéologiques, I: Les sites mégalithiques. *Mémoires de l'Institut Fondamental d'Afrique noire,* Dakar, IFAN, n° 91. 158p.

Thoma, A. (1984) Morphology and affinity of the Nazlet Khater man. *Journal of Human Evolution*, 13, 287-296.

Thomson, A. and Buxton, L.H.D. (1923) Man's nasal index in relation to certain climatic conditions. *Journal of the Royal Anthropological Institute,* 53, 92-122.

Times Atlas (1999) *The Times Comprehensive Atlas of the World.* London, Times Books & Bartholomew. 220p.

Tishkoff, S.A, Ruano, R., Kidd, J.R. and Kidd, K.K. (1996) Distribution and frequency of a polymorphic *Alu* insertion at the plasminogen activator locus in humans. *Human Genetics*, 97, 759–764.

Tishkoff, S.A., Goldman, A., Calafell, F., Speed, W.C., Deinard, A.S., Bonne-Tamir, B., Kidd, J.R., Pakstis, A.J., Jenkins, T. and Kidd, K.K. (1998) A global haplotype analysis of the myotonic dystrophy locus: implications for the evolution of modern humans and for the origin of myotonic dystrophy mutations. *American Journal of Physical Anthropology,* 62, 1389-1402.

Tishkoff, S.A. and Williams, S. (2002) Genetic analysis of African populations: human evolution and complex disease. *Nature Reviews, Genetics*, 3, 611-621.

Tishkoff, S.A., Reed, F.A., Friedlaender, F.R., Ehret, C., Ranciaro, A., Froment, A., Hirbo, J.B., Awomoyi, A.A., Bodo, J.-M., Doumbo, O., Ibrahim, M., Juma, A.T., Kotze, M.J., Lema, G., Moore, J.H., Mortensen, H., Nyambo, B.T., Omar, S.A., Powell, K., Pretorius, G.S., Smith, M.W., Thera, M.A., Wambebe, C., Weber, J.L. and Williams, S.M. (2009) The Genetic Structure and History of Africans and African Americans. *Science*, 324, 1035-1044.

Tobias, P.V. (1955) Physical Anthropology and somatic origins of the Hottentots. *African Studies,* 14, 1-22.

Treinen, F. (1975) La nécropole de Nemra, region de Koro-Toro, Tchad. *West African Journal of Archaeology*, 5, 53-79.

Ulijaszek, S. and Kerr, D.A. (1999) Anthropometric measurement error and the assessment of nutritional status. *British Journal of Nutrition*, 82, 165-177.

Underhill, P.A., Passarino, G., Lin, A.A., Shen, P., Lahr, M.M., Foley, R.A., Oefner, P.J. and Cavalli-Sforza, L.L. (2001) The phylogeography of Y chromosome binary haplotypes and the origins of modern human populations. *Annals of Human Genetics,* 65, 43-62.

Utermolhe, C.J. and Zegura, S. (1982) Intra-and interobserver error in craniometry: a cautionary tale. *American Journal of Physical Anthropology,* 57, 303-310.

Vallois, H.V. (1969) Les hommes de Cro-Magnon et les Guanches: les faits acquis et les hypothèses. *Anuario de Estudios Atlanticos,* 15, 97-118.

Van Bakel, M.A. (1981) The "Bantu" expansion: demographic models. *Current Anthropology,* 22(6), 688-691.

Van de Ginste, F. (1946) Anthropometric study of the Bapende and Basuku of the Belgian Congo. *American Journal of Physical Anthropology,* 4(2), 125-152.

Van Gerven, D.P. (1982) The contribution of time and local geography to craniofacial variation in Nubia's Batn el Hajar. *American Journal of Physical Anthropology,* 59, 307-316.

Van Noten, F. (1977) Excavations at Matupi cave.

Antiquity, LI, 35-40.

Van Noten, F. (1982) *The Archaeology of Central Africa.* Graz / Austria, Akademische Druck und Verlagsanstalt. 152p.

Vansina, J. (1990) *Paths in the rainforests, toward a history of political tradition in Equatorial Africa.* London, James Currey. 428p.

Vansina, J. (1995) New linguistic evidence and the Bantu expansion. *Journal of African History,* 36, 173-195.

Varela, H.H. and Cocilovo, J.A. (1999) Evaluation of the environmental component of the phenotypic variance in prehistoric populations. *Homo,* 50(1), 46-53.

Vecchi, F., Biondi, G., Passarello, P. and Priori, R. (1983) Distribuzione di alcune variabili morfologiche nell'Africa sub-sahariana. *Rivista di Antropologia,* LXII, 223-254.

Vecchi, F., Biondi, G., Passarello, P. and Priori, R. (1987) Distribution of several cephalic indexes in Sub-Saharan Africa. *Rivista di Antropologia,* LXV, 213-222.

Verdu, P., Austerlitz, F., Estoup, A., Vitalis, R., Georges, M., Théry, S., Froment, A., Le Bomin, S., Gessain, A., Hombert, J.-M., Van der Veen, L., Quintana-Murci, L., Bahuchet, S. and Heyer, E. (2009) Origins and genetic diversity of Pygmy hunter-gatherers from Western Central Africa. *Current Biology,* 19, 312–318.

Vermeersch, P. M., Gijselings, G. and Paulissen, E. (1984) Discovery of the Nazlet Khater man, Upper Egypt. *Journal of Human Evolution,* 13, 281-286.

Vidal, P. (1992) Au-delà des mégalithes: archéologie centrafricaine et histoire de l'Afrique centrale. *In:* J.-M. Essomba (ed.) *L'archéologie au Cameroun,* Actes du Ier Colloque International de Yaoundé (Jan. 1986). Paris, Karthala. p. 133-178.

Vidal, P., de Bayle des Hermens, R. and Menard, J. (1983) Le site archéologique de l'Ile de Toala sur la Haute Ouham (République Centrafricaine): néolithique et âge du fer. *L'Anthropologie,* 87(1), 113-133.

Viðarsdóttir, U.S, Higgins, P.O. and Stringer, C. (2002) A geometric morphometric study of regional differences in the ontogeny of the modern human facial skeleton. *Journal of Anatomy,* 201, 211-229.

Vigilant, L., Stoneking, M., Harpending, H., Hawkes, K. and Wilson, A.C. (1991) African populations and the evolution of human mitochondrial DNA. *Science,* 253, 1503-1507.

Vogel, J.O. (1997) Bantu expansion. *In:* J.O. Vogel (ed.) *Encyclopedia of Precolonial Africa,* London, Altamira Press. p. 435-438.

Wade, W.D. (1971) The skeletal biology of human remains from sites in the Lake Kainji Area of Nigeria. *West African Journal of Archaeology,* 1, 61-85.

Warnier, J.-P. (1984) Histoire du peuplement et genèse des paysages dans l'ouest camerounais. *Journal of African History,* 25, 395-410.

Watson, E., Bauer, K., Aman, R., Weiss, G. and von Haeseler, A. (1996) mtDNA sequence diversity in Africa. *American Journal of Human Genetics,* 59, 437-444.

Watson, E., Forster, P., Richards, M. and Bandelt, H.-J. (1997a) Mitochondrial footprints of human expansions in Africa. *Annals of Human Genetics,* 61, 691-704.

Watson, E., Forster, P., Richards, M. and Bandelt, H.-J. (1997b) Génétique, linguistique et préhistoire du peuplement sub-saharien. *Bulletin du Centre Genevois d'Anthropologie,* 3, 3-21.

Weber, W., Nash, D.J., Moutulsky, A.G., Henneberg, M., Crawford, M.H., Martin, S.K., Goldsmid, J.M., Spedini, G., Glidewell, S. and Schanfield, M.S. (2000) Phylogenetic relationships of human populations in sub-Saharan Africa. *Human Biology,* 72, 753–772.

Weiner, J.S. (1954) Nose shape and climate. *American Journal of Physical Anthropology,* 12, 615-618.

Weiss, K.M. (1988) In search of times past: gene flow and invasion in the generation of human diversity. *In:* C.G.N. Mascie-Taylor and G.W. Lasker (eds) *Biological Aspects of Human Migration.* Cambridge, Cambridge University Press. p. 130-166.

Wendorf, F., Schild, R. and Close, A.E. (1984) *Cattle-keepers of the Eastern Sahara: the Neolithic of Bir Kiseiba*. Dallas, Southern Methodist University Press. 452p.

Wendorf, F. and Schild, R. (1994) Are the Early Holocene cattle in the Eastern Sahara domestic or wild? *Evolutionary Anthropology,* 3(4), 118-128.

Williamson, K. (1993) Linguistic evidence for the use of some tree and tuber food plants in southern Nigeria. *In:* T. Shaw, P. Sinclair, B. Andah and A. Okpoke (eds) *The Archaeology of Africa.* London, Routledge. p. 139-153.

Wilson, M. (1969) The Sotho, Venda and Tsonga. *In:* M. Wilson and L. Thompson (eds) *The Oxford History of South Africa.* Oxford, Oxford University Press. p. 131-182.

Windenmann, A. (1896) Untersuchung von 30 Dschaggaschädeln. *Archiv für Anthropologie,* XXV, 361-396.

Wood, E.T., Stover, D.A., Ehret, C., Destro-Bisol, G., Spedini, G., McLeod, H., Louie, L., Bamshad, M., Strassmann, B.I., Soodyall, H. and Hammer, M.F. (2005) Contrasting patterns of Y chromosome and mtDNA variation in Africa: evidence for sex-biased demographic processes. *European Journal of Human Genetics*, 13, 867–876.

Yakan, M.Z. (1999) *Almanac of African Peoples and Nations.* New Brunswick, Transaction Publishers. 755p.

Zar, J.H. (1984) *Biostatistical Analysis.* New Jersey, Prentice-Hall, Englewood Cliffs. 718p.

Zegura, S.L. (1975) Taxonomic congruence in Eskimoid populations. *American Journal of Physical Anthropology*, 43, 271-28.

Appendix I.

Detailed list of the ethnic groups under study for the modern sample.

Region	References	Country	Other geographical information	Vegetation area[1]	Linguistic affiliation[2]	Ethnic affiliation	Subsistence strategy[3]	Murdock's classification[4]	Presence of archives	Place of deposit[5]	Total sample
WEST AFRICA (N = 531)	Reinecke (1896)	Cameroon	Yaounde	3	NK nc bc	?	1	BA eqb	-	AMNH	1
	Dronscthilow (1913)		Dschang	5	NK nc bc	Bana	1	BA camh	X	VM	5
			Dschang	5	NK nc bc	Bangwa	1	BA camh	X	VM	19
			Bamenda	5	NK nc bc	Bali	1	BA camh	X	VM	22
			"grassl.","waldl."	5, 1	NK nc bc	Ndu	1	BA camh	X	VM	4
			Bamenda	5	NK nc bc	Bafulero, Bamum, Bameka?	1	BA camh	X	VM	5
			Dschang	5	NK nc bc	Basu (Tuala?)	1	BA camh	X	VM	4
			Bamenda	5	NK nc bc	Babungo	1	BA camh	X	VM	2
			"waldland"	1	NK nc bc	Bakussi	1	BA nwb (north)	X	VM	2
			"waldland"	5	NK nc bc	Bakundi	1	BA nwb (north)	X	VM	3
			Mbo district	5	NK nc bc	Mbo	1	BA nwb	X	VM	4
			Jabassi, Mabi	1	NK nc bc	(Fondschanda?)	1	BA nwb	X	VM	1
			Kamalimpe	1	NK nc bc	Kabo	1	BA nwb (kabo)	X	VM	2
			Ossidinge	1	NK nc bc	Balondo	1	BA nwb (north)	X	VM	1
			Duala	1	NK nc bc	(Abo?)	1	BA nwb (north)	X	VM	1
			Bare	5	NK nc bc	?	1	(BA nwb north)	X	VM	2
			Bascho	5	NK nc bc	Anjang	1	YB south.niger.(bantoid)	X	VM	5
			Ossidinge	5	NK nc bc	Banjangi	1	YB south.niger.(bantoid)	X	VM	7
			Nkore	5	NK nc bc	Keaka	1	YB south.niger.(bantoid)	X	VM	1
			Bascho	5	NK nc bc	(Anjang?)	1	YB south.niger.(bantoid)	X	VM	1

1 Codes of vegetation zones: **1** = tropical evergreen rainforest &/or mangrove forests; **2** = category 1 mixed with various other areas; **3** = papyrus swamps; **4** = evergreen forest (mediterranean shrubs, bushes and woodland); **5** = tall short grass savanna, open woodland &/or thorn bush with grassland; **6** = temperate mountain grassland &/or evergreen shrubs; **7** = sand stone and mountain deserts &/or semi-desert (shrubs, succulents and some grassland);

2 Linguistic affiliation categories: **AA** = Afro-Asiatic including Semitic (**sem**), Cushitic (**cu**) and Chadic (**cha**); **NS** = Nilo-Saharan including Eastern Sudanic (**es**), Central Sudanic (**cs**); **NK** = Niger-Kordofanian including Niger-Congo (**nc**) that also includes West Atlantic (**wa**), Mande (**ma**), Kwa (**kw**), Adamawa Eastern (**ae**) and Benue-Congo or Bantu (**ba**); **K** = KhoiSan;

3 Subsistence strategy categories: **1** = agriculturalists with or without raising animal and/or fishing; **2** = pastoralists; **3** = hunter-gatherers; **4** = special case; **?** = unknown;

4 Categories of Murdock's classification: **WS** = Western Sudan; **YB** = Yam Belt; **BA** = Bantu Africa that includes Cameroon highlands (**camh**), North-western Bantu (**nwb**), Equatorial Bantu (**eqb**), Mongo (**mon**), Central Bantu (**cba**), Kenya highlands (**kenh**), Interlacustrine Bantu (**intlb**), Nguni (**ngu**); **EAP** = East African pastoralists; **NWAP** = North West African pastoralists; **K** = KhoiSan; **P** = Pygmy;

5 Place of deposit: **AM** = Anatomy Museum, Cambridge; **AMNH** = American Museum of Natural History, New York; **BD** = Barnard Davis' Collection, UK; **BM** = British Museum, London; **Duckw** = Duckworth Laboratory, Cambridge; **FA** = Faculty of Anatomy, Dakar; **Grav** = Gravenzande Collection, Leiden; **IFAN** = IFAN Cheikh Anta Diop Institute, Dakar; **IPH** = Institut de Paléontologie Humaine, Paris; **IRScNB** = Institut Royal des Sciences Naturelles de Belgique, Brussels; **MH** = Musée de l'Homme, Paris; **NHM** = Natural History Museum, London; **MHN** = Museum d'Histoire Naturelle, Paris; **MM** = Munich Museum; **NH** = Netley Hospital, Army Medical Museum, UK; **RCS** = Royal College of Surgeons, London; **ULB** = Université Libre de Bruxelles (Anthropology & Genetics Laboratory); **V** = Vesaliannum, Basle; **VM** = Volkenkunde Museum, Berlin; **W** = Wroclaw University, Poland.

Appendix I continued

Region	References	Country	Other geographical information	Vegetation area[1]	Linguistic affiliation[2]	Ethnic affiliation	Subsistence strategy[3]	Murdock's classification[4]	Presence of archives	Place of deposit[5]	Total sample
WEST AFRICA (N = 531)	Fieldwork in 2000	Cameroon	Bamum region	5	NK nc bc	Bamum	1	BA camh	-	AMNH	1
			Duala	1	NK nc bc	?	1	BA nwb	-	AMNH	1
			Ibea	5	NK nc bc	?	1	BA nwb (north)	-	AMNH	1
			Mabea	5	NK nc bc	?	1	BA nwb (north)	-	AMNH	3
			Edea	1	NK nc bc	?	1	(BA nwb north)	-	AMNH	4
			Yaounde	5	NK nc bc	Bawa	1	BA eqb	-	AMNH	3
			Yaounde	5	NK nc bc	?	1	BA eqb	-	AMNH	1
			Banyang region	1	NK nc bc	Banyang	1	YB south.niger.(bantoid)	-	AMNH	1
			Grassfields?	5	NK nc bc	Boki, Anyang	1	YB south.niger.(bantoid)	-	AMNH	23
			?	?	?	?	(1)	?	-	AMNH	1
			Kumabemba?	?	?	?	(1)	?	-	AMNH	1
	Keith (1911)	Nigeria	Ejagham, Oban?	1	NK nc	Ekoi	1	YB south.niger.(bantoid)	-	RCS	5
			Bonny	1	NK nc	Ogoni (Kana)	1	YB south.niger.	-	RCS	5
	Dronscthilow (1913)		south east	1	NK nc	Ekoi	1	YB south.niger.(bantoid)	-	VM	1
	Fieldwork in 1998		(Onatshi)?	1	NK nc kw	Ibo (Igbo)	1	YB south.niger.(centre)	-	NHM	35
	Fieldwork in 1999		Hausa region	1	AA cha	Haoussa (Hausa)	1	WS (Plateau Niger.)	-	IRScNB	1
	Fieldwork in 2000		Calabar	1	NK nc kw	?	1	YB south.niger.	-	AMNH	12
			Calabar, Creektown	1	NK nc kw	?	1	YB south.niger.	-	AMNH	2
			Calabar, Atakpa	1	NK nc kw	?	1	YB south.niger.	-	AMNH	7
			Calabar, Cross River	1	NK nc kw	?	1	YB south.niger.	-	AMNH	6
			(Ibadan)?	1	NK nc kw	Yoruba	1	YB south.niger.(Yoruba)	-	AMNH	1
	Fieldwork in 2000	Benin	Abome(y)?	1	NK nc kw	?	1	YB (Twi?)	-	AMNH	3
			?	?	NK nc kw	?	(1)	YB (Twi?)	-	AMNH	1
			Whydah (Ouidah)	1	NK nc kw	?	1	YB (Twi?)	-	AMNH	1
			?	?	NK nc kw	?	(1)	YB (Twi?)	-	AMNH	9
	Fieldwork in 2000	Togo	?	5	NK nc kw	Yendi (Mole-Dagbane)	1	YB (Twi?)	-	AMNH	2
			?	5	NK nc kw	Ewe	1	YB (Twi?)	-	AMNH	8
	Shrubsall (1899)	Ghana	Kumasi	1	NK nc kw	Ashanti (Asante)	1	YB (Twi, Akan)	-	NH	58
	Fieldwork in 2000		?	1	NK nc kw	Ashanti (Asante)	1	YB (Twi, Akan)	-	AMNH	38
			Gold Coast	1	NK nc kw	?	1	YB (Twi)	-	AMNH	3
			Gold C., Elmina	1	NK nc kw	?	1	YB (Twi)	-	AMNH	4
			Gold C., Aksim	1	NK nc kw	?	1	YB (Twi)	-	AMNH	2
			Gold C., Salaga	5	NK nc kw	?	1	YB (Twi)	-	AMNH	7
			Gold C., Tshi	1	NK nc kw	?	1	YB (Twi)	-	AMNH	1
			Gold C., Djago	1	NK nc kw	?	1	YB (Twi)	-	AMNH	6
			Gold C., Anyako	1	NK nc kw	?	1	YB (Twi)	-	AMNH	1
	Dronstchilow (1913)	Liberia	Monrovia	1	(NK nc ma)	?	1	YB (Kru & Periph.Ma.)	-	VM	1
	Fieldwork in 2000		Vai Camp	1	(NK nc ma)	?	1	YB (Kru & Periph.Ma.)	-	AMNH	1
			Monrovia	1	(NK nc ma)	?	1	YB (Kru & Periph.Ma.)	-	AMNH	2
	Dronscthilow (1913)	Guinea?	?	?	NK nc ma	Mandingo	1	WS (Mande)	-	AMNH	4
	Knip (1971)	Mali	Nokara	3	?	?	(1)	?	-	Grav	76
	Fieldwork in 2000		"	3	?	?	(1)	?	-	Grav	20

Appendix I continued

Region	References	Country	Other geographical information	Vegetation area[1]	Linguistic affiliation[2]	Ethnic affiliation	Subsistence strategy[3]	Murdock's classification[4]	Presence of archives	Place of deposit[5]	Total sample
WEST AFRICA (N = 531)	Fieldwork in 2000	Senegal	?	?	?	(Kere)?	?	?	-	FA	1
			north east	5	NK nc wa	Toucouleur (Tukulor)	1	NWAP	-	FA	1
			?	?	(NK nc wa)	(Moussa?)	(1)	(WS)	-	FA	2
			?	?	?	?	?	(WS)	-	FA	5
			Mande region	5	NK nc ma	Malinke	1	(WS)	-	FA	1
			?	?	?	?	?	?	-	FA	15
			Diola region	1	NK nc ma	Diola	1	WS (Senegambia)	-	IFAN	6
			Serer region	5	NK nc wa	Griots Serer	4	WS (Senegambia)	-	IFAN	15
			Lebu region	5	NK nc wa	Griots Lebu	4	WS (Senegambia)	-	IFAN	30
CENTRAL AFRICA (N = 447)	Shrubsall (1898)	D.R.C.	Low Congo	5	NK nc bc	?	1	BA cba (Kongo)	-	AM	3
	Benington (1912)		Makunji	1	NK nc bc	Batetela (Tetela)	1	BA cba (Mongo)	X	(2) destroyed	77
	Fieldwork in 1999		Low Congo	5	NK nc bc	?	1	BA cba (Kongo)	-	IRScNB	6
			Kabinda	5	NK nc bc	?	1	BA cba (Kongo)	-	IRScNB	2
			Ngudi Valley	5	NK nc bc	Basundi	1	BA cba (Kongo)	-	IRScNB	1
			Feshi	5	NK nc bc	Basuku (Batsamba)	1	BA cba (Kwango)	X	IRScNB	168
			Lusinga,W.Tang.	5	NK nc bc	Warua	1	BA cba (Bemba)	-	IRScNB	1
			Mpala, Maribu	5	NK nc bc	Warua	1	BA cba (Bemba)	-	IRScNB	1
			Sakania	5	NK nc bc	?	1	BA cba	-	IRScNB	1
			Kassai	5	NK nc bc	?	1	BA cba	-	IRScNB	1
			Costermanville	1	(NK nc bc)	(Kashusha)?	1	(BA egb)	-	IRScNB	1
			close to equator?	1	NK nc bc	Bangala	1	BA egb	-	IRScNB	1
			Befale?	1	NK nc bc	Mongo	1	BA egb	-	IRScNB	6
			Kisangani	1	NK nc bc	Bassoko (Basoko)	1	BA egb	-	IRScNB	3
			Songe (Uvira)	1	NK nc bc	Bafolero (Bafulero)	1	BA intlb (Rwanda)	-	IRScNB	1
			Beni	5	(NK nc bc)	?	1	BA (Mambuti)	-	IRScNB	1
			Mitombu	5	(NK nc bc)	?	1	(BA)	-	IRScNB	1
			Coquithatville	1	(NK nc bc)	?	1	(BA)	-	IRScNB	1
			Niaju	5	NS cs	Makere	1	YB Cent.Sud.(Mangbetu)	-	IRScNB	1
			Baka region	?	NS cs	Baka	1	YB Cent.Sud.(Madi)	-	IRScNB	1
			Dungu	5	NK nc bc	Mayanga	1	YB Cent.Sud.(Mangbetu)	-	IRScNB	1
	Fieldwork in 2000		Medje region	1	NS cs	Medje	1	YB Cent.Sud.(Mangbetu)	-	AMNH	1
	Fieldwork in 1999		Uele	5	NK nc bc	Azande	1	YB east.nigr.(Equator)		IRScNB	6
			Uele	5	NK nc bc	Mamvu	1	YB east.nigr.(Equator)	-	IRScNB	7
			(Ulese)?	?	?	?	?	?	-	IRScNB	1
			Buka Tsona	5	(NK nc bc)	?	1	?	-	IRScNB	1
			Goma	6	NK nc bc	?	?	?	-	IRScNB	1
	Shrubsall (1898)	Congo	Ubangui	1	NK nc bc	?	1	(BA)	-	AM	2
	Fieldwork in 1999		Alima N'Gunda	1	NK nc bc	Tegue (Teke)	1	BA nwb (south)	-	IRScNB	2
	Poutrin (1910)	Gabon	Gabon river mouth	1	NK nc bc	Kama	1	BA nwb (south)	X	MHN	3
			Gabon river mouth	1	NK nc bc	M'Pongwe	1	BA nwb (south)	X	MHN	2
			Campo, Coast	1	NK nc bc	Benga-Aka	1	BA nwb (north)	X	MHN	1
			Ikobaye Valley	1	NK nc bc	Ashango	1	BA nwb (south)	X	MHN	2

Appendix I continued

Region	References	Country	Other geographical information	Vegetation area[1]	Linguistic affiliation[2]	Ethnic affiliation	Subsistence strategy[3]	Murdock's classification[4]	Presence of archives	Place of deposit[5]	Total sample
CENTRAL AFRICA (N = 447)	Poutrin (1910)	**Gabon**	Ba-NJavi	1	NK nc bc	Adouma	1	BA nwb (south)	X	MHN	2
			Ogoue-Ngounie	1	NK nc bc	N'Javi	1	BA nwb (south)	X	MHN	2
	Benington (1912)		N'Komi region	1	NK nc bc	N'Komi	1	BA nwb (south)	X	NHM	133
	Fieldwork in 1999	**Angola**	Loango	5	NK nc bc	?	1	BA cba	-	IRScNB	1
		?	?	?	?	?	?	?	-	IRScNB	1
PYGMY (N = 28)	Poutrin (1910)	**Gabon**	High Ngounie	1	NK nc bc	Akoa	3	West-Central Pygmy	X	MHN	1
			High Ngounie	1	NK nc bc	O'Bongo	3	West-Central Pygmy	X	MHN	2
	Marquer (1972)		(inland)	1	NK nc bc	Babongo	3	West-Central Pygmy	-	MHN	10
		Congo	(central region)	5	NK nc bc	Babinga	3	West-Central Pygmy	-	MHN	10
		Cameroon	south east	1	NK nc bc	?	3	West-Central Pygmy	-	MHN	1
	Fieldwork in 1998	**D.R.C.**	Ituri bassin	5	NK nc bc	Efe (Irumu)	3	Eastern Pygmy	X	MHN	3
	Fieldwork in 1999		?	?	NK nc bc	(Twa)?	3	Eastern Pygmy	-	IRScNB	1
EAST AFRICA (N = 577)	Hiernaux (1972)	**Rwanda**	Gaseke	6	NK nc bc	Tutsi (king)	2	EAP (Nilotes)	-	?	1
	Fieldwork in 1999		Kivu L., Kibuye	6	NK nc bc	Tutsi	2	EAP (Nilotes)	-	IRScNB	1
			Kisenyi, Bugoye	6	NK nc bc	Bahutu (Hutu)	1	BA intlb (Rwanda)	X	IRScNB	15
			Ruhengeri (cave)	6	NK nc bc	Hutu	1	BA intlb (Rwanda)	X	IRScNB	88
			Kivu, Kisenyi	6	NK nc bc	?	?	(BA intlb Rwanda)	-	IRScNB	1
			Ngweshe	6	NK nc bc	?	1	(BA intlb Rwanda)	X	IRScNB	1
	Górnay (1957)	**Uganda**	(hospitals)?	9	NK nc bc	?	?	?	-	W	176
	Fieldwork in 1999	**Burundi**	Usumbura	6	NK nc bc	(Murundi)?	?	BA intlb (Rwanda)	-	IRScNB	1
	Hamy (1882)	**Somalia**	Medjeurtin	5	AA cu	Somali	2	EAP	-	MHN	2
	Sergi (1912)		?	5	AA cu	Somali	2	EAP	-	VM	5
	Lester (1927)		?	5	AA cu	Somali-Abgal	2	(EAP)	-	MHN	2
			Mogadiscio	5	AA cu	(Somali)?	2	EAP	-	MHN	3
	Fieldwork in 1998		(Diredawa)	6	AA cu	Somali (Beja, Galla)	2	EAP	X	Duckw	33
	Sergi (1912)	**Ethiopia**	Goba	5	AA cu	Galla	2	EAP	-	VM	5
	Lester (1928)		Goba, Kadjoa	6	AA cu	Galla-Arusi	2	EAP	-	MHN	20
			Oborrah	6	AA cu	Galla-Itu	2	EAP	-	MHN	1
			Goba	6	AA cu	Various Galla	2	EAP	-	MHN	1
	Reinecke (1896)	**Kenya**	?	5	NS es	Massai	2	EAP (Nilotes)	-	MM	8
			Pare mountains	5	NK nc bc	Wapare (Pare)	1	BA kenh	-	MM	4
			?	5	NK nc bc	Mschambaa (Shambaa)	1	BA kenh	-	MM	2
	Widenmann (1896)		Kilimandjaro	5	NK nc bc	Dschagga	1	BA kenh	-	VM	30
	Kitson (1931)		Dabida	5	NK nc bc	Teita (Taita)	1	BA kenh	-	Duckw	1
	Fieldwork in 2000		Dabida	5	NK nc bc	Teita (Taita)	1	BA kenh	-	Duckw	138
	Fieldwork in 1998	**Tanzania**	Musira Island	5	NK nc bc	Haya	1	BA intlb (Uganda)	X	Duckw	38
BANTU-SPEAK-ING SOUTH AFRICA (N = 125)	Shrubsall (1898)	**Mozambique**	Zambezi River	5	NK nc bc	Wahenga	1	(BA)	-	BM	2
			?	?	NK nc bc	(Urega)?	(1)	(BA)	-	BM	2
			Quiloa	5	NK nc bc	Wahiou (Wahiao, Yao)	1	BA cba (Yao)	-	RCS	10
			Zambezi River	5	NK nc bc	Manganja (Nyanja)	1	BA cba (Maravi)	-	RCS	6
			?	?	NK nc bc	(M'Kambo)?	1	?	-	BM	1

Appendix I continued

Region	References	Country	Other geographical information	Vegetation area[1]	Linguistic affiliation[2]	Ethnic affiliation	Subsistence strategy[3]	Murdock's classification[4]	Presence of archives	Place of deposit[5]	Total sample
BANTU-SPEAK-ING SOUTH AFRICA (N = 125)	Shrubsall (1898)	**South Africa**	Natal	6	NK nc bc	"Kaffirs" (Xhosa)	2	BA ngu	-	BD	42
			Natal	6	NK nc bc	Zulu	1	BA ngu	-	NHM	24
			Cape	4	NK nc bc	Angoni (Nguni, Ngoni)	1	BA ngu	-	BM	25
			?	?	NK nc bc	?	(1)	(BA)	-	Duckw	6
	Salmons (1925)		Tzitzikama	?	NK nc bc	?	(1)	(BA)		?	7
KHOISAN (N = 116)	Shrubsall (1897)	**South Africa**	Transvaal	5	K	San and/or Khoikhoi	3	KhoiSan	-	V; AM	4
			Port Elizabeth	4	K	San and/or Khoikhoi	3	KhoiSan	-	V; AM	3
			Orange River	7	K	San and/or Khoikhoi	3	KhoiSan	-	RCS	1
			Koranna	7	K	San and/or Khoikhoi	3	KhoiSan	-	RCS; AM	2
			Cape region	4	K	San and/or Khoikhoi	3	KhoiSan	-	BM; RCS; AM	30
			(Makapan)?	(5)	K	San and/or Khoikhoi	3	KhoiSan	-	RCS	6
			?	7	K	San and/or Khoikhoi	3	KhoiSan	-	NH	16
	Shrubsall (1907)		?	7	K	San and/or Khoikhoi	3	KhoiSan	-	?	25
	Salmons (1925)		?	?	K	San and/or Khoikhoi	3	KhoiSan	-	?	3
	Fieldwork in 1998		Knysna Cave	4	K	San and/or Khoikhoi	3	KhoiSan	-	Duckw	2
			Cape	4	K	San and/or Khoikhoi	3	KhoiSan	-	Duckw	2
			?	?	K	San and/or Khoikhoi	3	KhoiSan	-	Duckw	3
	Fieldwork in 2000		(Zeekoevlei)?	?	K	San and/or Khoikhoi	3	KhoiSan	-	AMNH	1
			Koranna	7	K	San and/or Khoikhoi	3	KhoiSan	-	AMNH	1
			Murraysburg	7	K	San and/or Khoikhoi	3	KhoiSan	-	AMNH	2
			Douglas	7	K	San and/or Khoikhoi	3	KhoiSan	-	AMNH	1
			Orange River	7	K	San and/or Khoikhoi	3	KhoiSan	-	AMNH	2
			Durban	6	K	San and/or Khoikhoi	3	KhoiSan	-	AMNH	1
			Cape	4	K	San and/or Khoikhoi	3	KhoiSan	-	AMNH	1
			Great Karoo	4	K	San and/or Khoikhoi	3	KhoiSan	-	AMNH	1
			?	?	K	San and/or Khoikhoi	3	KhoiSan	-	AMNH	2
	Shrubsall (1897)	**(Botswana)?**"	Kalahari	5	K	San and/or Khoikhoi	3	KhoiSan	-	BM	2
	Fieldwork in 1998		Northern Kalahari	5	K	San and/or Khoikhoi	3	KhoiSan	-	NHM	5
NORTH AFRICA (N = 82)	Fieldwork in 1998	**(Mauritania)?**	?	7	AA sem	Maure	2	NWAP	-	IFAN	1
	Storto (*pers.comm.*)	**4 countries**	(various)	(various)	(AA sem)	(Algeria, Tenerife, Morocco, Tunisia)	-	-	-	MH	81
	Howells (1996)	(various)	(various)	(various)	(AA sem)	(various)	-	-	-	(various)	10
WORLD-WIDE	Howells (1996)	**7 broad regions**	(various)	(various)	(various)	(Europe, Americas, Polynesia & West Pacific, Australia & Melanesia, India, Eskimos, Far East & South East Asia)	-	-	-	(various)	465
										TOTAL =	**2381**

Appendix II

Various grouping variables for testing morphological differentiation within modern Africa.

BROAD REGIONS

I	West Africa
II	Central Africa
III	Pygmies
IV	East Africa
V	South Africa
VI	KhoiSan

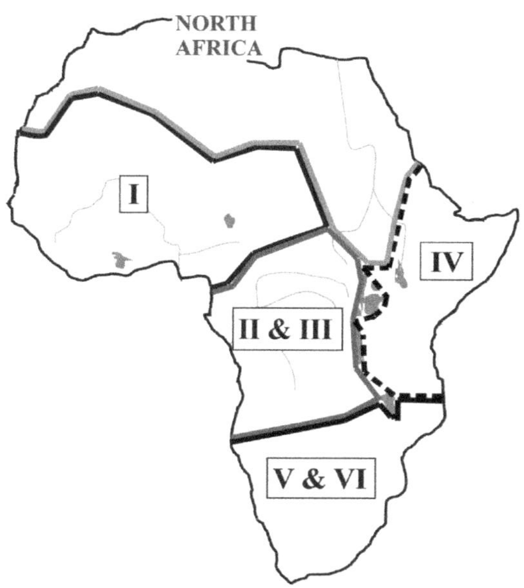

LATITUDE AREAS

1	Above 15.9°	5	4.9° to 1°
2	15.9° to 13°	6	0.9° to -2°
3	12.9° to 9°	7	-2.1° to -5°
4	8.9° to 5°	8	-5.1° to -8°
5	4.9° to 1°	9	Below -8°

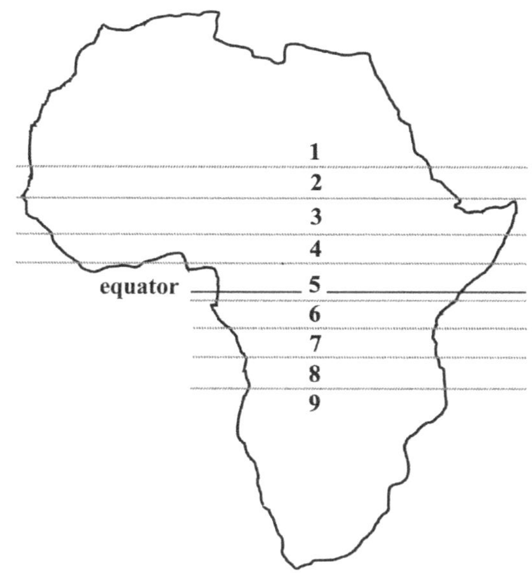

LONGITUDE AREAS

1	Above 36.9°	5	10.9° to 4°
2	36.9° to 29°	6	3.9° to -4°
3	28.9° to 20°	7	Below -4°
4	19.9° to 11°		

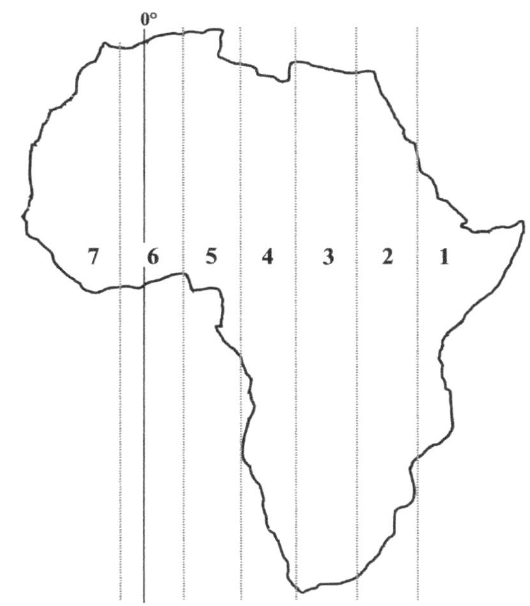

Appendix II continued

DISTANCE FROM 0° LATITUDE

1	0° to 3°
2	3.1° to 6°
3	6.1° to 9°
4	9.1° to 12°
5	12.1° to 16°
6	Above 16°

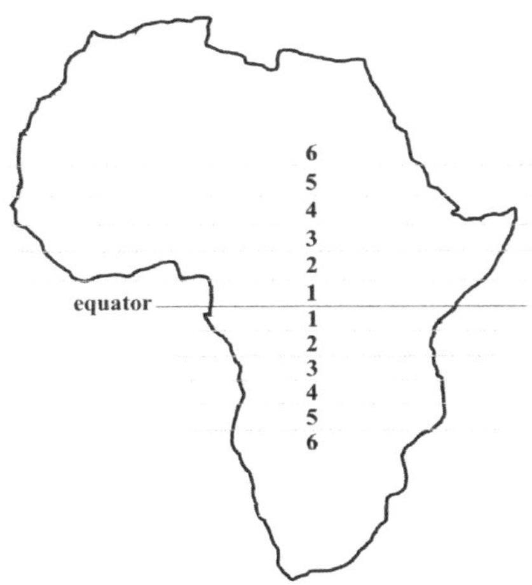

VEGETATION AREAS

1	tropical evergreen rainforest mangrove forest
2	category 1 mixed with various other areas
3	papyrus swamps
4	evergreen forest (mediterranean shrubs, bushes, woodland)
5	tall short grass savanna
	open woodland
	thorn bush with grassland
6	temperate mountain grassland
	Ethiopian evergreen shrubs
7	sand stone & mountain desert
	semi-desert
	riverine cultivation & oases
	mountain coniferous forest

(Philip, 1998)

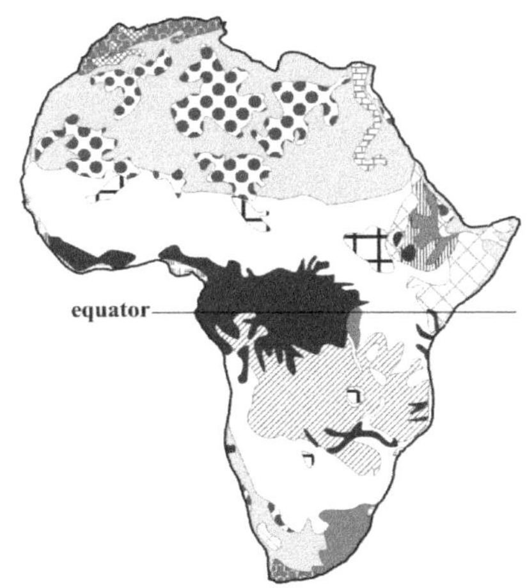

MEAN HUMIDITY

1	22.5 (gm/kgm)
2	20 (gm/kgm)
3	17.5 (gm/kgm)
4	15 (gm/kgm)
5	12.5 (gm/kgm)
6	10 (gm/kgm)
7	7.5 (gm/kgm)
8	5 (gm/kgm)
9	2.5 (gm/kgm)

(Jackson, 1961)

JANUARY

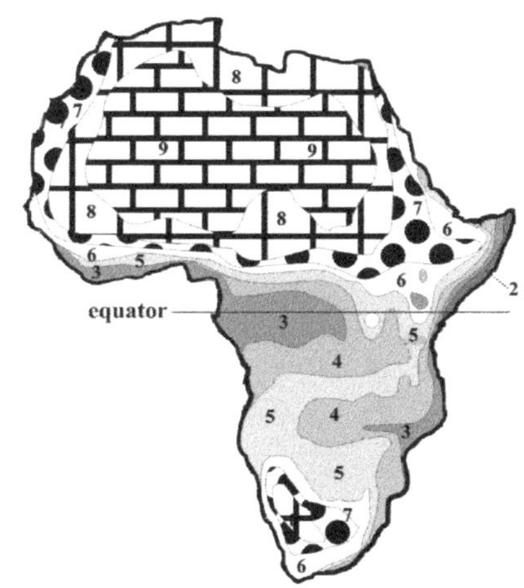

JULY

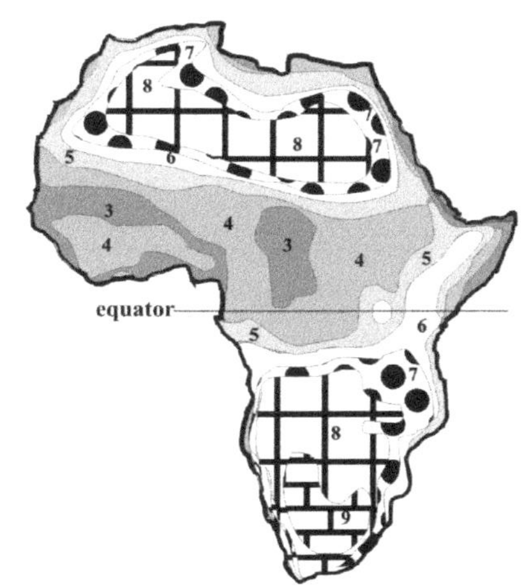

Appendix II continued

LANGUAGES

1 West Atlantic, Mande, Kwa
2 Adamawa Eastern
3 West & Central Bantu (Benue-Congo)
4 Eastern Bantu (Benue-Congo)
5 Southern Bantu (Benue-Congo)

Niger-Congo

6 Nilo-Saharan
7 Afro-Asiatic
8 KhoiSan +

(Greenberg, 1964; Holden, 2001)

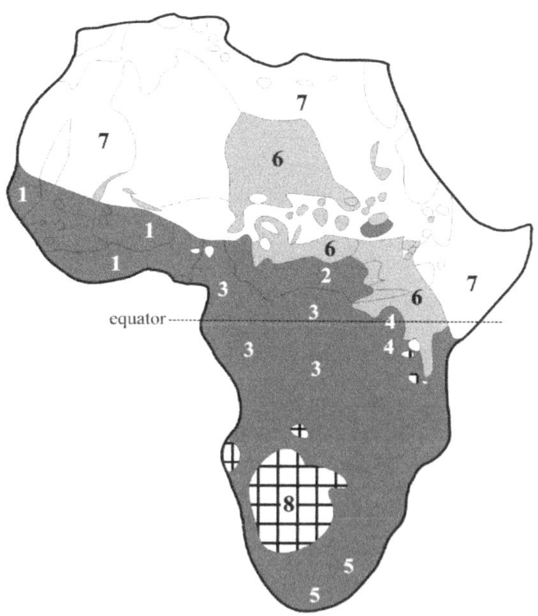

MURDOCK'S UNITS

1 Cameroon highlands
2 North Western
3 Equatorial
4 Mongo
5 Luba
6 Central group
7 Kenya highlands
8 Interlacustrine
9 Nguni

Bantu

10 Western Sudan +
11 Yam belt
12 East African pastoralists
13 North West African pastoralists
14 Pygmy
15 KhoiSan

non-Bantu

(Murdock, 1959)

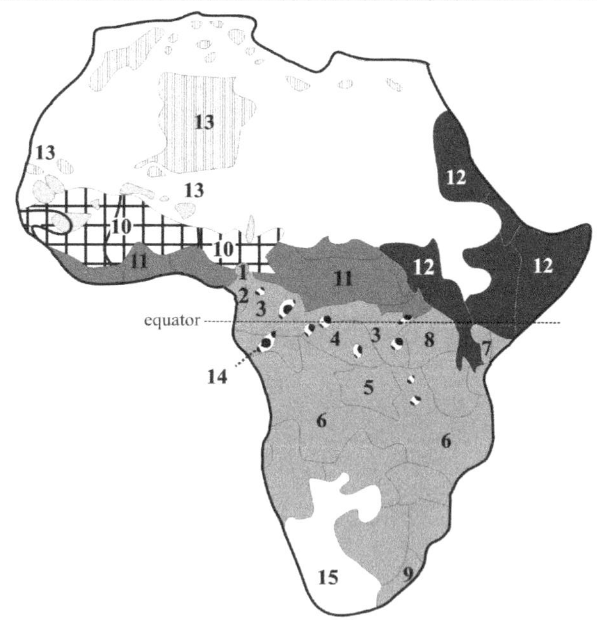

Appendix III

Detailed list of the sites under study for the archaeological sample.

Region	References	Country	Other geographical information	Date	Comments	Place of deposit[1]	Total sample
WEST AFRICA (N = 429)	Dutour (1989)	**Mali**	Hassi-el-Abiod	LSA-IA		?	27
	Georgeon *et al.* (1992)		Kobadi	Iron Age (2,415±120 BP)		?	5
	Fieldwork in 2000	**Senegal**	Walalde	Iron Age ($>6^{th}$ C. BC)	Abu Deme (*pers.comm.*)	IFAN	4
			Cayar	Iron Age (1st mill. BC)	Thilmans (*pers.comm.*)	IFAN	2
	Huizinga *et al.* (1967)	**Mali**	Tellem Cave A	Iron Age (390 BC)		Grav	71
	Treinen (1975)	**Central African Republic**	Koro Toro (Nemra)	Iron Age ($<5^{th}$ C. AD)		?	1
	Fieldwork in 2000	**Senegal**	Cap des Biches	Iron Age (1,300-1,900 BP)	Thilmans (*pers.comm.*)	IFAN	12
			Faboura	Iron Age(1,315-1,940±80 BP)	Descamps & Thilmans (1997)	IFAN	13
			Dioronboumak	Iron Age (1,140-1,580±80 BP)		IFAN	18
	Wade (1971)		Kainji	Late Iron Age (>1,500 AD)		?	10
	Huizinga *et al.* (1967)	**Mali**	Tellem Cave C	Iron Age (1,090-1,170 AD)		Grav	136
			Tellem Cave B	Iron Age (1,295 AD)		Grav	14
			Tellem Cave E	Iron Age (12th-13th C. AD)		Grav	18
			Tellem Cave F	Iron Age (1,560 AD)		Grav	34
			Tellem Cave H	Iron Age (12th-13th C. AD)		Grav	36
			Tellem Cave P	Iron Age (1,270 AD)		Grav	13
			Tellem Cave Z	Iron Age (12th-13th C. AD)		Grav	15
CENTRAL AFRICA (N = 26)	Fieldwork in 2000	**D.R.C.**	Ishango	Mesolithic (?) (7,000-750 BP)	Boaz *et al.* (1990)	IRScNB	6
	Ribot *et al.* (2001)	**Cameroon**	Shum Laka	LSA-IA (2,940±60 BP)	de Maret *et al.* (1993)	IRScNB	1
	Fieldwork in 2000		Sanga	Iron Age (8th-15th C. AD)	de Maret (1977, 1985)	IRScNB	6
	Gutierrez & Valentin (1995)	**Angola**	Kapanda	Iron Age (16th-18th C. AD)	Tumuli	?	6
EAST AFRICA (N = 13)	Fieldwork in 1998	**Kenya**	Gamble Cave	Later Stone Age (8,210±260 BP)	Prostch (1978) (C^{14} dating)	NHM	2
	Fieldwork in 2000		Bromhead (& various)	Elmenteita: LSA-IA (7,410±160 BP)	Prostch (1978) (C^{14} dating)	NHM	11
NORTH AFRICA (N = 131)	Lahr (*pers.comm.*)	**Morocco**	Taforalt	Iberomaurusian (13,750 BC)	Chamla (1978)	IPH	21
		Algeria	Afalou-bou-Rhummel	Iberomaurusian (10,500-8,500 BC)	Chamla (1978)	IPH	36
	Knip (1970)	**Sudan**	Duweishat	Iron Age (Christian period)		Grav	37
			Attiri	Iron Age (550-1,350 AD)		Grav	34
	Fieldwork in 2000		Attiri	Iron Age (550-1,350 AD)		Grav	2
						TOTAL =	**585**

[1] Place of deposit: **AM** = Anatomy Museum, Cambridge; **AMNH** = American Museum of Natural History, New York; **BD** = Barnard Davis' Collection, UK; **BM** = British Museum, London; **Duckw** = Duckworth Laboratory, Cambridge; **FA** = Faculty of Anatomy, Dakar; **Grav** = Gravenzande Collection, Leiden; **IFAN** = IFAN Cheikh Anta Diop Institute, Dakar; **IPH** = Institut de Paléontologie Humaine, Paris; **IRScNB** = Institut Royal des Sciences Naturelles de Belgique, Brussels; **MH** = Musée de l'Homme, Paris; **NHM** = Natural History Museum, London; **MHN** = Museum d'Histoire Naturelle, Paris; **MM** = Munich Museum; **NH** = Netley Hospital, Army Medical Museum, UK; **RCS** = Royal College of Surgeons, London; **ULB** = Université Libre de Bruxelles (Anthropology & Genetics Laboratory); **V** = Vesaliannum, Basle; **VM** = Volkenkunde Museum, Berlin; **W** = Wroclaw University, Poland.

Appendix IV

Detailed list of the metric traits under study for the cranium.

Instrument[1] used for measurement	Both compiled &/or collected	Only Compiled	Shrubsall, 1897 *Biometrika*	Martin & Saller, 1959	Howells, 1989	Lahr, 1996	Other references & comments	Anatomical definition	Code used
15 VARIABLES									
(compiled)		X	C				Górnay (1957)	cranial capacity (cm^3)	CC
SC	X		L	1	GOL	GOL	Górnay (1957)	maximum cranial length (glabella–opisthocranion or close)	GOL
SC	X			1d	NOL	NOL		nasion-opistocranion length	NOL
SC	X		B	8	XCB	XPB	Górnay (1957)	maximum cranial breadth (between parietals, perpendicular to sagittal plane)	XCB
SC	X		Biaster.B	12	ASB	ASB	Górnay (1957); Kitson (1931)	biasterionic breadth (occipital)	ASB
SC	X		S1'	29	FRC	FRC	Górnay (1957)	frontal chord (nasion-bregma)	FRC
SC	X		S2'	30	PAC	PAC	Górnay (1957)	parietal chord (bregma-lambda)	PAC
SC	X		S3'	31	OCC	OCC		occipital chord (lambda-opisthion)	OCC
T	X		Q'(bregma)	24a				transverse arch from each auricular passage (over bregma)	Q
T	X		Q'	24b				transverse arch from each auricular passage (perpendicular to horizontal plane)	Q1
T	X		S1	26		FARCH	Górnay (1957)	frontal arch (nasion-bregma)	FAR
T	X		S2	27		PARCH		parietal arch (bregma-lambda)	PAR
T	X		S3	28		OARCH		occipital arch (lambda-opisthion)	OAR
T	X							Sum of FAR, PAR & OAR	TOTARCH
SC	X		H'	17	BBH	BBH	Górnay (1957)	basion-bregma height (total height)	BBH
11 VARIABLES									
SC	X			13			Górnay (1957)	bimastoid breadth (mastoideale-mastoideale)	BIMAS.BR
SC	X					EAMEAM		breadth from each external auditory meatus	EAM/EAM
SC	X		fml	7	FOL	FOL		foramen magnum length	FOL
SLC	X		fmb	16		FOB		foramen magnum breadth	FOB
SLC	X						Giles & Elliot (1963)	mastoid breadth	MDB
SC	X			14	WCB	WCB		minimum cranial breadth	WCB
SLC	X					WBAB		minimum breadth of basisphenoid (perpendicular to long axis)	WBAB
CC	X					PALBA		chord from basion to posterior spine of palate	PALBA
CC	X					D1		subtense at hormion using previous measurement	D1
CC	X					PALOP		chord from opisthion to posterior spine of palate	PALOP
CC	X					D2		subtense at hormion using previous measurement	D2

[1] Instrument used for measurement: **SC** = spreading caliper; **SLC** = sliding caliper; **CC** = coordinate caliper; **T** = tape; **M** = mandibulometer.

Appendix IV continued

Instrument[1] used for measurement	Both compiled &/or collected	Only Compiled	Shrubsall, 1897 *Biometrika*	Martin & Saller, 1959	Howells, 1989	Lahr, 1996	Other references & comments	Anatomical definition	Code used
25 VARIABLES									
SC	X		LB	5	BNL	BNL	Górnay (1957)	basion-nasion length	BNL
SC	X		GL	40	BPL	BPL	Górnay (1957)	basion-prosthion length	BPL
SLC	X		G'H	48	NPH	NPH	Górnay (1957)	nasion-prosthion height (upper facial height)	NPH
SLC	X		GH	47			Górnay (1957)	nasion-gnathion (total facial height)	NASGNA
SC	X		J	45	ZYB	ZYB	Górnay (1957)	bizygomatic breadth (maximal breadth between zygomatic arches)	ZYB
SLC	X		(internal biorbital?)	44	EKB	EKB	Górnay (1957)	biorbital breadth (ectoconchion-ectoconchion)	EKB
SC	X			45.1	JUB	JUB	Górnay (1957)	bijugal breadth (jugale-jugale)	JUB
SC	X		St	10b	STB	STB	Górnay (1957)	bistephanic breadth (stephanion-stephanion)	STB
SC	X			10	XFB	XFB	Górnay (1957)	maximum frontal breadth (on coronal suture but not on stephanion)	XFB
SLC	X		B'	9		WFB	Górnay (1957)	minimal frontal breadth (frontotemporalia)	WFB
SLC	X			43.1	FMB	FMB		bifrontal breadth (using fronto-malare anterior)	FMB
CC	X			46	ZMB	ZMB	Górnay (1957)	bimaxillary breadth (zma-zma)	ZMB
SLC	X				IML			inferior malar length (from zygomaxillare anterior to lowest point of zygotemporal suture on external surface)	IML
SLC	X				XML			maximum malar length (from zygo-orbitale to lowest point of zygotemporal suture on external surface)	XML
SLC	X			61	MAB	MAB	Górnay (1957)	maximum maxillo-alveolar breadth or external palate breadth (taken at level of ectomolare or M2 in general)	MAB
SLC	X			60			Górnay (1957)	maxillo-alveolar length (prosthion-alveolon)	MAL
SLC	X		G1	62a				internal palate length (orale-alveolon)	PAL
SLC	X		G2	63			Górnay (1957)	palate breadth between endomolare (M2)	PABR
SLC	X		PH					premaxillary height (from prosthion to anterior nasal spine)	PMH
SLC	X		NH (NH')	55	NLH	NLH	Górnay (1957)	nasal height (from nasion to inferior margin or lowest point of nasal roof); in some cases taken from nasion to nasospinale (NH')	NLH
SLC	X		NB	54	NLB	NLB	Górnay (1957)	maximal nasal breadth (alare-alare)	NLB
CC	X		SC	57	WNB	WNB		minimum nasal breadth or simotic chord	WNB
SLC	X		DC	49a	DKB	WIOB	Kitson (1931)	interorbital breadth (dacryon-dacryon)	DKB
CC	X		O1'	51a	OBB	OBB		orbital breadth (ectoconchion-dacryon)	OBB
SLC	X		O2	52	OBH	OBH	Górnay (1957)	orbital height (perpendicularly to OBB); internal measurement	OBH

Appendix V

Detailed list of the metric traits under study for the mandible.

Instrument[1] used for measurement	Both compiled &/or collected	Only Compiled	Shrubsall, 1897 *Biometrika*	Martin & Saller, 1959	Howells, 1989	Lahr, 1996	Other references & comments	Anatomical definition	Code used
10 VARIABLES									
SLC	X		W1	65		MAXB	Górnay (1957)	bicondylar breadth (lateral point of condyle)	bb
SLC	X		GoGo	66		BIGO	Górnay (1957)	bigonial width (gonion-gonion)	bw
M	X		ml	68.1		MAX= MANDL		maximum projective length	mpl
SLC	X		h1	69		ANTH	Górnay (1957)	symphyseal height (infradentale-gnathion)	sh
SLC	X		m2h	69.2				height of body at level of M2	bmb2
SLC	X			70				not projective ascending ramus height (gonion-condylion)	har
M	X		rl	70a		CONDH		similar to previous, but projective chord	phar
SLC	X		rb'	71a		RAMUSB	Górnay (1957)	minimal width of ascending ramus	war
SLC	X							maximal width of ascending ramus	mwar
M	X		M<	79			Górnay (1957)	mandibular angle	ma

[1] Instrument used for measurement: **SC** = spreading caliper; **SLC** = sliding caliper; **CC** = coordinate caliper; **T** = tape; **M** = mandibulometer.

Appendix VI

Descriptive data (in mm) with paired t-tests and correlation coefficients for the various intra-observer data sets.

Kenyan cranial sample

Variable code	N	Mean 1a	Mean 1b	Mean difference	t[1]	r[2]
GOL	15	182.23	182.57	0.34	-	0.98
XCB	15	133.00	133.40	0.40	-	0.99
BBH	15	134.30	133.80	0.50	*	0.98
BNL	15	99.93	99.43	0.50	-	0.98
BPL	15	95.27	95.41	0.14	-	0.99
ZYB	15	124.63	125.27	0.64	-	0.89
NPH	15	69.16	69.39	0.23	-	0.92
NLB	15	25.18	25.22	0.04	-	0.92
OBB	15	39.26	39.36	0.10	-	0.92
EKB	15	95.07	95.43	0.36	-	0.98
DKB	15	22.40	22.74	0.34	*	0.96

Rwandese cranial sample

Variable code	N	Mean 2a	Mean 2b	Mean difference	t[1]	r[2]
GOL	28	183.29	183.30	0.01	-	0.99
NOL	28	181.69	181.86	0.17	-	0.99
XCB	29	133.07	132.95	0.12	-	0.98
ASB	29	107.50	107.00	0.50	-	0.97
FRC	30	106.85	106.37	0.48	-	0.90
PAC	29	111.40	111.00	0.40	-	0.95
OCC	26	91.60	91.10	0.50	-	0.96
Q	27	284.86	284.00	0.90	-	0.97
Q1	26	286.93	288.50	1.57	*	0.96
FAR	30	124.40	124.60	0.20	-	0.99
PAR	29	123.70	123.30	0.40	-	0.96
OAR	26	112.30	111.40	0.90	-	0.93
BBH	23	123.70	123.90	0.20	-	0.93
BIMAS.BR	38	99.50	100.30	0.80	-	0.97
EAM/EAM	26	99.70	99.80	0.10	-	0.92
FOL	25	36.00	36.20	0.20	*	0.98
FOB	25	28.90	29.10	0.20	-	0.99
MDB	17	20.90	19.85	1.05	**	0.81
WCB	24	65.90	67.50	1.60	**	0.90
WBAB	28	22.86	23.20	0.34	-	0.82
PALBA	22	44.90	44.60	0.30	-	0.92
D1	22	15.00	15.40	0.40	-	0.81
PALOP	22	81.40	81.30	0.10	-	0.90
D2	22	15.30	15.90	0.60	-	0.90
BNL	26	98.83	98.60	0.23	-	0.99
BPL	24	100.70	100.50	0.20	-	0.97
NPH	26	67.90	67.60	0.30	-	0.99
ZYB	16	129.50	129.30	0.20	*	0.95
EKB	24	99.30	98.80	0.50	-	0.89
JUB	17	115.30	115.90	0.60	-	0.98
STB	28	108.70	108.80	0.10	-	0.96
XFB	28	113.10	113.90	0.80	-	0.97
WFB	27	95.20	96.00	0.80	**	0.99
FMB	27	101.20	101.40	0.20	-	0.86

Variable code	N	Mean 2a	Mean 2b	Mean difference	t[1]	r[2]
ZMB	25	93.33	94.06	0.70	-	0.96
IML	11	38.15	38.60	0.45	-	0.94
XML	10	54.60	55.85	1.25	-	0.98
MAB	28	62.60	62.70	0.10	-	0.99
MAL	28	55.90	56.30	0.40	*	0.97
PAL	28	49.80	50.60	0.80	*	0.87
PABR	29	38.90	38.80	0.10	-	0.95
PMH	28	20.20	18.90	1.30	-	0.99
NLH	26	47.80	47.85	0.05	-	0.99
NLB	28	26.40	26.30	0.10	-	0.92
WNB	25	9.30	9.10	0.20	-	0.96
DKB	29	24.40	24.70	0.30	*	0.95
OBB	20	39.70	39.65	0.05	-	0.93
OBH	20	34.28	34.35	0.07	-	0.97

[1] Level of significance of paired t-test ($P<0.05$*; $P<0.01$**; $P<0.001$***);
[2] Pearson correlation coefficient.

Appendix VI continued

Various Congolese sample of mandibles

Variable code	N	Mean 3a	Mean 3b	Mean difference	t^1	r^2
bb	23	111.28	112.05	0.77	-	0.99
bw	29	91.91	92.14	0.23	-	0.99
mpl	26	103.71	104.25	0.54	-	0.98
sh	29	31.93	32.00	0.07	-	0.98
hmb2	30	24.05	24.30	0.25	-	0.91
har	22	52.85	53.31	0.46	-	0.99
phar	27	47.23	48.52	1.29	-	0.91
war	27	34.11	34.34	0.23	***	0.99
mwar	24	41.55	41.59	0.04	-	0.95
ma	29	120.67	119.48	1.19	*	0.96

Kenyan sample of mandibles

Variable code	N	Mean 4a	Mean 4b	Mean difference	t^1	r^2
bb	9	114.29	114.31	0.02	-	0.99
bw	15	90.56	90.28	0.28	-	0.99
sh	20	28.05	28.16	0.11	-	0.99
har	14	57.27	57.78	0.51	**	0.98
phar	21	46.94	46.64	0.30	-	0.95
war	22	39.54	39.27	0.27	**	0.93
ma	17	114.50	114.76	0.26	-	0.97

Appendix VII

Descriptive data (in mm) with paired t-tests and correlation coefficients for the various inter-observer data sets.

Gabonese cranial sample

Variable code	N	Mean 5a	Mean 5b	Mean difference	t[1]	r[2]
GOL	33	180.33	178.71	1.62	***	0.99
XCB	34	134.71	134.27	0.44	*	0.95
BBH	35	136.46	135.33	1.13	***	0.98
BNL	34	100.12	98.93	1.19	***	0.93
BPL	33	101.82	100.61	1.21	***	0.95
NPH	35	67.94	67.99	0.05	-	0.91
ZYB	31	128.97	127.90	0.07	***	0.96
NLB	35	26.49	27.10	0.61	***	0.94
DKB	35	24.20	25.30	1.10	***	0.72
OBB	33	41.34	39.80	1.54	***	0.85

Kenyan cranial sample

Variable code	N	Mean 6a	Mean 6b	Mean difference	t[1]	r[2]
GOL	38	185.03	183.37	1.66	***	0.99
XCB	37	129.69	128.88	0.81	***	0.99
BBH	32	130.70	129.39	1.31	***	0.99
BNL	31	102.39	101.00	1.39	***	0.99
BPL	24	101.41	101.82	0.41	-	0.99
NPH	25	69.18	68.25	0.93	***	0.98
ZYB	20	130.95	129.90	1.05	***	0.98
NLB	34	27.60	27.38	0.22	*	0.96
DKB	15	25.08	25.81	0.73	**	0.92
OBB	16	40.11	39.57	0.54	*	0.88
WFB	38	96.57	96.22	0.35	***	0.99
PAL	19	52.23	49.75	2.48	***	0.87

Malian sample of crania & mandibles

Variable code	N	Mean 7a	Mean 7b	Mean difference	t[1]	r[2]
GOL	16	175.56	175.25	0.31	-	0.98
XCB	16	130.56	129.69	0.87	*	0.94
BBH	16	133.00	132.22	0.78	***	0.99
NPH	16	65.69	65.67	0.02	-	0.98
ZYB	15	126.87	126.81	0.06	-	0.98
NLB	16	26.69	28.35	1.66	-	0.95
FRC	16	106.69	107.16	0.47	**	0.99
PAC	16	113.06	113.00	0.06	-	0.99
OCC	16	94.00	94.22	0.22	-	0.99
FAR	16	122.19	122.28	0.09	-	0.94
PAR	16	126.50	126.50	0.00	-	0.98
OAR	16	108.75	109.84	1.09	**	0.99
MDH	15	29.87	30.05	0.18	-	0.94
MAB	14	62.79	64.87	2.08	*	0.95
NLH	16	46.31	45.95	0.36	-	0.98
Variable code	**N**	**Mean 8a**	**Mean 8b**	**Mean difference**	**t[1]**	**r[2]**
bb	9	113.70	112.33	1.37	**	0.92
bw	9	92.43	92.20	0.23	-	0.94
sh	8	34.54	34.13	0.41	-	0.88
hmb2	10	32.05	31.20	0.85	-	0.88
war	10	34.53	33.50	1.03	-	0.82
mwar	9	43.20	41.00	2.20	-	0.84

1 Level of significance of paired t-test (P<0.05*; P<0.01**; P<0.001***);
2 Pearson correlation coefficient.

Appendix VIII

Summary of descriptive data (in mm) for worldwide comparisons of Africa when pooled sexes.

	1. Europe			2. Asia (various)			3. Australia, Melanesia & Polynesia			4. Americas (including Arctic & Eskimos)			5. West Africa		
	N	**Mean**	**S.D.**	**N**	**Mean**	**S.D.**	**N**	**Mean**	**S.D.**	**N**	**Mean**	**S.D.**	**N**	**Mean**	**S.D.**
VAULT															
GOL	75	179.75	8.68	126	178.59	8.22	140	181.17	9.19	100	178.07	8.29	426	179.31	7.10
NOL	75	177.61	8.29	126	176.48	8.01	140	178.09	8.54	100	176.01	7.84	2	180.50	9.19
XCB	75	141.57	8.46	126	138.64	6.78	140	135.71	8.21	100	136.93	5.31	421	133.42	6.85
PAC	75	110.17	5.75	126	113.32	6.46	140	113.16	7.76	100	109.47	7.71	351	113.04	6.32
OCC	75	95.77	5.63	126	96.16	5.84	137	95.62	5.47	100	96.83	5.21	333	95.65	5.93
BBH	75	131.39	6.59	126	134.33	6.32	140	134.69	7.83	100	130.41	6.57	383	133.20	6.09
BASE															
WCB	75	73.71	5.35	126	73.28	5.04	140	69.83	4.42	100	71.36	4.42	217	67.63	5.08
FACE															
BNL	72	99.19	4.31	126	99.64	5.52	140	101.24	5.98	100	92.92	5.85	404	99.28	4.99
BPL	75	98.51	5.14	126	97.64	6.24	140	101.61	5.69	100	95.56	6.55	376	101.12	5.99
NPH	75	68.41	6.15	126	66.02	5.66	140	66.59	5.73	100	68.12	4.77	395	66.32	5.17
ZYB	75	133.09	7.62	126	133.65	8.92	140	133.40	7.90	100	132.41	6.50	384	128.94	7.05
JUB	75	116.76	6.32	126	117.99	6.97	140	117.41	6.39	100	114.91	5.94	210	113.96	5.60
STB	75	112.48	7.42	126	111.03	6.60	140	106.81	9.09	99	109.73	7.46	221	110.40	6.62
XFB	75	116.57	6.69	126	114.36	6.07	138	111.22	6.64	100	114.47	6.43	259	114.51	6.53
FMB	75	98.49	4.81	126	97.33	4.56	140	98.89	4.68	100	96.47	4.97	230	101.66	4.49
ZMB	75	96.61	6.09	126	98.45	5.90	140	95.92	5.92	100	94.98	5.37	216	94.93	5.36
IML	75	35.75	4.14	126	35.41	4.21	140	38.42	3.54	100	35.82	3.75	205	37.10	4.28
XML	75	53.56	4.34	126	53.02	4.87	140	54.47	4.04	100	52.83	4.75	205	53.04	4.31
NLH	75	51.59	4.29	126	50.20	3.97	140	50.67	4.18	100	50.24	3.49	250	48.19	3.50
NLB	75	25.81	2.26	126	27.06	2.20	140	26.69	2.00	100	24.98	2.35	402	27.51	2.24
DKB	75	20.77	1.96	126	22.11	2.49	140	21.72	2.25	98	21.08	2.70	267	24.84	2.63
OBB	75	40.35	2.37	126	39.36	1.87	140	40.37	2.17	100	39.25	2.22	407	39.72	2.29
OBH	75	32.91	2.47	126	33.44	2.31	140	34.11	2.52	100	34.45	1.86	410	33.70	2.17

Appendix VIII continued

	6.			7.			8.			9.			10.		
	Central Africa						East Africa			South Africa					
	Bantu-speaking			Pygmy						Bantu-speaking			KhoiSan		
	N	**Mean**	**S.D.**	**N**	**Mean**	**S.D.**	**N**	**Mean**	**S.D.**	**N**	**Mean**	**S.D.**	**N**	**Mean**	**S.D.**
VAULT															
GOL	409	174.79	7.59	41	173.43	6.90	543	180.30	7.24	116	185.57	7.33	97	178.74	6.66
NOL	198	172.14	7.53	3	167.00	6.25	102	180.01	6.73	-	-	-	-	-	-
XCB	410	130.99	6.71	41	131.79	5.78	543	130.97	5.45	113	134.88	6.11	96	134.73	5.27
PAC	272	109.77	6.80	26	105.78	5.89	485	110.84	6.63	-	-	-	12	109.15	7.99
OCC	405	93.85	6.67	26	93.22	4.48	445	92.83	5.12	-	-	-	11	92.71	3.85
BBH	324	129.95	6.56	40	129.76	5.58	476	127.16	6.40	113	135.85	6.09	91	126.91	5.03
BASE															
WCB	199	65.43	4.53	3	67.44	1.34	93	67.68	4.23	-	-	-	-	-	-
FACE															
BNL	393	96.30	5.12	35	97.67	4.87	431	97.50	5.12	114	102.52	4.99	89	94.94	4.95
BPL	373	97.55	6.14	23	101.00	5.82	388	98.19	6.25	103	103.45	5.47	81	94.34	6.68
NPH	383	62.88	5.58	31	63.60	4.93	428	66.73	5.27	108	68.50	5.32	80	61.81	4.61
ZYB	345	123.65	7.05	41	126.20	6.08	334	126.32	6.35	83	130.80	6.36	71	121.50	6.61
JUB	255	110.83	5.83	3	109.17	1.44	193	111.65	6.70	93	117.67	6.45	55	110.69	6.51
STB	413	109.98	6.57	3	107.67	2.52	104	108.37	5.59	112	111.55	7.86	54	110.07	4.74
XFB	273	111.33	5.58	39	111.64	5.40	422	110.71	5.46	113	116.42	5.88	59	112.75	4.35
FMB	273	99.14	5.37	3	100.17	2.47	101	101.51	4.31	-	-	-	-	-	-
ZMB	390	93.11	5.89	3	92.83	6.93	166	94.94	6.01	101	95.49	5.85	66	91.93	6.36
IML	239	36.15	3.34	3	33.78	2.53	60	37.00	3.76	-	-	-	-	-	-
XML	240	51.57	4.14	3	50.59	1.33	58	53.38	3.72	-	-	-	-	-	-
NLH	261	45.74	3.34	15	46.94	4.15	259	48.35	3.47	-	-	-	11	48.06	5.46
NLB	396	26.21	2.15	36	27.14	2.21	472	25.99	2.42	114	27.16	2.17	83	25.54	2.12
DKB	416	24.16	2.57	36	22.06	3.14	261	24.26	2.38	112	23.84	2.87	85	22.17	2.08
OBB	394	39.51	2.24	36	38.27	2.00	296	39.26	1.99	110	38.65	1.93	85	37.75	1.96
OBH	393	33.68	2.12	36	32.58	2.37	480	33.78	2.08	112	33.64	1.96	86	31.82	2.34

Appendix IX

One-way analyses of variance for worldwide comparisons of Africa.

	pooled sexes				only males				only females			
	Compared groups[1]	F	Sig.[2]	L.[3]	Compared groups[1]	F	Sig.[2]	L.[3]	Compared groups[1]	F	Sig.[2]	L.[3]
VAULT												
GOL	all **(1.-10.)**	29.63	***	**	all **(1.-10.)**	15.76	***	***	all **(2.-10.)** but not 1.	12.27	***	-
NOL	**1., 2., 3., 4., 6., 8.**	17.66	***	-	**1., 2., 3., 4., 6., 8.**	6.49	***	*	**2., 3., 4., 6., 8.**	11.69	***	-
XCB	all **(1.-10.)**	44.25	***	***	all **(1.-10.)**	20.42	***	***	all **(2.-10.)** but not 1.	20.07	***	***
PAC	**1., 2., 3., 4., 5., 6., 8.**	10.05	***	*	**1., 2., 3., 4., 5., 6., 8.**	2.58	*	-	**2., 3., 4., 5., 6., 8.**	9.43	***	-
OCC	**1., 2., 3., 4., 5., 6., 8.**	11.58	***	*	**1., 2., 3., 4., 5., 6., 8.**	8.99	***	-	**2., 3., 4., 5., 6., 8.**	8.29	***	-
BBH	all **(1.-10.)**	45.51	***	***	all **(1.-10.)**	30.23	***	*	all **(2.-10.)** but not 1.	18.03	***	-
BASE												
WCB	**1., 2., 3., 4., 5., 6., 8.**	48.86	***	-	**1., 2., 3., 4., 5., 6., 8.**	22.57	***	-	**2., 3., 4., 5., 6., 8.**	16.02	***	-
FACE												
BNL	all **(1.-10.)**	28.80	***	-	all **(1.-10.)**	18.14	***	-	all **(2.-10.)** but not 1.	6.21	***	-
BPL	**1.-10.** but not 7.	27.10	***	-	**1.-10.** but not 7.	13.35	***	-	**2.-10.** but not 1. & 7.	8.71	***	-
NPH	all **(1.-10.)**	27.22	***	-	all **(1.-10.)**	16.61	***	-	all **(2.-10.)** but not 1.	13.12	***	*
ZYB	all **(1.-10.)**	52.00	***	*	all **(1.-10.)**	33.53	***	***	all **(2.-10.)** but not 1.	22.35	***	**
JUB	**1.-10.** but not 7.	28.10	***	-	**1.-10.** but not 7.	14.81	***	-	**2.-10.** but not 1. & 7.	11.42	***	-
STB	**1.-10.** but not 7.	6.53	***	*	**1.-10.** but not 7.	7.67	***	***	**2.-10.** but not 1. & 7.	2.13	-	-
XFB	all **(1.-10.)**	21.38	***	*	all **(1.-10.)**	13.40	***	-	all **(2.-10.)** but not 1.	4.02	***	-
FMB	**1., 2., 3., 4., 5., 6., 8.**	22.76	***	*	**1., 2., 3., 4., 5., 6., 8.**	23.90	***	**	**2., 3., 4., 5., 6., 8.**	24.20	***	-
ZMB	**1.-10.** but not 7.	12.56	***	-	**1.-10.** but not 7.	8.48	***	-	**2.-10.** but not 1. & 7.	4.42	***	-
IML	**1., 2., 3., 4., 5., 6., 8.**	10.08	***	**	**1., 2., 3., 4., 5., 6., 8.**	8.67	***	-	**2., 3., 4., 5., 6., 8.**	7.45	***	**
XML	**1., 2., 3., 4., 5., 6., 8.**	7.34	***	-	**1., 2., 3., 4., 5., 6., 8.**	2.07	-	-	**2., 3., 4., 5., 6., 8.**	4.12	***	-
NLH	**1., 2., 3., 4., 5., 6., 8.**	33.64	***	**	**1., 2., 3., 4., 5., 6., 8.**	28.28	***	-	**2., 3., 4., 5., 6., 8.**	12.34	***	-
NLB	all **(1.-10.)**	21.97	***	-	all **(1.-10.)**	14.41	***	-	all **(2.-10.)** but not 1.	16.40	***	**
DKB	all **(1.-10.)**	49.65	***	***	all **(1.-10.)**	37.41	***	***	all **(2.-10.)** but not 1.	31.48	***	*
OBB	all **(1.-10.)**	14.22	***	-	all **(1.-10.)**	17.65	***	-	all **(2.-10.)** but not 1.	6.88	***	**
OBH	all **(1.-10.)**	11.27	***	*	all **(1.-10.)**	11.34	***	*	all **(2.-10.)** but not 1.	3.84	***	-

[1] Compared groups on a worldwide geographical scale (only with N≥30): **1.** Europe, **2.** Various Asia, **3.** Australia, Melanesia and Polynesia, **4.** Americas including Eskimos, **5.** West Africa, **6.** Bantu-speaking Central Africa, **7.** Pygmy Central Africa, **8.** East Africa, **9.** Bantu-speaking South Africa, **10.** KhoiSan South Africa;

[2] Level of significance (P<0.05*; P<0.01**; P<0.001***);

[3] Levene's test of homogeneity of variances.

Appendix X

Summary of descriptive data (in mm) for inter-regional comparisons in Africa.

	I. West Africa			Central Africa						IV. East Africa			South Africa					
				II. Bantu-speaking			III. Pygmy						V. Bantu-speaking			VI. KhoiSan		
	N	Mean	S.D.	N	Mean	S.D.	N	Mean	S.D.	N	Mean	S.D.	N	Mean	S.D.	N	Mean	S.D.
VAULT																		
CC	96	1408.07	117.06	206	1318.08	139.83	23	1277.17	124.59	294	1287.34	136.45	-	-	-	40	1319.90	127.96
ASB	252	103.89	5.22	292	101.98	4.94	3	98.96	4.25	403	105.13	5.17	111	107.79	5.65	59	105.52	4.90
Q	33	293.52	12.47	414	295.36	15.09	24	293.29	12.79	230	288.30	11.84	-	-	-	-	-	-
Q1	191	298.54	14.97	298	294.99	13.02	6	291.83	9.52	425	291.96	11.76	105	300.96	13.32	51	290.00	9.80
FAR	404	125.00	7.17	445	123.61	7.09	27	121.07	5.39	404	123.89	6.30	108	130.30	6.73	84	127.73	7.34
PAR	404	126.63	8.23	424	124.14	8.51	24	119.08	8.27	417	122.49	8.33	107	128.25	7.13	83	121.95	6.45
OAR	387	110.77	7.32	417	109.13	7.36	24	109.38	5.63	218	111.79	7.15	105	113.29	8.90	85	110.92	6.95
TOTARCH	380	362.60	15.40	414	356.87	14.22	24	349.29	13.34	207	359.62	14.72	107	371.54	14.78	82	360.38	13.50
BASE																		
BIMAS.BR	119	98.63	5.36	211	93.84	5.40	5	95.51	6.45	271	98.85	5.92	-	-	-	-	-	-
EAM/EAM	62	98.12	5.86	211	94.32	6.12	3	100.23	4.93	98	100.24	5.76	-	-	-	-	-	-
FOL	305	36.58	2.69	419	35.46	2.89	28	34.84	2.69	45	36.09	2.74	110	36.97	2.70	69	36.31	2.47
FOB	270	29.50	2.28	418	28.82	2.64	24	28.55	2.71	392	28.82	2.29	109	29.58	2.29	60	28.64	2.09
WBAB	79	23.05	2.08	208	22.54	1.87	3	22.06	0.66	98	22.82	1.62	-	-	-	-	-	-
PALBA	152	45.91	3.78	238	43.48	3.94	3	43.17	3.25	73	44.06	3.95	-	-	-	-	-	-
D1	152	14.65	2.10	238	13.83	2.07	3	11.83	3.33	72	14.79	2.05	-	-	-	-	-	-
PALOP	152	82.40	4.44	238	79.51	5.15	3	76.50	3.04	72	81.14	5.27	-	-	-	-	-	-
D2	151	15.48	2.33	238	14.35	2.34	3	12.67	2.52	73	15.80	2.35	-	-	-	-	-	-
FACE																		
NASGNA	201	114.48	6.58	71	111.46	9.15	12	104.42	6.52	62	112.86	7.98	39	114.72	8.15	34	103.21	7.78
EKB	269	99.37	4.30	279	97.02	5.27	26	97.94	4.75	271	96.95	4.80	104	101.09	5.23	93	96.08	4.17
WFB	329	94.47	5.50	443	93.77	4.93	27	92.76	4.70	485	94.18	5.21	113	98.01	5.36	74	94.21	4.28
MAB	356	63.79	4.34	260	61.34	3.81	23	59.16	2.64	316	61.07	4.08	107	63.42	3.85	74	59.84	4.28
MAL	310	56.06	3.78	201	54.38	4.17	19	54.39	3.69	233	54.02	4.06	101	56.83	3.73	65	52.33	4.38
PABR	345	39.59	3.95	358	37.85	3.48	18	35.79	2.30	327	38.38	3.02	110	36.36	3.41	66	35.00	3.70
PMH	222	17.54	3.13	236	16.42	3.40	3	14.54	3.14	135	19.52	2.85	-	-	-	11	17.52	2.33
WNB	109	9.08	2.48	198	8.94	2.72	3	8.17	1.13	185	9.60	2.38	-	-	-	-	-	-
MANDIBLE																		
bb	176	112.12	6.23	102	11.68	5.90	3	112.06	3.37	52	112.43	6.70	39	116.13	5.89	32	106.97	6.37
bw	183	92.13	6.47	105	92.00	6.83	3	82.68	2.92	69	91.42	6.57	39	95.03	6.96	33	87.49	7.38
mpl	160	105.12	5.35	105	103.37	6.00	3	96.67	1.15	6	106.00	6.26	-	-	-	10	101.75	7.28
sh	171	33.77	3.61	97	31.29	3.87	-	-	-	68	32.48	3.78	38	34.41	3.32	32	30.77	3.83
phar	143	54.16	8.06	107	50.31	7.51	2	44.75	4.60	6	52.50	6.41	2	58.50	6.36	10	46.60	4.72
har	186	53.43	5.37	107	53.89	5.89	3	51.41	3.87	61	55.97	5.85	33	46.61	4.27	31	40.08	6.64
war	188	34.47	3.42	110	33.05	2.97	3	32.20	4.60	65	34.53	3.47	40	36.88	4.03	34	34.07	2.96
mwar	186	42.54	3.70	107	41.51	3.02	3	40.65	4.14	6	43.62	1.76	8	40.50	4.20	12	42.38	3.90
ma	163	119.65	8.13	109	121.99	8.22	3	112.33	8.74	66	123.89	7.28	20	113.30	5.53	9	123.06	3.54

Appendix XI

Various one-way analyses of variance (pooled sexes): re-analysis of 23 variables within Africa.

	Compared groups[1]	F	Sig.[2]	L.[3]
VAULT				
GOL	I, II, III, IV, V, VI	51.79	***	-
NOL	II, IV	78.13	***	-
XCB	I, II, III, IV, V, VI	15.03	***	***
PAC	I, II, IV	20.04	***	-
OCC	I, II, IV	21.38	***	**
BBH	I, II, III, IV, V, VI	68.48	***	-
BASE				
WCB	I, II, IV	13.27	***	-
FACE				
BNL	I, II, III, IV, V, VI	37.55	***	-
BPL	I, II, III, IV, V, VI	37.97	***	-
NPH	I, II, III, IV, V, VI	39.17	***	-
ZYB	I, II, III, IV, V, VI	36.67	***	-
JUB	I, II, IV, V, VI	25.72	***	-
STB	I, II, IV, V, VI	4.04	**	***
XFB	I, II, III, IV, V, VI	23.22	***	**
FMB	I, II, IV	19.14	***	***
ZMB	I, II, III, IV, V, VI	6.36	***	-
IML	I, II, IV	9.79	***	-
XML	I, II, IV	9.58	***	-
NLH	I, II, IV	42.01	***	-
NLB	I, II, III, IV, V, VI	32.50	***	-
DKB	I, II, III, IV, V, VI	20.08	***	**
OBB	I, II, III, IV, V, VI	15.64	***	-
OBH	I, II, III, IV, V, VI	22.01	***	-

[1] Compared groups on a geographical scale within Africa (only with N≥20): **I.** West Africa, **II.** Central Africa (only Bantu-speaking), **III.** Central Africa (only Pygmy), **IV.** East Africa, **V.** South Africa (only Bantu-speaking), **VI.** South Africa (only KhoiSan);

[2] Level of significance (P<0.05*; P<0.01**; P<0.001***);

[3] Levene's test of homogeneity of variances.

Appendix XII

One-way analyses of variance within Africa for both pooled and separated sexes using 35 additional variables.

	pooled sexes				only males				only females			
	Compared groups[1]	F	Sig.[2]	L.[3]	Compared groups[1]	F	Sig.[2]	L.[3]	Compared groups[1]	F	Sig.[2]	L.[3]
VAULT												
CC	I, II, III, IV, VI	15.19	***	-	I, II, IV	8.27	***	-	II, IV, VI	5.79	**	-
ASB	I, II, IV, V, VI	31.27	***	-	I, II, IV, VI	7.24	***	-	I, II, IV, VI	14.62	***	-
Q	I - IV	12.73	***	***	II, IV	39.66	***	*	II, IV	14.83	***	**
Q1	I, II, IV, V, VI	21.75	***	***	I, II, IV, V, VI	15.25	***	*	I, II, IV, VI	4.88	**	**
FAR	I - VI	22.66	***	-	I, II, IV, V, VI	12.07	***	-	I, II, IV, VI	5.66	***	*
PAR	I - VI	18.92	***	*	I, II, IV, V, VI	8.77	***	*	I, II, IV, VI	8.67	***	*
OAR	I - VI	7.40	***	*	I, II, IV, V, VI	4.99	***	**	I, II, IV, VI	2.57	-	-
TOTARCH	I - VI	21.47	***	-	I, II, IV, V, VI	8.93	***	-	I, II, IV, VI	3.25	*	-
BASE												
BIMAS.BR	I, II, IV	52.79	***	-	II, IV	0.13	-	-	II, IV	0.31	-	**
EAM/EAM	I, II, IV	35.40	***	-	II, IV	0.69	-	-	II, IV	11.36	***	-
FOL	I - VI	9.73	***	-	I, II, IV, V, VI	2.85	*	-	I, II, IV, VI	6.35	***	-
FOB	I - VI	4.90	***	*	I, II, IV, V, VI	3.07	*	**	I, II, IV, VI	0.21	-	-
WBAB	I, II, IV	2.38	-	*	II, IV	0.35	-	**	II, IV	0.01	-	-
PALBA	I, II, IV	18.48	***	-	II, IV	8.83	**	-	II, IV	1.66	-	-
D1	I, II, IV	9.56	***	-	II, IV	0.17	-	-	II, IV	0.01	-	-
PALOP	I, II, IV	16.07	***	-	II, IV	2.96	-	-	II, IV	0.48	-	-
D2	I, II, IV	16.55	***	-	II, IV	1.96	-	-	II, IV	0.03	-	-
FACE												
NASGNA	I, II, IV, V, VI	17.37	***	*	I, II, IV, V	0.71	-	-	I, II	15.48	***	-
EKB	I - VI	20.56	***	*	I, II, IV, V, VI	13.68	***	-	I, II, IV, VI	13.19	***	*
WFB	I - VI	13.15	***	*	I, II, IV, V, VI	7.36	***	-	I, II, IV, VI	1.48	-	-
MAB	I – VI	27.15	***	-	I, II, IV, V, VI	13.59	***	-	I, II, IV, VI	15.12	***	-
MAL	I, II, IV, V, VI	22.81	***	-	I, II, IV, V, VI	10.26	***	-	I, II, IV, VI	10.37	***	-
PABR	I, II, IV, V, VI	35.85	***	***	I, II, IV, V, VI	40.24	***	-	I, II, IV, VI	26.82	***	-
PMH	I, II, IV	40.58	***	*	I, II, IV	13.15	***	-	I, II, IV	28.03	***	-
WNB	I, II, IV	3.47	*	-	I, II, IV	0.03	-	-	I, II, IV	4.41	*	-
MANDIBLE												
bb	I, II, IV, V, VI	9.85	***	-	I, II, IV, V	2.00	-	-	I, II	0.76	-	-
bw	I, II, IV, V, VI	5.88	***	-	I, II, IV, V	1.37	-	-	I, II, IV	0.82	-	-
mpl	I, II	5.91	*	-	I, II	2.04	-	-	I, II	3.84	-	-
sh	I, II, IV, V, VI	10.83	***	-	I, II, IV, V	4.64	**	-	I, II, IV	6.58	**	-
phar	I, II	14.83	***	-	I, II	7.48	**	-	I, II	9.72	**	-
har	I, II, IV, V, VI	57.56	***	-	I, II, IV, V	36.66	***	-	I, II, IV	2.56	-	-
war	I, II, IV, V, VI	9.99	***	-	I, II, IV, V	8.41	***	-	I, II, IV	2.19	-	-
mwar	I, II	6.00	*	*	I, II	6.09	*	-	I, II	3.89	-	-
ma	I, II, IV, V	11.37	***	-	I, II, IV	7.95	***	-	I, II, IV	4.25	*	-

[1] Compared groups on a geographical scale within Africa (only with N≥20): **I.** West Africa, **II.** Central Africa (only Bantu-speaking), **III.** Central Africa (only Pygmy), **IV.** East Africa, **V.** South Africa (only Bantu-speaking), **VI.** South Africa (only KhoiSan);

[2] Level of significance (P<0.05*; P<0.01**; P<0.001***);

[3] Levene's test of homogeneity of variances.

Appendix XIII

One-way analyses of variance within Africa using cranial variables with matching mandibles.

	Compared groups[1]	F	Sig.[2]	L.[3]
VAULT				
CC	I, II, III, IV	14.75	***	-
ASB	I, II, IV, V	12.37	***	-
Q	II, III	8.30	**	-
Q1	-	-	-	-
FAR	I, II, III, IV, V, VI	6.53	***	-
PAR	I, II, III, IV, V, VI	8.35	***	*
OAR	I, II, III, V, VI	5.16	***	-
TOTARCH	I, II, III, V, VI	15.17	***	-
BASE				
BIMAS.BR	I, IV	0.88	-	-
EAM/EAM	-	-	-	-
FOL	I, II, III, IV	6.86	***	-
FOB	I, II, III, IV	6.02	***	*
WBAB	-	-	-	-
PALBA	-	-	-	-
D1	-	-	-	-
PALOP	-	-	-	-
D2	-	-	-	-
FACE				
NASGNA	I, II, IV, V, VI	15.09	***	-
EKB	I, II, III, IV, V, VI	10.30	***	-
WFB	I, II, III, IV, V, VI	4.55	***	**
MAB	I, II, III, IV, V, VI	19.92	***	-
MAL	I, II, IV, V, VI	12.59	***	-
PAL	I, II, III, V	11.10	***	-
PABR	I, II, IV, V	23.88	***	-
PMH	I, II	2.58	-	-
WNB	I, II	2.22	-	-

	Compared groups[1]	F	Sig.[2]	L.[3]
MANDIBLE				
bb	I, II, IV, V, VI	10.62	***	-
bw	I, II, IV, V, VI	7.86	***	-
mpl	I, II	5.09	*	-
sh	I, II, IV, V, VI	9.54	***	-
phar	I, II	12.28	***	-
har	I, II, IV, V, VI	54.85	***	-
war	I, II, IV, V, VI	11.34	***	-
mwar	I, II	6.60	*	-
ma	I, II, IV, V, VI	9.77	***	-

1 Compared groups on a geographical scale within Africa (only with N≥20): **I.** West Africa, **II.** Central Africa (only Bantu-speaking), **III.** Central Africa (only Pygmy), **IV.** East Africa, **V.** South Africa (only Bantu-speaking), **VI.** South Africa (only KhoiSan);
2 Level of significance (P<0.05*; P<0.01**; P<0.001***);
3 Levene's test of homogeneity of variances.

Appendix XIV

Summary of various multiple discriminant analyses using skull variables and geographical groupings within Africa.

	Pooled sexes					Only males					Only females				
Entered variables	**Groups compared & sample size**	**Total rate of correct classification**	**Highest rate of mis-classification (original with other group)**	**N° of factors extracted**	**Highest standard canonical discrimin. function coefficients**	**Groups compared**	**Total rate of correct classification**	**Highest rate of mis-classification (original with other group)**	**N° of factors extracted**	**Highest standard canonical discrimin. function coefficients**	**Groups compared**	**Total rate of correct classification**	**Highest rate of mis-classification (original with other group)**	**N° of factors extracted**	**Highest standard canonical discrimin. function coefficients**
VAULT &/or BASE															
Analysis 1: CC, ASB, Q1, FAR, PAR	II, IV, VI (N=336)	69.0%	82.4% VI → IV	2	Q1 (1) FAR (2)	II, IV, VI (N=190)	64.7%	100% II → IV 73.3% VI → IV	2	Q1 (1) CC (2)	IV, VI (N=240)	54.2%	95.7% VI → IV	2	CC (1)
Analysis 2: GOL, XCB, PAC, OCC, BBH,WCB, BNL	I, II, IV, VI (N=1389)	42.6%	100% VI → II, IV, I	3	BBH (1) BNL (2) XCB (3)	I, II, IV, VI (N=789)	38.0%	94.9% VI → I	3	BBH (1) XCB (2) GOL (3)	I, II, IV, VI (N=543)	43.1%	100% VI → IV, II, I	3	BBH (1) GOL (2) XCB (3)
BASE															
Analysis 3: BIMAS.BR, EAM/EAM, FOL, FOB	I, II, IV (N=347)	64.0%	96.7% I → II, IV	2	BIMAS.BR (1) FOL (2)	I, II, IV (N=144)	62.5%	94.9% VI → I	1	EAM/EAM (1)	I, II, IV (N=189)	68.3%	100% VI → IV, II, I	2	BIMAS.BR (1) EAM/EAM (2)
FACE															
Analysis 4: BPL, ZYB, JUB, STB, XFB, FMB, ZMB	I, II, IV (N=596)	44.5%	87.7% IV → II, I	2	STB (1) XFB (2)	I, II, IV (N=309)	35.3%	80.8% IV → I, II	2	STB (1, 2)	I, II, IV (N=425)	42.6%	80.8% IV → I, II	2	ZYB (1) BPL (2)
Analysis 5: IML, XML, NLH, NLB, DKB, OBB, OBH	I, II, IV (N=909)	39.3%	80.7% IV → I, II	2	NLB (1, 2)	I, II, IV (N=806)	33.0%	100% IV → I, II	1	NLB (1)	I, II, IV (N=404)	44.3%	85.2% IV → II, I	2	OBB (1) NLB (2)
MANDIBLE															
Analysis 6: war, bw, sh, ma, bb, har	I, II, IV (N=294)	60.9%	72.9% IV → I, II	2	ma (1) war (2)	I, II, IV (N=146)	63.7%	69.7% IV → II	2	ma (1) war (2)	I, II, IV (N=133)	54.1%	100% IV → I	1	sh (1)
Analysis 7: war, bw, sh, ma, bb, har	I, II (N=245)	71.8%	50.5% II → I	1	sh (1)	I, II (N=119)	79.0%	38.5% II → I	1	ma (1)	I, II (N=109)	70.6%	59.0% II → I	1	sh (1)
VARIOUS (vault, face &/or base)															
Analysis 8: BBH, GOL, BNL, ZYB, XCB, BPL, NPH	I, II, IV-VI (N=1086)	54.7%	100% III → I, II 85.5% V → I	4	BBH (1) ZYB (2, 4) XCB (3)	I, II, IV-VI (N=608)	55.9%	72.1% II → I	4	BBH (1) NPH (2) ZYB (3) XCB (4)	I, II, IV-VI (N=399)	62.9%	66.7% V → I, II	5	BBH (1) XCB (2, 3) GOL (4)
Analysis 9: BBH, OBB, GOL, DKB, EKB, ZYB, FMB	I, II, IV-VI (N=401)	72.1%	88.9% I → II, V	4	OBB (1) BBH (2) ZYB (3) EKB (4)	I, II, IV-VI (N=404)	61.4%	51.5% II → I	4	EKB (1) BBH (2) ZYB (3) ZYB (4)	I, II, IV-VI (N=310)	60.3%	60.5% VI → II	4	GOL (1) ZYB (2) OBB (3) ZYB (4)
Analysis 10: BBH, GOL, ZYB, XCB, BPL, NPH, BNL, WFB	I, II, IV-VI (N=934)	56.7%	76.0% V → I	4	BBH (1) ZYB (2, 3) XCB (4) BPL (5)	I, II, IV-VI (N=535)	55.6%	68.7% VI → IV	4	BBH (1) GOL (2) ZYB (3) BPL (4)	I, II, IV-VI (N=399)	59.6%	88.9% V → II	4	BBH (1) XCB (2,3,4)
Analysis 11: FMB, OBH, Q1, DKB, NLB	I, II, IV (N=453)	60.3%	52.5% IV → II, I	2	NLB (1) FMB (2)	I, II, IV (N=516)	34.7%	71.6% IV → II, I	2	Q1 (1) OBH (2)	I, II, IV (N=232)	62.1%	47.4% IV → II	2	NLB (1) FMB (2)

Appendix XV

Details for six multiple discriminant analyses on the skull: predicted geographical group membership within Africa with pooled sexes.

Variables entered, total sample size & groups analysed	N	Classification results						Total correct rate	Selected variables	Standardized canonical discriminant function coefficients			
		I.	II.	III.	IV.	V.	VI.			function 1	function 2	function 3	function 4
Analysis 1: CC, ASB, Q1, FAR, PAR (N=336)													
II. Central Africa (only Bantu-speaking)	73	-	**0**	-	73	-	0		CC	0.877	-0.847	-	-
IV. East Africa	229	-	0	-	**226**	-	3	**69.0%**	**Q1**	**-1.355**	0.637	-	-
VI. South Africa (only KhoiSan)	34	-	0	-	28	-	**6**		**FAR**	0.751	**0.926**	-	-
Analysis 2: GOL, XCB, PAC, OCC, BBH, WCB, BNL (N=1474)													
I. West Africa	418	**256**	139	0	23	-	0						
II. Central Africa (only Bantu-speaking)	346	145	**184**	0	17	-	0		**GOL**	0.704	**0.480**	-0.833	-
III. Central Africa (only Pygmy)	28	7	18	**1**	2	-	0	**40.1%**	**XCB**	0.491	0.363	**0.959**	-
IV. East Africa	467	124	192	0	**150**	-	1		**BBH**	**-1.152**	0.412	-0.007	-
VI. South Africa (only KhoiSan)	105	27	37	0	41	-	**0**						
Analysis 2 bis : idem previous but without Pygmy (N=1389)													
I. West Africa	415	**274**	118	-	23	-	0		GOL	0.656	0.137	-0.243	-
II. Central Africa (only Bantu-speaking)	339	148	**168**	-	22	-	1	**42.6%**	**XCB**	0.506	0.385	**0.820**	-
IV. East Africa	423	118	155	-	**150**	-	0		**BBH**	**-1.172**	0.285	0.221	-
VI. South Africa (only KhoiSan)	102	32	37	-	33	-	**0**		**BNL**	0.109	**0.521**	-0.665	-
Analysis 3: BIMAS.BR, EAM/EAM, FOL, FOB (N=347)													
I. West Africa	60	**2**	41	-	17	-	-		**BIMAS.BR**	**0.755**	-0.768	-	-
II. Central Africa (only Bantu-speaking)	206	1	**185**	-	20	-	-	**64.0%**	EAM/EAM	0.396	0.767	-	-
IV. East Africa	81	0	46	-	**35**	-	-		**FOL**	-0.296	**0.821**	-	-
Analysis 4: BPL, ZYB, JUB, STB, XFB, FMB, ZMB (N=596)													
I. West Africa	165	**71**	87	-	7	-	-		BPL	0.181	0.735	-	-
II. Central Africa (only Bantu-speaking)	247	50	**187**	-	10	-	-	**44.5%**	JUB	0.874	-0.015	-	-
IV. East Africa	57	19	31	-	**7**	-	-		**STB**	**-0.981**	1.518	-	-
									XFB	0.579	**-1.626**	-	-
Analysis 6: war, bw, sh, ma, bb, har (N=294)													
I. West Africa	142	**120**	19	-	3	-	-		**war**	0.103	**0.763**	-	-
II. Central Africa (only Bantu-speaking)	93	43	**43**	-	7	-	-	**60.9%**	sh	-0.824	0.414	-	-
IV. East Africa	59	25	18	-	**16**	-	-		**ma**	**1.034**	0.518	-	-
									har	0.911	0.244	-	-

Appendix XVI

One-way analyses of variance for regression factor scores within Africa based on various geographical groupings: broad regions, latitude and longitude areas.

			Broad regions[1]				Latitude areas[2]				Longitude areas[3]			
Variables with high factor loadings (>0.6)	Type	Analysis & factors	Compared groups[4]	F	Sig.[5]	L.[6]	Compared groups[4]	F	Sig.[5]	L.[6]	Compared groups[4]	F	Sig.[5]	L.[6]
BPL, BNL	only	**VIb** 1	I, II, IV-VI	18.92	***	-	2-6, 8, 9	5.73	***	-	1-7	8.75	***	***
XCB	males	2		15.10	***	-		15.05	***	-		15.63	***	-
BPL, BNL	only	**VIc** 1	I, II, IV, VI	41.86	***	-	2, 4-6, 8, 9	17.21	***	-	1-6	12.73	***	-
XCB	females	2		6.85	***	-		14.07	***	***		16.66	***	-
BPL, XML, GOL	pooled	**VIIa** 1	I, II, IV	16.75	***	-	2, 4-6, 8	10.29	***	-	2-7	7.89	***	-
OBH, NLH	sexes	2		8.29	***	-		4.35	**	-		4.85	***	-
BPL, XML	only	**VIIb** 1	I, II, IV	2.27	-	-	2, 4-6, 8	2.09	-	-	2-7	1.80	-	-
OBH, NLH	males	2		0.96	-	-		0.99	-	-		0.82	-	**
BPL, GOL, XML	only	**VIIc** 1	I, II, IV	13.38	***	-	2, 4-6, 8	9.10	***	-	2-7	6.30	***	-
OBH, NLH	females	2		8.26	***	-		4.34	**	-		5.06	***	-
BBH, OCC, FRC	only	**VIIIb** 1	I, II, IV	25.95	***	-	2, 4-6, 8	9.58	***	*	1, 2, 4-7	13.15	***	-
PAC, NLB	males	3		1.65	-	*		2.97	*	*		9.30	***	-
BBH, FRC, OCC, PAC	only	**VIIIc** 1	I, II, IV	2.93	-	-	2, 4-6, 8	3.83	*	-	1-6	8.74	***	-
OBB, NPH, NLB	females	2		27.71	***	-		15.62	***	-		8.22	***	-
BPL, JUB, NLB, WCB	pooled	**IXa** 1	I, II, IV	26.34	***	-	2, 4-6, 8	14.53	**	-	2-7	11.92	***	-
OBH, NLH	sexes	2		8.10	***	-		4.47	**	-		5.13	***	-
BPL, JUB	only	**IXb** 1	I, II, IV	5.12	**	-	2, 4-6, 8	3.50	***	-	2, 4, 5-7	3.35	*	-
OBH, NLH	males	2		1.80	-	*		1.22	-	-		1.72	-	-
bw, sh, bb	pooled	**Xa** 1	I, II, IV	2.08	-	-	2, 4-6, 9	1.27	-	-	2, 3, 5-6	3.09	*	-
ma, war	sexes	2		2.31	-	-		3.05	*	-		0.15	-	-
ma, war	only	**Xb** 1	I, II, IV	3.67	*	-	4-6, 9	2.00	-	-	1-3, 5-6	1.87	-	-
bw, sh	males	2		0.86	-	-		0.74	-	-		1.64	-	-

1 Compared groups for broad geographical regions: **I** (West Africa), **II** (Central Africa, only Bantu-speaking), **III** (Central Africa, only Pygmy), **IV** (East Africa), **V** (South Africa, only Bantu-speaking), and **VI** (South Africa, only KhoiSan);

2 Compared groups for latitude areas: (≥16°), **2** (15.9° to 13°), **3** (12.9° to 9°), **4** (8.9° to 5°), **5** (4.9° to 1°), **6** (0.9° to -2°), **7** (-2.1° to -5°), **8** (-5.1° to –8°) and **9** (<-8°);

3 Compared groups for longitude areas: **1** (≥37°), **2** (36.9° to 29°), **3** (28.9° to 20°), **4** (19.9° to 11°), **5** (10.9° to 4°), **6** (3.9° to –4°) and **7** (<-4°);

4 Compared groups within Africa (only with N≥15);

5 Level of significance (P<0.05*; P<0.01**; P<0.001***);

6 Levene's test of homogeneity of variances.

Appendix XVII

Spearman Rank correlation coefficients between metrical variables of the skull and various geographical and ecological variables.

Geographical and ecological continuous variables	Individual metric variables of the skull[1]											
FIRST SERIES OF VARIABLES USED ON A WORLDWIDE SCALE												
	GOL	*NOL*	*XCB*	*FRC*	*PAC*	*OCC*	*BBH*	*WCB*	*BNL*	*BPL*	*NPH*	*ZYB*
Latitude[2]	0.036	-0.318**	0.040	0.091**	0.142**	0.104**	0.281**	-0.178**	0.139**	0.059*	-0.039	-0.012
Longitude[3]	0.146**	0.370**	-0.056*	-0.059*	-0.084**	-0.139**	-0.262**	-0.005**	-0.085**	-0.161**	0.053*	-0.065*
Mean annual temperature	-0.201**	-0.342**	-0.021	0.063*	0.093**	0.139**	0.185**	0.122**	0.021	0.055*	-0.079**	0.017
Mean total rainfall	-0.135**	-0.285**	0.146**	0.103**	-0.025	0.161**	0.105**	0.071	-0.056*	0.050	-0.045	0.137**
	JUB	*STB*	*XFB*	*FMB*	*ZMB*	*IML*	*XML*	*NLH*	*NLB*	*DKB*	*OBB*	*OBH*
Latitude[2]	0.122**	-0.016	0.135**	-0.077	-0.030	-0.054	-0.120**	-0.017	0.182**	-0.094**	-0.203**	-0.158**
Longitude[3]	-0.034	0.064	-0.146**	-0.083*	0.040	-0.055	-0.012	0.062	-0.225**	-0.051	-0.099**	-0.021
Mean annual temperature	-0.032	-0.010	-0.046	0.142**	0.003	0.067	0.065	0.024	0.126**	0.103**	0.150**	0.058*
Mean total rainfall	-0.010	0.060	0.127**	0.070	-0.028	0.107*	0.140**	-0.088*	0.008	0.096**	0.259**	0.176**
SECOND SERIES OF VARIABLES USED ONLY WITHIN AFRICA												
	CC	*ASB*	*Q*	*Q1*	*FAR*	*PAR*	*OAR*	*TOTARCH*	*BIMASBR*	*EAM/EAM*	*FOL*	*FOB*
Latitude[2]	0.168**	-0.027	-0.113**	0.167**	0.127**	0.181**	0.045	0.142**	-0.071	-0.191**	0.113**	0.053
Longitude[3]	-0.169**	0.102**	-0.195**	-0.051	0.017	-0.148**	0.067*	0.006	0.138**	0.212**	0.023	-0.029
Mean annual temperature	0.123**	-0.252**	0.424**	0.061*	-0.092**	0.101**	-0.114**	-0.069*	-0.162**	-0 .187**	-0.078**	-0.007
Mean total rainfall	0.160**	-0.067*	0.198**	-0.007	-0.024	0.006	-0.041	-0.033	-0.293**	-0.332**	-0.050	0.022
	WBAB	*PALBA*	*D1*	*PALOP*	*D2*	*NASGNA*	*EKB*	*WFB*	*MAB*	*MAL*	*PABR*	
Latitude[2]	-0.011	0.090	-0.084	0.035	-0.105*	-0.016	0.095**	0.005	0.125**	0.090**	-0.114**	
Longitude[3]	0.017	-0.247**	0.050	-0.170**	-0.042	-0.068	-0.165**	0.018	-0.217**	-0.154**	-0.145**	
Mean annual temperature	0.005	0.240**	0.035	0.187**	0.029	0.064	0.058	-0.090**	0.123**	0.064	0.183**	
Mean total rainfall	-0.113	-0.065	-0.021	-0.043	-0.046	0.138**	0.078*	0.070**	0.091**	0.109**	0.168**	
	PMH	*WNB*	*bb*	*bw*	*mpl*	*sh*	*phar*	*har*	*war*	*mwar*	*ma*	
Latitude[2]	-0.185**	-0.085	-0.022	-0.009	0.116	0.133**	0.075	-0.314**	0.109*	0.080	-0.178*	
Longitude[3]	0.186**	0.114*	0.055	0.027	-0.119*	-0.113*	-0.131*	-0.088	0.056	-0.104	0.008	
Mean annual temperature	-0.120**	-0.061	-0.071	-0.019	0.103	0.085	0.114	0.254**	-0.153**	-0.062	0.102	
Mean total rainfall	-0.195**	-0.124**	0.113*	0.120*	-0.040	-0.036	0.176**	0.167**	-0.024	0.051	-0.108*	

1 Level of significance: P<0.05*, P<0.01** and P<0.001***;
2 Latitude expresses here the distance from the equator (a south-north or north-south trend starting from 0°);
3 Longitude expresses here the distance from 0° which reflects mainly a west-east trend.

Appendix XVIII

One-way analyses of variance within Africa (including only Bantu-speaking populations).

	pooled sexes				only males				only females			
	Compared groups[1]	F	Sig.[2]	L.[3]	Compared groups[1]	F	Sig.[2]	L.[3]	Compared groups[1]	F	Sig.[2]	L.[3]
VAULT												
CC	I, II, III, IV	21.10	***	-	I, II, IV	6.73	***	-	II, IV	3.88	-	-
ASB	I, II, IV, V	38.51	***	-	I, II, IV	8.17	***	-	II, IV	23.67	***	-
Q	I - IV	19.29	***	***	II, IV	36.27	***	*	II, IV	15.18	***	**
Q1	I, II, IV, V	26.24	***	-	II, IV, V	4.70	*	-	II, IV	1.66	-	-
FAR	I - V	26.61	***	-	I, II, IV, V	12.76	***	-	II, IV	4.88	*	-
PAR	I - V	11.17	***	-	I, II, IV, V	1.84	-	-	II, IV	0.35	-	-
OAR	I - V	9.21	***	*	I, II, IV, V	6.16	***	*	II, IV	5.81	*	-
TOTARCH	I - V	29.67	***	-	I, II, IV, V	9.70	***	-	II, IV	3.66	-	-
BASE												
BIMAS.BR	I, II, IV	82.91	***	-	II, IV	29.94	***	-	II, IV	49.76	***	-
EAM/EAM	I, II, IV	63.12	***	-	II, IV	27.12	***	-	I, IV	47.08	***	-
FOL	I - V	13.37	***	-	II, IV	6.21	*	-	II, IV	7.76	**	-
FOB	I - V	7.46	***	*	II, IV	0.22	-	**	II, IV	1.60	-	-
WBAB	I, II, IV	1.21	-	-	II, IV	0.01	-	-	II, IV	3.27	-	-
PALBA	I, II, IV	1.39	-	-	II, IV	0.02	-	-	II, IV	1.31	-	-
D1	I, II, IV	9.21	***	-	II, IV	2.35	-	-	II, IV	4.73	*	-
PALOP	I, II, IV	3.96	*	-	II, IV	0.44	-	-	II, IV	4.76	*	-
D2	I, II, IV	15.31	***	-	II, IV	12.25	***	-	II, IV	5.60	*	-
FACE												
NASGNA	I, II, IV, V	3.56	*	*	I, II, V	1.26	-	-	-	-	-	-
EKB	I - V	13.80	***	-	I, II, IV, V	3.23	*	*	II, IV	5.73	*	-
WFB	I - V	21.81	***	-	I, II, IV, V	6.24	***	-	II, IV	10.42	***	-
MAB	I – V	14.83	***	-	I, II, IV, V	2.00	-	-	II, IV	2.44	-	-
MAL	I, II, IV, V	11.93	***	-	I, II, IV, V	0.89	-	-	II, IV	8.25	**	-
PAL	I - V	6.68	***	*	I, II, IV, V	9.96	***	*	II, IV	0.01	-	-
PABR	I, II, IV, V	30.62	***	-	I, II, IV, V	25.12	***	-	II, IV	1.44	-	-
PMH	I, II, IV	42.23	***	**	I, II, IV	13.27	***	-	II, IV	52.41	***	*
WNB	I, II, IV	7.00	**	-	II, IV	0.01	-	-	II, IV	7.27	***	-
MANDIBLE												
bb	I, II, IV	7.63	***	-	II, V	4.08	*	-	-	-	-	-
bw	I, II, IV	3.59	*	-	II, V	1.50	-	-	-	-	-	-
mpl	I, II	0.61	-	-	I, II	-	-	-	-	-	-	-
sh	I, II, IV	8.65	***	-	II, V	3.66	-	-	-	-	-	-
phar	I, II	23.87	***	-	I, II	-	-	-	-	-	-	-
har	I, II, IV	23.02	***	-	II, V	75.78	***	-	-	-	-	-
war	I, II	11.44	***	*	II, V	23.55	***	-	-	-	-	-
mwar	I, II	4.54	*	-	-	-	-	-	-	-	-	-
ma	I, II	21.41	***	-	-	-	-	-	-	-	-	-

1 Compared groups on a geographical scale within Bantu-speaking Africa (only with N≥20): **I.** West Africa, **II.** Central Africa (only non-Pygmy), **III.** Central Africa (only Pygmy), **IV.** East Africa, **V.** South Africa;
2 Level of significance (P<0.05*; P<0.01**; P<0.001***);
3 Levene's test of homogeneity of variances.

Appendix XIX

One-way analyses of variance for regression factor scores within Africa based on various cultural groupings: languages and Murdock's ethno-linguistic groupings.

				Languages[1]				Murdock's ethno-linguistic units[2]			
Variables with high factor loadings (>0.6)	**Type**	**Analysis & factors**		**Compared groups**	**F**	**Sig.[3]**	**L.[4]**	**Compared groups**	**F**	**Sig.[3]**	**L.[4]**
BPL, BNL	only	**VIb**	1	1, 3-5, 7, 8	21.43	***	-	1, 2, 6-9, 11,	10.15	***	-
XCB	males		2		10.58	***	-	15	23.86	***	-
BPL, BNL	only	**VIc**	1	1, 3, 4, 8	40.64	***	-	2, 6, 8, 11, 15	8.12	***	-
XCB	females		2		5.08	**	-		8.75	***	*
BPL, XML, GOL	pooled	**VIIa**	1	1, 3, 4	7.63	***	-	6, 8, 11, 13	1.78	-	-
OBH, NLH	sexes		2		3.08	*	-		0.82	-	-
BPL, XML	only	**VIIb**	1	1, 3, 4	1.01	-	-	6, 8, 11	2.54	-	-
OBH, NLH	males		2		0.68	-	-		1.18	-	-
BPL, GOL, XML	only	**VIIc**	1	1, 3, 4	8.77	-	-	6, 8, 11	13.53	***	-
OBH, NLH	females		2		5.05	**	-		9.36	***	-
BBH, OCC, FRC	only	**VIIIb**	1	1, 3, 4	22.23	***	-	1, 2, 6-8, 11	17.37	***	-
PAC, NLB	males		3		16.37	***	-		7.52	***	*
BBH, FRC, OCC,PAC	only	**VIIIc**	**1**	1, 3, 4	2.86	-	*	2, 6, 8, 11	6.68	***	-
OBB, NPH, NLB	females		**2**		15.40	***	-		11.70	***	-
BPL, JUB, NLB,WCB	pooled	**IXa**	**1**	1, 3, 4	11.69	***	-	6, 8, 11, 13	16.58	***	-
OBH, NLH	sexes		**2**		3.08	*	*		5.12	**	-
BPL, JUB	only	**IXb**	**1**	1, 3, 4	2.06	-	-	6, 8, 11	4.43	*	-
OBH, NLH	males		**2**		0.31	-	-		1.30	-	-
bw, sh, bb	pooled	**Xa**	**1**	1, 3	2.73	-	-	2, 6, 11	0.40	-	-
ma, war	sexes		**2**		0.31	-	-		4.83	**	-
ma, war	only	**Xb**	**1**	1, 3	3.74	-	-	2, 11	10.27	**	-
bw, sh	males		**2**		2.11	-	-		1.96	-	-

1 Compared groups for languages (at least N≥20 individuals): **1** (Niger-Congo: West Atlantic, Mande & Kwa), **2** (Niger-Congo: Adamawa Eastern), **3** (Niger-Congo & Benue-Congo: West & Central Bantu), **4** (Niger-Congo & Benue-Congo: Eastern Bantu), **5** (Niger-Congo & Benue-Congo: Southern Bantu), **6** (Nilo-Saharan), **7** (Afro-Asiatic), **8** (KhoiSan);

2 Compared groups for Murdock's ethno-linguistic units: **1** (Cameroon highlands), **2** (North West Bantu), **3** (Equatorial Bantu), **4** (Mongo), **5** (Luba), **6** (Central Bantu), **7** (Kenya highland Bantu), **8** (Interlacustrine Bantu), **9** (Nguni), **10** (Western Sudan), **11** (Yam belt), **12** (East African pastoralists), **13** (North West African pastoralists), **14** (Pygmy), **15** (KhoiSan);

3 Level of significance (P<0.05*; P<0.01**; P<0.001***);

4 Levene's test of homogeneity of variances.

Appendix XX.

Descriptive data (pooled sexes) for four main groups of both known (Bahutu, Basuku, Ashanti) and unknown (Ghana, Gold Coast) ethnic affiliation.

	Bahutu			**Basuku**			**Ashanti**			**? (Ghana, Gold Coast)**		
	N	**Mean**	**S.D.**	**N**	**Mean**	**S.D.**	**N**	**Mean**	**S.D.**	**N**	**Mean**	**S.D.**
GOL	50	179.69	7.38	120	173.19	6.40	31	176.52	6.58	15	178.33	8.80
XCB	50	132.67	4.83	120	126.70	4.89	31	127.86	6.13	15	126.80	7.11
PAC	50	109.69	6.83	120	108.98	6.87	31	113.23	7.29	15	113.87	7.55
OCC	50	91.70	4.52	120	92.13	5.99	31	94.34	6.65	15	92.90	6.57
BBH	50	123.32	5.27	120	126.95	5.92	31	132.08	5.40	15	131.40	8.94
WCB	50	67.29	3.97	120	64.49	4.24	31	66.57	4.09	15	67.03	5.81
BNL	50	96.54	4.87	120	93.77	4.70	31	97.94	4.43	15	97.97	5.16
BPL	50	99.23	6.18	120	96.12	6.09	31	99.94	5.28	15	102.45	6.93
ZYB	50	127.83	6.84	120	120.49	6.76	31	124.77	5.41	15	124.50	8.32
JUB	50	115.19	5.44	120	109.41	5.77	31	113.26	3.90	15	113.20	7.26
STB	50	108.25	5.02	120	107.09	5.25	31	107.19	5.18	15	106.20	7.27
XFB	50	112.52	4.78	120	109.78	4.72	31	109.55	5.45	15	108.87	7.05
FMB	50	101.77	3.82	120	97.31	6.02	31	99.79	3.55	15	100.13	5.42
ZMB	50	94.71	4.62	120	92.08	6.19	31	95.03	4.09	15	96.60	5.56
NLB	50	25.61	2.01	120	26.77	2.13	31	27.85	1.88	15	28.26	2.60
OBH	50	34.09	1.47	120	32.83	1.88	31	33.60	1.89	15	33.39	1.86

Appendix XXI

Summary of descriptive data (in mm) when pooled sexes for various West (and North) African archaeological and modern samples: Dioronboumak, Faboura, Early Tellem, Late Tellem, Griots Lebu and Griots Serer, Nokara and North Africa (various countries).

	Dioronboumak (Senegal)			Faboura (Senegal)			Griots Lebu (Senegal)			Griots Serer (Senegal)		
	N	Mean	S.D.	N	Mean	S.D.	N	Mean	S.D.	N	Mean	S.D.
VAULT												
GOL	17	177.00	6.80	8	186.50	6.05	22	178.30	5.23	10	173.60	5.28
XCB	17	134.35	7.87	10	137.40	6.47	22	128.39	4.50	10	127.90	4.82
PAC	17	109.94	5.09	10	115.50	7.62	22	111.55	5.01	10	110.00	7.29
OCC	15	95.07	8.51	8	98.19	3.61	20	93.40	2.95	8	90.13	4.04
ASB	17	104.29	5.32	10	109.16	5.93	9	100.19	4.01	10	101.84	2.19
QI	-	-	-	-	-	-	21	291.19	9.46	10	286.10	15.67
BBH	15	132.20	4.44	6	136.00	6.72	22	129.16	5.03	8	126.63	7.15
TOTARCH	15	358.00	15.56	7	374.00	9.27	19	356.13	13.58	8	347.19	16.47
BASE												
WCB	14	69.79	3.56	8	71.75	4.13	22	65.56	3.57	9	62.55	3.43
FACE												
BNL	15	100.40	4.56	6	104.67	4.72	21	98.67	4.56	8	96.25	2.54
BPL	11	99.73	7.50	3	105.33	7.77	19	99.10	4.94	8	96.96	4.81
NPH	12	66.08	4.27	6	67.50	3.15	19	64.70	4.21	9	64.04	4.97
ZYB	15	126.00	9.27	7	132.86	6.28	17	123.79	5.26	9	119.89	5.18
JUB	12	112.75	7.61	6	117.83	6.01	18	109.28	5.44	9	108.00	5.68
STB	16	112.25	8.99	11	110.82	7.64	21	107.26	4.18	10	104.20	6.56
XFB	17	115.59	8.20	11	113.73	7.68	21	110.07	3.44	10	107.85	4.60
EKB	12	97.00	4.31	6	101.00	2.76	20	98.28	2.67	9	94.33	2.93
WFB	16	94.19	6.71	10	96.10	3.73	23	92.61	3.21	10	91.53	5.40
FMB	16	98.69	4.77	10	101.60	4.14	22	100.77	3.45	10	97.20	4.12
ZMB	12	94.83	6.42	6	97.33	2.81	20	94.08	6.32	9	90.56	3.62
XML	11	51.00	3.82	4	53.64	1.98	18	52.36	3.78	9	50.21	3.56
PABR	-	-	-	-	-	-	15	39.87	2.77	8	36.26	2.52
NLH	13	48.85	2.64	6	49.00	2.83	19	48.59	3.39	9	46.62	2.36
NLB	12	28.42	1.83	7	28.00	2.24	20	27.52	1.42	9	26.16	2.43
OBH	13	34.08	1.85	6	33.83	1.60	20	32.87	1.92	9	33.42	1.79
MANDIBLE												
bb	-	-	-	2	106.94	10.08	6	109.36	4.70	2	113.16	2.51
sh	-	-	-	7	35.33	1.38	10	34.12	3.06	1	36.31	-
ma	-	-	-	3	114.33	5.51	7	127.00	5.89	2	116.50	4.95

Appendix XXI continued

	Tellem (early period) (Mali)			Tellem (late period) (Mali)			Nokara (Mali)			North Africa (various)		
	N	Mean	S.D.	N	Mean	S.D.	N	Mean	S.D.	N	Mean	S.D.
VAULT												
GOL	25	173.32	4.34	107	175.35	6.45	57	174.90	8.06	71	180.54	7.69
XCB	26	133.23	4.97	110	136.79	5.85	56	130.55	5.49	70	135.76	5.22
PAC	-	-	-	23	109.35	7.26	57	112.10	6.70	70	113.53	6.50
OCC	-	-	-	23	93.26	6.28	57	93.43	6.32	70	94.40	5.45
ASB	1	110.00	-	2	108.00	-	30	101.86	5.18	71	105.58	4.63
Q1	1	300.00	-	2	302.00	-	30	286.53	12.03	-	-	-
BBH	22	128.68	3.93	109	131.98	5.80	54	131.58	6.10	70	132.37	6.52
TOTARCH	1	355.00	-	25	350.20	13.99	57	354.72	16.67	70	365.23	20.53
BASE												
WCB	-	-	-	-	-	-	-	-	-	69	72.06	4.29
FACE												
BNL	22	95.82	4.48	107	97.46	4.18	54	99.34	5.32	70	99.77	5.54
BPL	19	96.00	5.33	99	97.43	5.05	42	100.31	5.33	70	96.79	5.26
NPH	22	63.28	3.78	100	65.99	4.40	42	66.21	4.95	71	65.81	4.47
ZYB	24	123.67	6.91	105	126.46	6.48	44	126.21	5.79	70	126.79	6.50
JUB	1	119.00	-	2	118.00	-	26	113.25	5.14	67	111.75	5.34
STB	1	108.00	-	2	116.00	-	30	106.08	4.45	71	111.42	4.87
XFB	1	111.00	-	2	119.00	-	30	110.42	4.72	70	114.44	4.44
EKB	1	101.00	-	2	98.00	-	30	98.83	4.45	70	96.19	3.94
WFB	1	95.00	-	2	92.00	-	30	90.06	4.03	70	94.40	4.60
FMB	-	-	-	-	-	-	30	101.55	5.22	71	99.42	4.55
ZMB	1	102.00	-	2	96.00	-	26	94.65	5.66	70	93.84	6.38
XML	-	-	-	-	-	-	26	52.54	3.38	70	52.18	4.39
PABR	1	37.00	-	11	36.64	4.43	40	40.22	3.67	-	-	-
NLH	-	-	-	-	-	-	30	46.80	3.47	71	49.05	3.33
NLB	23	27.17	2.15	107	27.65	1.92	53	28.49	2.27	71	24.10	1.97
OBH	1	34.00	-	3	32.00	1.73	30	33.04	1.62	71	32.66	1.95
MANDIBLE												
bb	20	106.50	7.13	183	106.28	7.38	39	109.98	5.82	29	113.52	6.05
sh	34	31.38	5.43	179	32.64	4.71	41	33.55	3.79	-	-	-
ma	41	123.44	6.62	253	123.13	6.96	23	124.00	5.04	28	124.01	7.98

www.ingramcontent.com/pod-product-compliance
Ingram Content Group UK Ltd.
Pitfield, Milton Keynes, MK11 3LW, UK
UKHW061847041026
15591UKWH00015B/957

9 781407 307732